P9-CKY-260

The Elementary Functions

Margaret R. Hutchinson
University of St. Thomas

Charles E. Merrill Publishing Company
A Bell & Howell Company
Columbus, Ohio

Merrill Mathematics Series

Erwin Kleinfeld, *Editor*

Published by
Charles E. Merrill Publishing Co.
A Bell & Howell Company
Columbus, Ohio 43216

International Standard Book Number: 0-675-08855-0
Library of Congress Catalog Number: 73-89294
1 2 3 4 5 6—79 78 77 76 75 74

Printed in the United States of America

To my mother

Preface

At one time the precalculus sequence consisted of three separate and distinct courses, college algebra, trigonometry, and analytic geometry. Then, in 1965, the Committee on the Undergraduate Program in Mathematics (CUPM) recommended that the first two of these be combined into a single course, organized around the central theme of the function, to be called elementary functions. To the surprise of many, this combination worked very well. Not only did it unify the two topics, but it was found that the two could be covered in a one-term course by leaving out those topics (such as solving triangles) not essential to the calculus. In fact, by reorganizing the material, many topics not previously covered in the two separate courses, such as functions and their properties, and exponential functions, could now be included.

This textbook is written for such a course. Experience has shown that Chapters 2 through 5, which contain the essentials of the elementary functions, can be covered in a one-term course by students not exceptionally well prepared. Chapter 1, Preliminaries, is included by way of review and may be skipped or covered very quickly. Chapter 6 contains material on computational trigonometry. Chapters 7 and 8 contain a rather thorough treatment of plane and solid analytic geometry. There is sufficient material here for the book to be used as a two-term elementary functions-analytic geometry text.

The starred sections indicate optional material. With a well-prepared class the instructor may wish to cover some of these sections; with a weak class they can be left out. Starred problems in the Review Exercises correspond to these optional sections. A Review Exercise follows each chapter.

Particular attention has been paid throughout to those topics which will be needed later in the calculus. For example, in the sections on trigonometric identities, problems are included which involve those identities needed to integrate trigonometric functions.

A great deal of effort has been spent to make this a text the student can read. Historical comments and discussion of applications (for example, Section 4-3 on applications of the exponential function, and Section 5-7 on mathematics and music) have been included to make the text more interesting and enjoyable to read.

I would like to express my appreciation to Professor Edward L. Hutton of the University of St. Thomas, who used a preliminary version of this text and made many helpful suggestions; to my husband, Professor J. D. Hutchinson of the University of Houston, for his help and encouragement; and finally to my students whose frank and uninhibited comments contributed considerably to the clarity of the text.

Contents

1

Preliminaries

1-1 Sets and Subsets

The ideas of set theory are basic in mathematics. Set notation is used in practically every branch of mathematics, including calculus. Accordingly, we begin our study of precalculus mathematics with some of the elementary notions of set theory. A set is simply a collection of objects. The only requirement we will make is that a set be sufficiently well defined that we can always decide whether a given object is in the set or not.

The objects that make up a set are called its *elements* or its *members*. Thus, if A is the set of all the letters in the alphabet, then a is an element of the set A. In symbols we write

$$a \in A$$

where $\in$ stands for the phrase "is an element of" or "is a member of." The number 2 is not an element of set A, and we write

$$2 \notin A$$

where $\notin$ means "is not an element of." (In mathematics a diagonal slash through a symbol denotes negation. Thus, $\neq$ means "is not equal to.")

The simplest way of describing a set is by listing its elements, separated by commas, and enclosing the list in braces. Thus, if B is the set of vowels in the alphabet,

$$B = \{a, e, i, o, u\}$$

If a set is too large to list all of its elements, then sometimes a partial listing may be sufficient. We can describe the set of all counting numbers from 1 to 100, for example, as

$$C = \{1, 2, 3, 4, \ldots, 100\}$$

The three dots, called an ellipsis, can be thought of as indicating that we are to continue counting according to the pattern indicated until we reach 100. An infinite set can be described in a similar manner. The set of all counting numbers is

$$N = \{1, 2, 3, \ldots\}$$

In this case, of course, there is no last element.

A set may have just one element. This is called a *singleton set.* Note that $\{a\}$ is the *set* whose only member is the letter a. This is not the same as the *element* a. A set may even have no members at all. This set is called the *empty set* and is denoted by the symbol ϕ or $\{\ \}$.

Another way of describing a set is by the "rule" method or set builder notation. In this method we describe a set by giving a rule by which membership can be determined. For example,

$$N = \{x \mid x \text{ is a counting number}\}$$

Here x is a symbol standing for any member of the set. It should not be interpreted as saying that the *letter* x is in the set. The bar is read "such that." In this method, N is described as the set of all objects x such that x is a counting number. To decide whether or not an object belongs to the set N, we apply the rule. Is the object a counting number? If so, it is in the set. We can describe the set C mentioned above by the rule method also.

$$C = \{x \mid x \text{ is a counting number less than or equal to } 100\}$$

Two sets A and B are *equal,* $A = B$, if they contain exactly the same

7

members. If $A = \{1, 2, 3, 4\}$ and $B = \{3, 4, 1, 2\}$, then $A = B$. Note that the order in which the elements are listed is not important since a set is simply a collection of objects.

If every element in a set A is also found in a set B, then we say that A is a *subset* of B, and describe this relationship by

$$A \subseteq B$$

The set C described above is a subset of N, since every element of C is a counting number, and we write $C \subseteq N$. Note that every set is a subset of itself. In fact, if A and B are equal, then $A \subseteq B$ and $B \subseteq A$. If A is a subset of B, but A is not equal to B, then we say A is a *proper* subset of B, and write

$$A \subset B$$

For example, if $A = \{1, 2, 3, 4\}$, $B = \{3, 4, 1, 2\}$, $C = \{1, 2, 3\}$, then

$$A \subseteq B \quad \text{but} \quad C \subset A$$

Notice that we write $A \subseteq B$ if we mean that either A is a proper subset of B or perhaps A is equal to B, while $A \subset B$ is reserved for the case in which A is a subset of B but A is not equal to B. (Compare these symbols to the symbols $\leq$ and $<$ used with real numbers. See Section 1-4.)

A slash through the symbol $\subseteq$ indicates negation. Thus,

$$A \nsubseteq C$$

since there is an element in A, 4, which is not found in C. The empty set is considered to be a subset of every set.

Note that the symbols $\subset$, $\subseteq$, $\not\subset$, $\nsubseteq$ describe a relation between sets, while the symbols $\in$ and $\notin$ describe a relation between an element and a set. Thus, if $A = \{1, 2, 3, 4\}$, it is correct to write

$$1 \in A$$

since 1 is an element. However, we write

$$\{1\} \subset A$$

since $\{1\}$ is a set.

1-2 Operations on Sets

Just as two numbers can be combined to give a third number by the operations of addition or multiplication, two sets can be combined in several ways to get a third set.

One such operation is called *set union* and we use the symbol $\cup$ for this operation. If A and B are sets, then $A \cup B$ (read "A union B") is the set containing all the elements that are in A or in B or in both. In set notation

$$A \cup B = \{x \mid x \in A \text{ or } x \in B \text{ or both}\}$$

Example 1-1. If $A = \{1, 2, 3, 4\}$ and $B = \{3, 4, 5\}$, then $A \cup B = \{1, 2, 3, 4, 5\}$.

Example 1-2. If $C = \{x \mid x \text{ is a counting number between 1 and 10}\}$ and $D = \{x \mid x \text{ is a counting number between 5 and 25}\}$, then $C \cup D = \{x \mid x \text{ is a counting number between 1 and 25}\}$.

Another operation on sets is called *set intersection* and is denoted by the symbol $\cap$. If A and B are two sets, then $A \cap B$ (read "A intersect B") is a set containing those elements that are found in both A and B. In symbols,

$$A \cap B = \{x \mid x \in A \text{ and } x \in B\}$$

Example 1-3. Referring to the sets of Examples 1-1 and 1-2, $A \cap B = \{3, 4\}$ and $C \cap D = \{x \mid x \text{ is a counting number between 5 and 10}\}$.

If two sets have no elements in common then their intersection will be the empty set, and we say that the sets are *disjoint*.

Example 1-4. The following two sets are disjoint, since $E \cap O = \phi$.

$$E = \{2, 4, 6, 8, \ldots\}$$
$$O = \{1, 3, 5, 7, \ldots\}$$

It is often helpful to draw diagrams to illustrate the relations between sets and the operations on sets. These pictures are called *Venn diagrams*. In Figure 1-1 for example, the set A is represented by the region inside the circle labeled A. If two sets A and B are disjoint, then we picture them as in Figure 1-1 (a), having no region in common. If A is a proper subset of B, then we draw diagram (b), showing the region representing A to be a part of the region representing B. If the two sets have some elements in common, but neither is a subset of the other, we draw diagram (c), and finally, if the two sets are equal, we draw diagram (d).

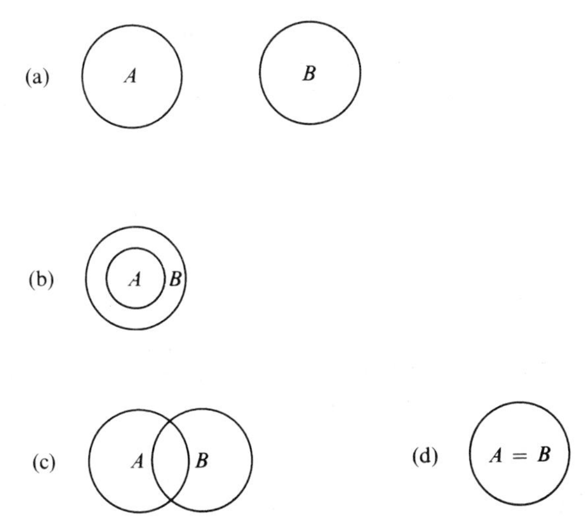

Figure 1-1

Once we have pictured the relationship between two sets A and B, then the new set $A \cup B$ or $A \cap B$ can be illustrated by shading the appropriate region—the region found in either set A or set B for $A \cup B$, and the region common to both for $A \cap B$. (See Figure 1-2.)

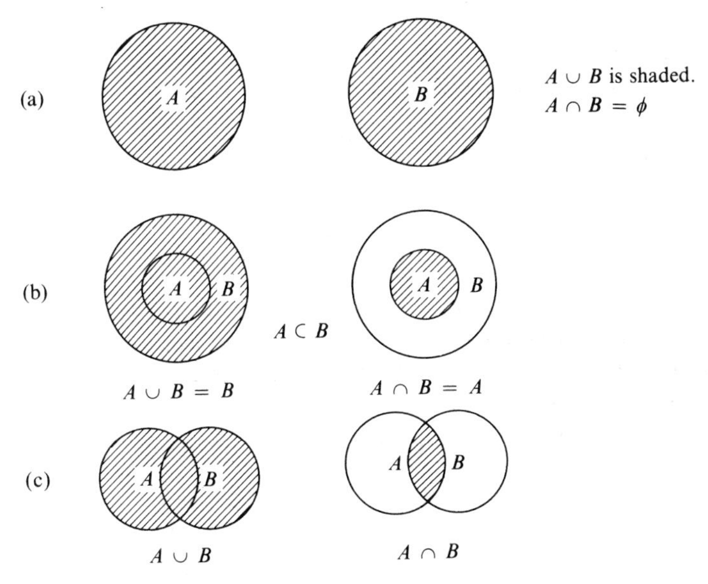

Figure 1-2

In any discussion of sets, all of the sets involved will be subsets of some larger set which is called the *universal set,* usually called U. This universal set may vary from problem to problem. If the sets we are discussing are all sets of counting numbers, as in Examples 1-1 and 1-2, then the universal set can be N, the set of all counting numbers. In a different problem the universal set might be the set of all the letters in our alphabet or the set of all the students in the university.

If A is a subset of some universal set U, then the complement of A, written $\overline{A}$, is the set of all elements in U that are *not* in A. In set notation

$$\overline{A} = \{x \mid x \in U \text{ and } x \notin A\}$$

or if U is understood

$$\overline{A} = \{x \mid x \notin A\}$$

Example 1-5. If $U = N = \{1, 2, 3, \ldots\}$ and $E = \{2, 4, 6, \ldots\}$, then $\overline{E} = \{1, 3, 5, 7, \ldots\}$.

Example 1-6. If $U = \{x \mid x \text{ is a student at this university}\}$ and $A = \{x \mid x \text{ is a female student at this university}\}$, then $\overline{A} = \{x \mid x \text{ is a male student at this university}\}$.

We can picture $\overline{A}$ in a Venn diagram by representing U by the region inside a rectangle. If A is a subset of U, then $\overline{A}$ will be represented by all the region in U which is outside of A. (See Figure 1-3.)

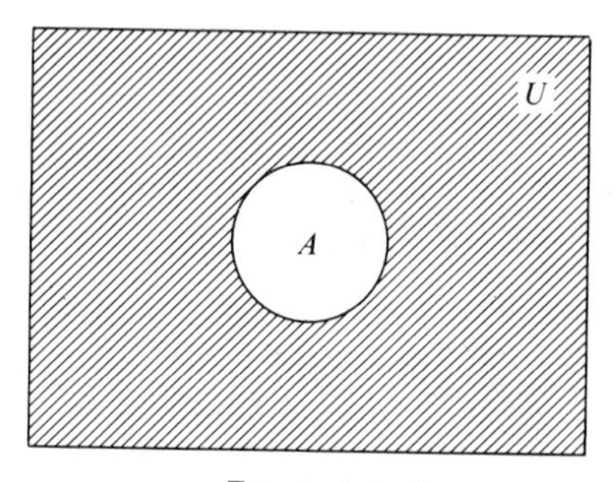

$\overline{A}$ is shaded.

Figure 1-3

Exercise 1-2

1. If $A = \{1, 2, 3, 4, 5\}$, $B = \{1, 2, 3\}$, $C = \{2, 3, 4\}$, fill in the blanks with the appropriate symbol ($\subset$, $\subseteq$, $\in$, $\not\subset$, $\not\subseteq$, $\notin$).

   (a) B ______ A (b) B ______ C
   (c) ϕ ______ B (d) 1 ______ C
   (e) 2 ______ B (f) $\{2\}$ ______ B
   (g) ϕ ______ $\{\phi\}$ (h) A ______ A

2. Describe each of the following sets by the listing method:
   (a) $\{x \mid x$ is a living former president of the United States$\}$
   (b) $\{x \mid x$ is a member of this class who is ten feet tall$\}$
   (c) $\{n \mid n$ is a counting number between 5 and 7$\}$
   (d) $\{x \mid x$ is a letter in the word *mississippi*$\}$
   (e) $\{y \mid y$ is a two-digit counting number whose second digit is one$\}$

3. Describe each of the following sets by the rule method:
   (a) {Alaska, Hawaii} (b) $\{1, 2, 3, 4\}$
   (c) {red, white, blue} (d) $\{5, 6, 7, \ldots, 100\}$
   (e) $\{3, 6, 9, 12, \ldots\}$

4. Find the union and intersection of each of the following pairs of sets:
   (a) $A = \{1, 3, 5, 7, 9\}$, $B = \{2, 4, 6, 8, 10\}$
   (b) $C = \{x \mid x$ is a counting number less than 100$\}$
   $D = \{x \mid x$ is a counting number greater than 50$\}$
   (c) $X = \{x \mid x$ is a student at this university$\}$
   $Y = \{x \mid x$ is an English major at this university$\}$
   (d) $R = \{a, b, c, d\}$, $S = \phi$

5. Let A be any subset of a universal set U. Complete the following:
   (a) $A \cup \overline{A} =$ (b) $A \cap \overline{A} =$
   (c) $A \cup U =$ (d) $A \cap U =$
   (e) $A \cup \phi =$ (f) $A \cap \phi =$
   (g) $\overline{\phi} =$ (h) $\overline{U} =$
   (i) $\overline{\overline{A}} =$

6. If $U = \{1, 2, 3, 4, 5, 6, 7, 8, 9, 10\}$, $A = \{1, 2, 3, 4\}$, $B = \{4, 5, 6\}$, $C = \{9, 10\}$, find (a) $\overline{A}$; (b) $\overline{B}$; (c) $\overline{A} \cup \overline{B}$; (d) $\overline{A \cup B}$; (e) $\overline{A \cap B}$; (f) $\overline{B} \cup \overline{C}$; (g) $A \cup \overline{C}$; (h) $\overline{A} \cap C$; (i) $\overline{A \cap C}$.

7. If $U = \{x \mid x$ is a letter in our alphabet$\}$, $A = \{a, e, i, o, u\}$, $B = \{a, b, c, d, e\}$, $C = \{x, y, z\}$, find (a) $\overline{A}$; (b) $B \cap C$; (c) $\overline{A} \cap C$; (d) $\overline{A \cap C}$; (e) $\overline{B \cup C}$.

8. Let U be the set of all living things, F the set of living things that can fly, G the set of living things that are green in color, and S the set of living things that sing.
   (a) Describe in words the members of the sets: $F \cap S$; $F \cup G \cup S$; $F \cap G \cap S$; $F \cap G \cap \overline{S}$; $\overline{F \cup G}$; $\overline{F \cap G}$.

(b) Describe in symbols the set of green flying things, the set of living things that either fly or sing, the set of living things that can neither fly nor sing, the set of living things that are green and sing but cannot fly.

9. For each of the following, copy the Venn diagram shown in Figure 1-4 and shade the region that represents each given set:
(a) $A \cap B$ (b) $\overline{A \cap B}$
(c) $A \cup B$ (d) $\overline{A \cup B}$
(e) $\overline{A} \cup B$ (f) $\overline{A} \cap B$
(g) $\overline{A} \cup \overline{B}$ (h) $\overline{A} \cap \overline{B}$

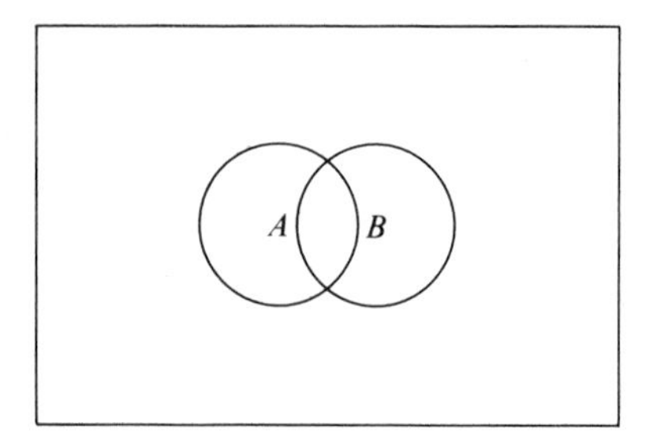

Figure 1-4

10. For each of the following, copy the Venn diagram shown in Figure 1-5 and shade the region that represents each given set:
(a) $A \cap B \cap C$ (b) $A \cup B \cup C$
(c) $\overline{A \cup B \cup C}$ (d) $\overline{A \cap B \cap C}$
(e) $\overline{A} \cup \overline{B} \cup \overline{C}$ (f) $\overline{A} \cap \overline{B} \cap \overline{C}$
(g) $\overline{A} \cup B \cup C$ (h) $\overline{A} \cap \overline{B} \cap C$

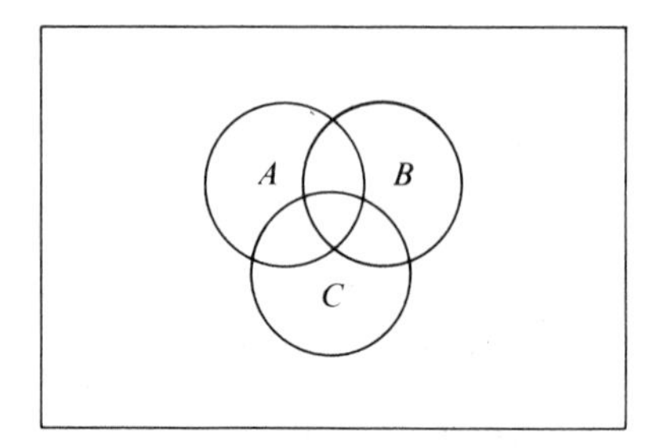

Figure 1-5

11. The De Morgan Laws state that
(a) $\overline{A \cup B} = \overline{A} \cap \overline{B}$.
(b) $\overline{A \cap B} = \overline{A} \cup \overline{B}$.
Verify these two laws by making four copies of the Venn diagram of problem 9 and shading $\overline{A \cup B}$, $\overline{A} \cap \overline{B}$, $\overline{A \cap B}$, and $\overline{A} \cup \overline{B}$.

1-3 Subsets of Real Numbers and Complex Numbers

The *counting numbers* are the numbers you count with, 1, 2, 3, ... Another name for this set of numbers is the *natural numbers*. We will call this set N.

$$N = \{1, 2, 3, \ldots\}$$

If we add the number zero to this set, we have the set of *whole numbers*.

$$W = \{0, 1, 2, 3, \ldots\}$$

The set of *integers* contains zero, the counting numbers, and the negative of every counting number. We will call this set J.

$$J = \{\ldots -3, -2, -1, 0, 1, 2, 3, \ldots\}$$

A *rational number* is a number that can be written as the ratio of two integers, p/q, where q is not zero. Thus, $1/2$, $-3/2$, and $5/-1$ are rational numbers. You are probably accustomed to calling rational numbers fractions. Two rational numbers a/b and c/d are *equivalent* provided $ad = bc$. The numbers $1/2$ and $2/4$ are equivalent since $1 \cdot 4 = 2 \cdot 2$. We can think of an integer as a rational number in which the second integer is one; for example, we can write the integer 2 as $2/1$. If we do this, then the set J of integers is a subset of the set Q of rational numbers. Note that $N \subset W$, $W \subset J$, and $J \subset Q$.

$$Q = \{p/q \mid p \text{ and } q \text{ are integers and } q \neq 0\}$$

All of these sets of numbers are subsets of a larger set R, called the set of *real numbers*. A real number is a number that can be written as a decimal. The student no doubt knows that every rational number can be written as a decimal by dividing the denominator into the numerator. Sometimes this division "comes out even," that is, we eventually get a zero remainder. In this case we say the decimal is finite. An example is $1/8 = .125$. It may happen, however, that we never get a zero remainder, in which case the decimal will eventually repeat a block of digits over and over. We call these infinite repeating decimals. An example is $20/11 = 1.818181 \ldots$.

In addition to the rational numbers, R contains numbers which are not rational. These are called the *irrational numbers*. Since they are real numbers, they can be written as decimals, however their decimal representations are not finite and they do not repeat as do the rational numbers. An example of an irrational number is $\sqrt{2}$, the number which when multiplied by itself gives $2 (\sqrt{2} \cdot \sqrt{2} = 2)$.

$$\sqrt{2} = 1.41421 \ldots$$

Another example is the number π which is the ratio of the circumference of a circle to its diameter.

$$\pi = 3.14159 \ldots$$

These decimals never terminate and never repeat.

The irrationals are a very large set. In fact, if b is a counting number which is not the square of a counting number, then $\sqrt{b}$ is irrational. Thus, $\sqrt{3}$, $\sqrt{5}$, $\sqrt{6}$, $\sqrt{7}$ are all irrational numbers, but $\sqrt{4}$ is not irrational, because $\sqrt{4} = 2$. Another example of an irrational number is the cube root of 2, written $\sqrt[3]{2}$. This number when taken as a factor three times gives 2 ($\sqrt[3]{2} \cdot \sqrt[3]{2} \cdot \sqrt[3]{2} = 2$). Indeed, if b is any integer which is not the cube of some integer, then $\sqrt[3]{b}$ is an irrational number. In addition to these, the sum or product of a rational and an irrational number is irrational. Hence, $2 - \sqrt{2}$, $2\sqrt[3]{7}$, 2π, and $(2 + \sqrt{3})/2$ are all irrational numbers.

Every real number is either rational or irrational and the two sets do not overlap. We can picture the relationship between these subsets of the reals in a Venn diagram. (See Figure 1-6.)

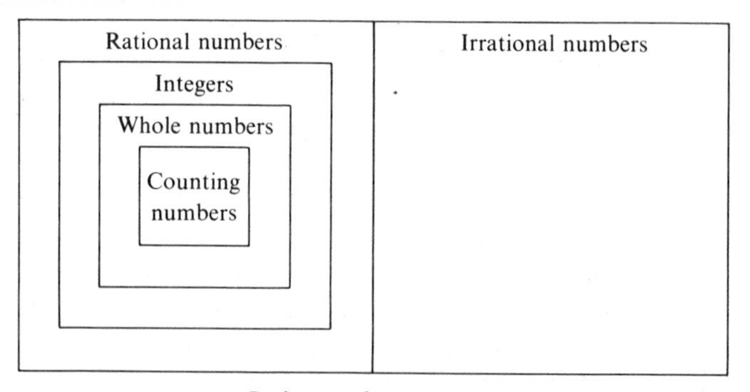

Figure 1-6

One of the fundamental assumptions we make in mathematics is that there is a one-to-one correspondence between the set of real numbers and the points on a line. If we choose a point on the line and assign to it the number zero, then pick some fixed length as our unit, we can set up this correspondence. (See Figure 1-7.)

The number one will be assigned to the point one unit to the right of the point labeled zero, the number two will be assigned to the point two units to the right of zero, and so on. Negative numbers will be

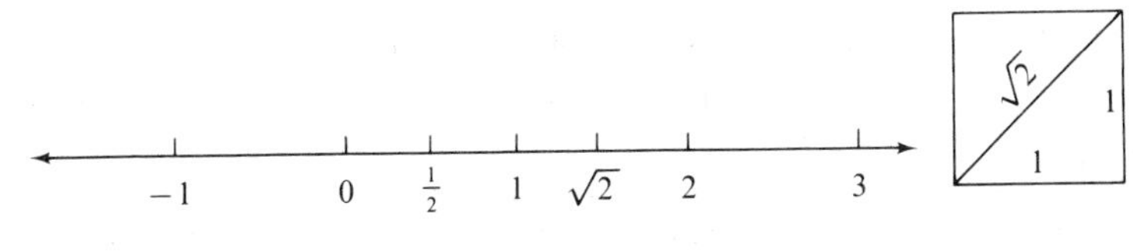

Figure 1-7

assigned to points to the left of zero. Each rational number can also be assigned a point. For example, the number 1/2 will be assigned to the point halfway between 0 and 1. Irrational numbers are also assigned points. To find the point labeled $\sqrt{2}$, for example, we could construct the diagonal of a square each of whose sides was one unit in length. By the Pythagorean Theorem, this diagonal has length $\sqrt{2}$. The point labeled $\sqrt{2}$ will be this distance to the right of zero. Because of this one-to-one correspondence between points on the line and real numbers, we often think of the two concepts as being identical and refer to "the point 2" instead of "the point to which is assigned the number 2." The line is usually called the *number line* or sometimes the *real line*.

This assignment of numbers to points on the line gives a very good illustration of the *order property* of the real numbers. If a and b are real numbers, then to say that a is less than b (written $a < b$) is equivalent to saying that on the number line point a lies to the left of point b. To put this another way, saying b is greater than a (written $b > a$) is equivalent to saying that point b lies to the right of point a. Thus, for example, -5 is less than -1, and every negative number is less than zero.

If x is any real number, then exactly *one* of the following holds:

(1) $x > 0$
(2) $-x > 0$ (in which case $x < 0$)
(3) $x = 0$

This is known as the Trichotomy Principle.

Since the product of two positive or two negative real numbers is always positive, it follows that if x is any real number whatsoever, then x^2 is either positive or zero—it can never be negative. Thus, a number whose square is negative cannot be a real number, but must be some entirely different sort of number. In order to define this new class of numbers, called *complex numbers*, first let us agree to designate by the letter i that number whose square is -1. Then $i^2 = -1$ or $i = \sqrt{-1}$.

A complex number is a number of the form $a + bi$, where a and b

are real numbers. Examples of complex numbers are $1/2 + 3i$, $2i$ ($= 0 + 2i$), and $4 - i\sqrt{2}$. The square root of any negative number is a complex number. For example, $\sqrt{-7} = \sqrt{(-1)(7)} = \sqrt{-1}\sqrt{7} = i\sqrt{7}$. The real numbers are a subset of the set of complex numbers, since any real number, such as 2, can be written as $2 + 0i$; and zero, of course is a real number. Thus, every real number is a complex number. However in actual practice when we refer to a complex number, we usually mean one which is not real, that is, a number of the form $a + bi$, where b is not zero.

Exercise 1-3

1. Write each of the following as a decimal: $2/3$, $1/7$, $1/4$, $3/16$, 0, $\pi/2$, $2\sqrt{2}$, $\sqrt{2} - 1$, 2π.

2. Classify each of the following numbers as rational or irrational.
   (a) $\sqrt{7} - 1$
   (b) $1/2 + 3/5$
   (c) $-\sqrt{9}$
   (d) $1/\sqrt{2}$
   (e) $.35802802802\ldots$
   (f) $(1 + 2\sqrt{3})/2$
   (g) $\sqrt[3]{3}$
   (h) $\sqrt[3]{8}$
   (i) $.5$
   (j) $.01001000100001\ldots$
   (k) 0
   (l) $\pi/2$

3. Describe each of the following sets by the listing method:
   (a) $A = \{x \mid x \text{ is a counting number and } x \text{ is less than } 4\}$
   (b) $B = \{x \mid x \text{ is a whole number and } x \text{ is less than } 4\}$
   (c) $D = \{x \mid x \text{ is an integer and } x \text{ is less than } 4\}$

4. Tell whether each of the following numbers is or is not an element of each of the sets N, W, J, Q, R. Use set symbols, i.e., $1/2 \in R$, $-5 \notin N$, etc.
   (a) $1/2$
   (b) -2
   (c) 0
   (d) $2 + \sqrt{2}$
   (e) $1{,}000$
   (f) $-2/3$
   (g) $-\sqrt{2}$
   (h) $.151515\ldots$
   (i) -4.23
   (j) $\sqrt{-2}$

5. Give an example to show that the sum of two irrational numbers may be a nonzero rational number. Give an example to show that the product of two irrationals may be rational.

6. Classify each of the following numbers as real or nonreal:
   (a) $2i$
   (b) $1 + \sqrt{-7}$
   (c) $1 - i$
   (d) $1 - \sqrt{7}$
   (e) $i^2 + 1$
   (f) $1/i$
   (g) $i + 4$
   (h) 0

7. If $i^2 = -1$, then $i^3 = i^2 \cdot i = (-1)i = -i$. Find i^4, i^5, i^6, i^7, i^8. Use the pattern you observe to predict the values of i^{29} and i^{107}.

8. Given two infinite decimals

$$x.a_1a_2a_3 \ldots \quad \text{and} \quad y.b_1b_2b_3 \ldots$$

state a rule for determining which number is the smaller.

1-4 Inequalities

If a and b are real numbers, then $a < b$ is read "a is less than b." We say that $a < b$ if there is a positive real number k such that $a + k = b$. Equivalently, if a and b are represented by points on the number line, the point a lies to the left of the point b. Thus, $0 < 2$ and $-2 < -1$. This can be restated b is greater than a, which is written $b > a$. Note that in each case the arrow ($<$ or $>$) points to the smaller number. For example, we can write $2 < 4$ or, equivalently, $4 > 2$.

The notation $a \leq b$ means *either* a is less than b *or* a is equal to b. Hence, $2 \leq 4$ and $2 \leq 2$ are both true statements. Equivalently we can write $b \geq a$, which is read "b is greater than or equal to a."

Statements involving the symbols $<$, $>$, $\leq$, or $\geq$ are called *inequalities*. If an inequality contains a variable such as x, then we often ask what values of the variable will make the inequality a true statement.* Such a value is called a solution and the set of all solutions is called the solution set of the inequality. Thus, zero is a solution of the inequality $x + 1 < 2$, since $0 + 1 < 2$ is a true statement; but one is not a solution since it is not true that $1 + 1 < 2$. The solution set of this inequality is the set of all real numbers less than one. In set notation we describe this solution set as $\{x \mid x < 1\}$, where it is understood that the universal set is the set of real numbers.

There are rules for solving inequalities just as there are rules for solving equations.

Let a, b, and c be real numbers.

Rule 1. If $a < b$, then $a + c < b + c$.

Rule 2. If $a < b$ and $c > 0$, then $ac < bc$.

Rule 3. If $a < b$ and $c < 0$, then $ac > bc$.

Rule 1 says that we can add (or subtract, since c can be negative) the same number to both sides of an inequality, just as we did with equations.

---

* A variable is a symbol which can assume different (usually numerical) values.

Example 1-7. If $x - 1 \leq 2$, then adding 1 to both sides of the inequality gives $x \leq 3$. If $x + 2 > 1$, then adding -2 to both sides gives $x > -1$.

We can sketch these solution sets on the number line. (See Figure 1-8.) The half bracket indicates that the number 3 is included in the solution set. The half parenthesis indicates that the number -1 is not included in the solution set.

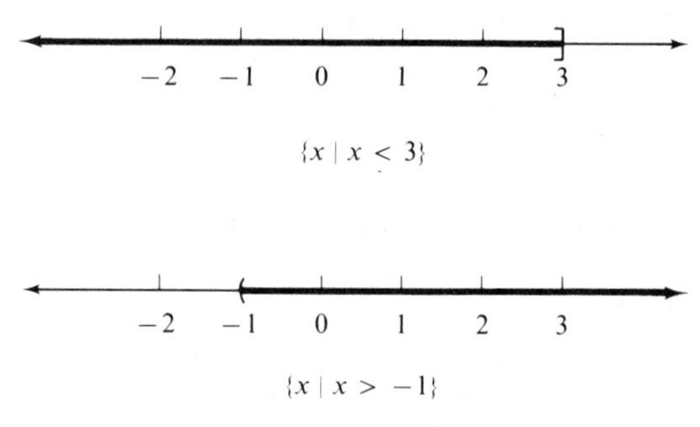

Figure 1-8

Rule 2 tells us that we can multiply both sides of an inequality by the same number without affecting the inequality *as long as the multiplier is positive.* By Rule 3, if the multiplier is negative, then the direction of the inequality is reversed.

Example 1-8. If $x/2 < 1$, multiplying both sides of the inequality by 2 gives $x < 2$. If $-2x < 4$, multiplying both sides by $-1/2$ reverses the inequality and we have $x > -2$.

Example 1-9. Solve the inequality $4 - x > 1$.

Solution: Adding -4 to both sides we get $-x > -3$. Multiplying both sides by -1 gives $x < 3$.

If $a < b$ and x is a variable standing for any real number between a and b, then the two inequalities $a < x$ and $x < b$ are true simultaneously and we can combine these two into one statement.

$$a < x < b \quad \text{(or equivalently } b > x > a\text{)}$$

This is read "x is between a and b" (where $a < b$) or "x is greater than a and less than b." Note that the arrows all point in the same direction. We *never* write, for example,

$$2 < x > 5 \qquad 5 > x < -1 \qquad 5 < x < 1$$

Example 1-10. Find the solution set of $x < 2x - 1 < x + 5$.

Solution: This is really two inequalities in one, $x < 2x - 1$ and $2x - 1 < x + 5$. Solving the first gives $1 < x$, while from the second we get $x < 6$. Thus, we have $1 < x$ *and* $x < 6$, or $1 < x < 6$. We can sketch this solution set on the number line. (See Figure 1-9.)

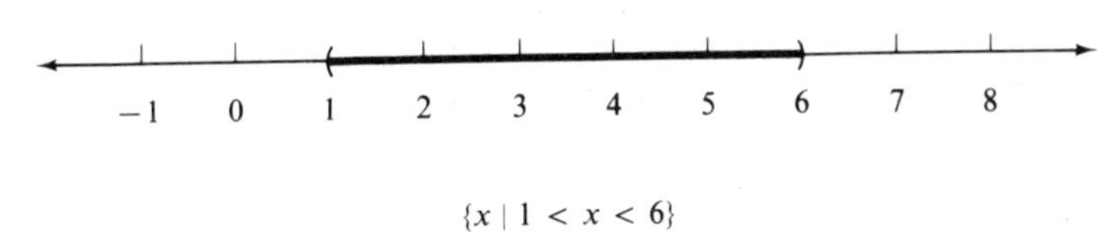

$\{x \mid 1 < x < 6\}$

Figure 1-9

It is interesting to note that the complex numbers are not ordered. In fact, it is impossible to assign an ordering to the complex numbers so that Rules 2 and 3 and the Trichotomy Principle hold. Consider the complex number i, for instance. If $i > 0$, then by Rule 2, $i^2 > 0$, and $i \cdot i^2 > 0$. But $i \cdot i^2 = -i$, thus we have $-i > 0$. This violates the Trichotomy Principle, since both i and $-i$ are positive.

If we assume that $i < 0$, we again reach a contradiction, for then $i^2 > 0$ by Rule 3 and $i \cdot i^2 < 0$ by Rule 3, thus $-i < 0$, and both i and $-i$ are negative.

Exercise 1-4

1. Fill in the blanks with the appropriate symbol: $<, >, \leq, \geq$.
   (a) -5 ____ 0 (b) 2 ____ 1
   (c) 0 ____ 0 (d) -2 ____ -5
   (e) 7 ____ 0 (f) -2 ____ $3 - 4$
   (g) $3 - 5$ ____ $-8 + 7$ (h) $0 - 1$ ____ $7 - 9$
   (i) $.005$ ____ $3/10{,}000$ (j) $3/24$ ____ $6/43$

2. Find the solution set of each of the following inequalities. Sketch each solution set on the number line.
   (a) $2x + 1 \leq 5$ (b) $7 - x > 4$
   (c) $x + 1 < 2x - 3$ (d) $2x + 7 \leq x - 4$
   (e) $\frac{x}{2} - 2 > \frac{2x}{3} + 5$ (f) $\frac{x - 1}{2} < \frac{x + 3}{4}$
   (g) $2(x - 3) \geq 3x - 1$ (h) $8x - 2 \leq 10x + 9$
   (i) $\frac{2x - 1}{3} + \frac{x}{2} < \frac{3x}{5}$ (j) $1 - x < x - 1$
   (k) $x + 2 > x + 1$ (l) $\frac{3x + 1}{-2} < 0$

(m) $-1 < x + 2 < 8$
(n) $0 < 2x - 1 < 10$
(o) $x - 1 < 2x < x + 1$
(p) $x - 1 < x - 5$

3. Express the following in symbols:
   (a) x is between 3 and 10
   (b) y is between 5 and -1
   (c) x is greater than 10 and less than 100
   (d) z is greater than or equal to zero and less than or equal to 1
   (e) n is greater than -10 and less than or equal to 14
   (f) x is positive and less than 10
   (g) x is negative and greater than -5
   (h) x is not positive and greater than -2

4. If the universal set is the set of real numbers, give the complement of each of the following sets:
   (a) $\{x \mid x < 1\}$
   (b) $\{x \mid x \geq 4\}$
   (c) $\{x \mid x > -1\}$
   (d) $\{x \mid 0 \leq x\}$
   (e) $\{x \mid 0 < x < 1\}$
   (f) $\{x \mid -1 < x < 5\}$

5. Use set notation to describe each of the following sets:
   (a) $\{x \mid x < 1\} \cap \{x \mid x > -4\}$
   (b) $\{x \mid x < 0\} \cap \{x \mid x > 10\}$
   (c) $\{x \mid x \leq -1\} \cup \{x \mid x \leq 10\}$
   (d) $\{x \mid x > 0\} \cup \{x \mid x > -4\}$
   (e) $\{x \mid x > 0\} \cap \{x \mid x > -4\}$

6. There are three more rules for inequalities. Let a, b, c, and d be real numbers.

   Rule 4. If $a < b$ and a and b have the same sign, and neither is zero, then $1/a > 1/b$.

   Rule 5. If $a < b$ and a and b have different signs and neither is zero, then $1/a < 1/b$.

   Rule 6. If $a < b$ and $c < d$, then $a + c < b + d$.

   Give several numerical examples to illustrate each of these rules.

7. Suppose a and b are located on the number line as in Figure 1-10. On the number line locate $1/a$, $1/b$, $-a$, $-b$, $(a + b)/2$, $1 + a$, $1 + b$.

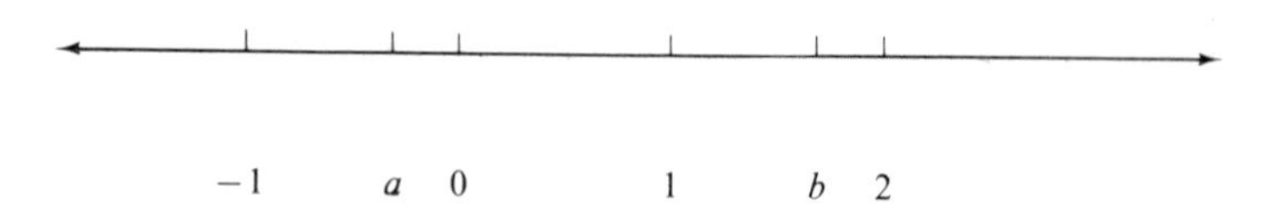

Figure 1-10

8. Solve the inequality $(1/x) < 2$. (Since x can be either positive or negative, you must consider two cases, $x < 0$ and $x > 0$.)

1-5 Absolute Value

If we consider real numbers as points on a number line, then every real number except zero has associated with it a magnitude and a direction. Its direction—either to the right or to the left of the point zero on the number line—is given by its sign, positive or negative. Its magnitude can be described as its distance from the point zero. Since distance is never negative, then magnitude is also nonnegative.

The magnitude of a real number is called its *absolute value*. Absolute value is denoted by two vertical lines; the absolute value of a real number x is written $|x|$. Since magnitude is never negative, then $|x|$ is never negative. If x is positive or zero, clearly $|x| = x$. On the other hand, if x is negative, then $-x$ is a positive number and $|x| = -x$. For example, $|3| = 3$ and $|-3| = -(-3) = 3$.

Definition 1-1. For *any real number x,* the absolute value of x, written $|x|$, is defined as follows:

$$|x| = x \text{ if } x \geq 0$$

$$|x| = -x \text{ if } x < 0$$

We can use absolute value to describe the distance between two points on the number line. The distance between the point 3 and the point 5, for example, is 2 units, and we found this by subtracting the smaller number from the larger one. In general, if a and b are real numbers, then the distance between point a and point b is $|a - b|$ since

(1) $|a - b| = a - b$ if a is greater than or equal to b (i.e., $a - b \geq 0$).
(2) $|a - b| = -(a - b) = b - a$ if b is the larger of the two (i.e., $a - b < 0$).

Thus, $|5 - 3| = |2| = 2$ and $|3 - 5| = |-2| = 2$. We can see from this that the distance from a to b is the same as the distance from b to a. Interpreted in this way, $|a| = |a - 0|$ is the distance from the point a to the point 0.

A solution to an equation involving an absolute value, such as $|x - 2| = 3$, is a real number x whose distance from the point 2 is 3 units. A sketch (Figure 1-11) shows that there are two solutions to this equation, the point 5 and the point -1.

We can find these solutions algebraically by solving two equations: $x - 2 = 3$ if $x - 2 > 0$ and $x - 2 = -3$ if $x - 2 < 0$.

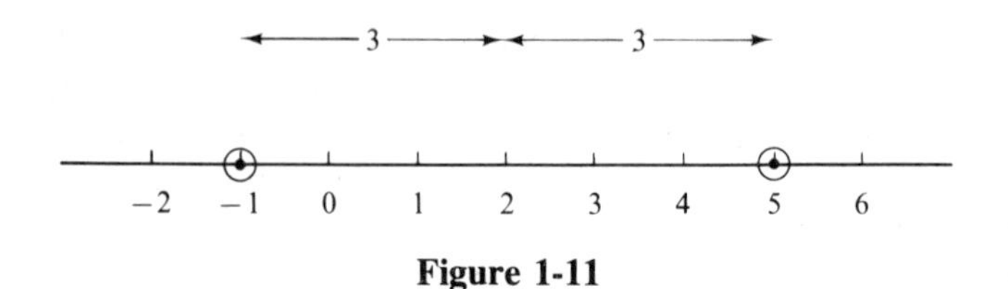

Figure 1-11

Example 1-11. Solve the equation $|x + 4| = 2$ (a) geometrically and (b) algebraically.

Solution: (a) Since $|x + 4| = |x - (-4)| = 2$, a solution to this equation will be a number x whose distance from the point -4 is 2 units. From Figure 1-12 we can see that the two solutions are $x = -2$ and $x = -6$.

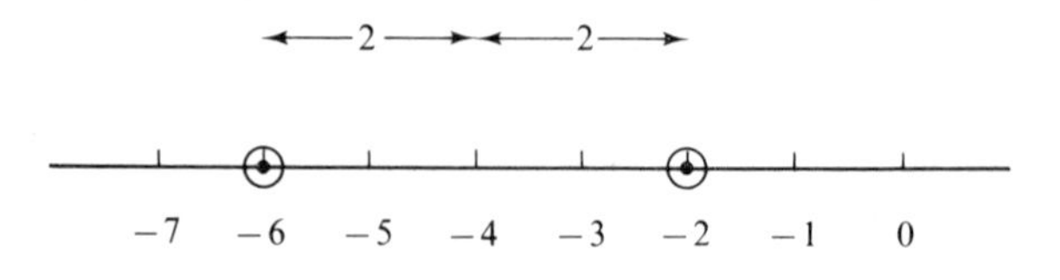

Figure 1-12

(b) Algebraically, we solve the two equations $x + 4 = 2$ and $x + 4 = -2$, getting $x = -2$ and $x = -6$.

The solution set of an inequality involving an absolute value, such as $|x - 2| < 3$ contains all real numbers x whose distance from the point 2 is less than three units. A sketch (Figure 1-13) shows that this set contains all real numbers between -1 and 5.

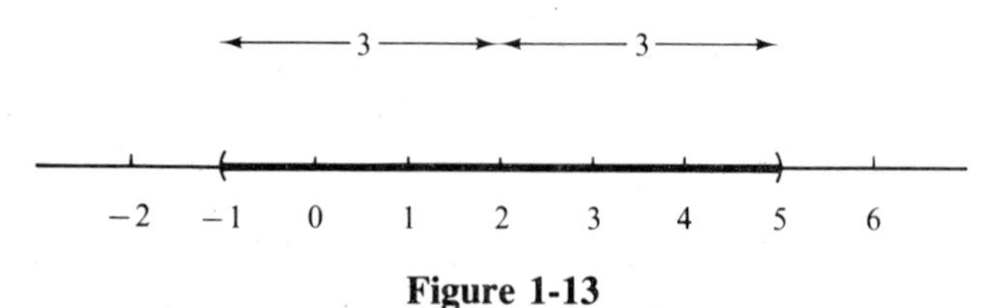

Figure 1-13

Algebraically, we can solve this inequality by using the following theorem:

Theorem 1-1. If $|x - a| < r$ and $r > 0$, then $-r < x - a < r$. Conversely, if $-r < x - a < r$ and $r > 0$, then $|x - a| < r$.

Proof: If $|x - a| < r$, then we must consider two cases, either $x - a \geq 0$ or $x - a < 0$.

Case 1. If $x - a \geq 0$, then $|x - a| = x - a$, and the inequality becomes $x - a < r$. Since r is positive, $-r$ is negative, and $-r < x - a$. These two inequalities may be written $-r < x - a < r$.

Case 2. If $x - a < 0$, then $|x - a| = -(x - a)$ and the inequality becomes $-(x - a) < r$. Multiplying both sides of the inequality by -1 gives $x - a > -r$ or $-r < x - a$. Since $x - a$ is negative and r is positive, $x - a < r$, and the two inequalities can be combined to give $-r < x - a < r$. ■

The proof of the converse is left for the student.

Example 1-12. Solve $|x - 2| < 3$ algebraically.

Solution: By the theorem this is equivalent to $-3 < x - 2 < 3$. This is two inequalities, $-3 < x - 2$ and $x - 2 < 3$. Solving both inequalities gives $-1 < x$ and $x < 5$, or $-1 < x < 5$.

Example 1-13. Solve $|x + 1| \leq 2$ (a) geometrically and (b) algebraically.

Solution: (a) Since $|x + 1| = |x - (-1)|$ the solution set of this inequality is the set of real numbers x whose distance from the point -1 is less than or equal to two units. A sketch (Figure 1-14) shows that this is the set $\{x \mid -3 \leq x \leq 1\}$.

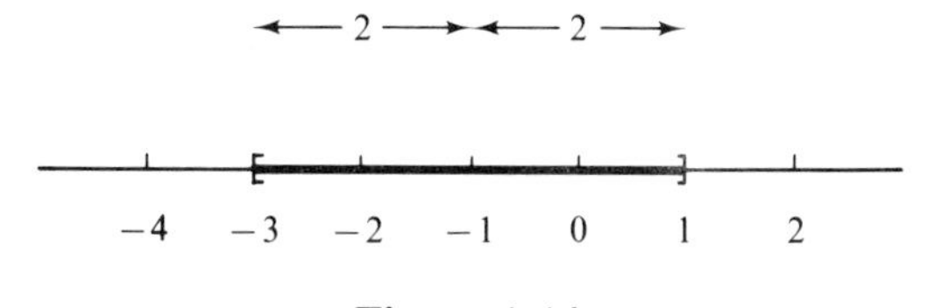

Figure 1-14

(b) Using Theorem 1-1 we see that $|x + 1| \leq 2$ is equivalent to $-2 \leq x + 1 \leq 2$. Solving these two inequalities gives $-3 \leq x \leq 1$.

Solutions to the inequality $|x - 2| > 3$ will be real numbers x whose distance from the point 2 is greater than 3 units. A sketch (Figure 1-15) shows that these solutions will be either greater than 5 or less than -1.

In set notation we describe this set as $\{x \mid x > 5 \text{ or } x < -1\}$. Note that we do not combine these two inequalities into one expression. It is *not* correct to write $5 < x < -1$, since this would imply that the

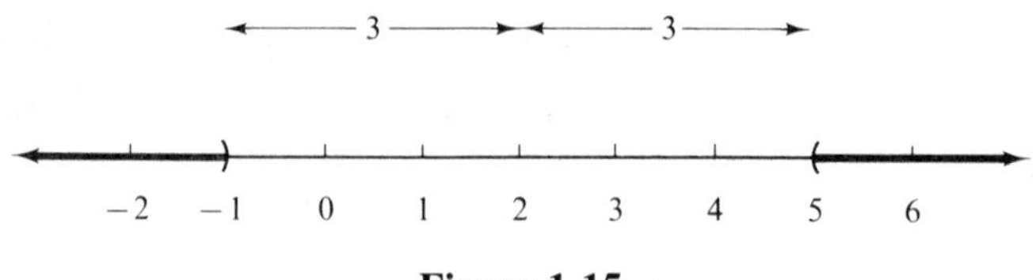

Figure 1-15

two inequalities are true simultaneously—that the number x is greater than 5 and *at the same time* less than -1, which is of course not the case.

To solve this inequality algebraically we need the following theorem:

Theorem 1-2. If $|x - a| > r$ and $r \geq 0$, then either $x - a > r$ or $x - a < -r$. Conversely, if $x - a > r$ or $x - a < -r$ and $r \geq 0$, then $|x - a| > r$.

Proof: If $x - a > 0$, then $|x - a| = x - a$ and the given inequality becomes $x - a > r$. On the other hand, if $x - a < 0$, then $|x - a| = -(x - a)$ and the inequality becomes $-(x - a) > r$ or $x - a < -r$. Thus, either $x - a > r$ or $x - a < -r$. ■

Proof of the converse is left for the student.

Example 1-14. Solve $|x - 2| > 3$ algebraically.

Solution: By Theorem 1-2, either $x - 2 > 3$ or $x - 2 < -3$. Solving these two inequalities we find that either $x > 5$ or $x < -1$.

Exercise 1-5

1. Evaluate: $|-8|$, $|4 - 5|$, $|5 - 4|$, $|2 - 2|$, $|-1 - 4|$, $|(-1/2) + 1|$, $|(-1/2) - (1/2)|$.
2. Write an expression using absolute value for the distance from the point 3 to the point 1, for the distance from the point 3 to the point -2, for the distance from the point 3 to the point 0.
3. Solve each of the following equations (i) geometrically and (ii) algebraically.

   (a) $|x - 4| = 1$ (b) $|x| = 7$
   (c) $|x + 3| = 2$ (d) $|2 - x| = 1/2$
   (e) $|-1 - x| = 4$ (f) $|-x| = 2$
   (g) $|(1/2) - x| = 2$ (h) $|x - \sqrt{2}| = 2$
   (i) $|x + (1/2)| = 3/4$ (j) $|2x + 1| = 3$
4. Is $|a + b| = |a| + |b|$? Give several examples to justify your answer.

5. Is $|a - b| = |a| - |b|$? Give several examples to justify your answer.

6. Solve each of the following inequalities (i) geometrically and (ii) algebraically.
   (a) $|x - 3| < 2$ (b) $|x + 1| \leq 5$
   (c) $|x| < 4$ (d) $|1 - x| < 4$
   (e) $|x + 5| \leq 1$ (f) $|2 - x| < 3$
   (g) $|x - (1/2)| < 1/2$ (h) $|x - 1| < .01$

7. Solve each of the following inequalities (i) geometrically and (ii) algebraically.
   (a) $|x - 5| > 1$ (b) $|x + 2| \geq 2$
   (c) $|x| > 1$ (d) $|x + 5| > 2$
   (e) $|1 - x| \geq 4$ (f) $|x - 1| \geq 3$
   (g) $|x - (1/2)| > 3/4$ (h) $|x + 1| > 100$

8. Solve each of the following:
   (a) $|2x - 1| \leq 3$ (b) $|x + 3| = |x - 7|$
   (c) $|x| = |x - 2|$ (d) $|x + 1| < -1$

9. (a) Show that if $|x - a| < r$, then $a - r < x < a + r$.
   (b) Show that if $|x - a| > r$, then either $x > a + r$ or $x < a - r$.

10. Show that $|a \cdot b| = |a|\,|b|$ by considering three cases: (i) a and b are both positive, (ii) a and b are both negative, (iii) a is positive and b is negative. Give a numerical example to illustrate each of these cases.

11. Show that $|a| = |-a|$ (i) by considering the cases in which a is positive, negative, or zero and (ii) by using problem 10. Give a geometric interpretation of this equation.

12. Prove the converse of Theorem 1-1.

13. Prove the converse of Theorem 1-2.

14. In Theorem 1-1 we have the restriction $r > 0$. What would be the solution set of $|x - a| < r$ if r were negative or zero?

15. In Theorem 1-2 we have the restriction $r \geq 0$. What would be the solution set of $|x - a| > r$ if r were negative; if r were zero?

1-6 Cartesian Coordinates and the Distance Formula

If we take two number lines which are perpendicular, then we can set up a one-to-one correspondence between the set of points in the plane and the set of ordered pairs of real numbers. Let the point of intersection be the point labeled zero on both number lines, and let the positive direction on the horizontal line be to the right and on the vertical line be upward. (See Figure 1-16.) We call the horizontal number line the x-axis and the vertical number line the y-axis.

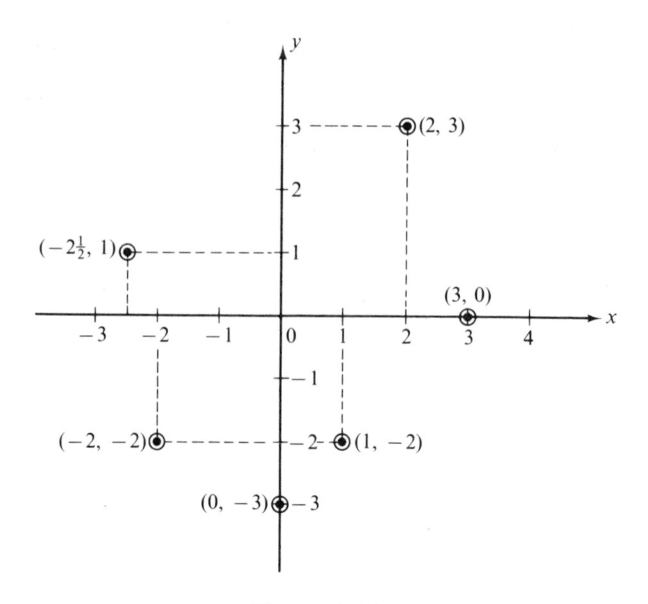

Figure 1-16

To every ordered pair of real numbers, (x_0, y_0), we can assign a point of the plane in the following way: Erect a perpendicular to the x-axis at the point on the horizontal number line labeled x_0 and a perpendicular to the y-axis at the point on the vertical line labeled y_0. The intersection of these two perpendiculars will be the point labeled (x_0, y_0). Conversely, given any point in the plane we can find a unique ordered pair of real numbers by dropping perpendiculars to the x-axis and y-axis from this point. The number assigned to the point at which the perpendicular intersects the x-axis is called the x-coordinate of the point and is the first number in the ordered pair. The number assigned to the point at which the perpendicular intersects the y-axis is called the y-coordinate of the point and is the second number in the ordered pair. A few examples (Figure 1-16) will make the correspondence clear. Points on the x-axis will have zero for their y-coordinate and points on the y-axis will have zero for their x-coordinate. The intersection of the two axes is called the origin and its coordinates are (0,0). This system of assigning an ordered pair of real numbers to every point in the plane is called a *Cartesian coordinate system* or sometimes a *rectangular coordinate system*. Because of this one-to-one correspondence between points and ordered pairs of real numbers, we will use the two notions interchangeably and refer to "the point (a, b)" when we mean "the point whose coordinates are (a, b)."

The *graph* of an algebraic sentence involving x and y is the set of all points (x, y) that make the sentence a true statement. An algebraic sentence might be an equality, an inequality, or a combination of these. For example,

$$x + 2y = 3$$
$$x \geq 0$$
$$y = x^2 - 2x + 1$$
$$y = |x|$$

are all algebraic sentences. Their graphs are pictured in Figure 1-17.

The distance between two points $A(x_0, y_0)$ and $B(x_1, y_1)$ in the plane can be found by using the Pythagorean Theorem. This theorem states that in any right triangle with sides of length a, b, and c, where c is the length of the side opposite the right angle, $a^2 + b^2 = c^2$. If we draw lines

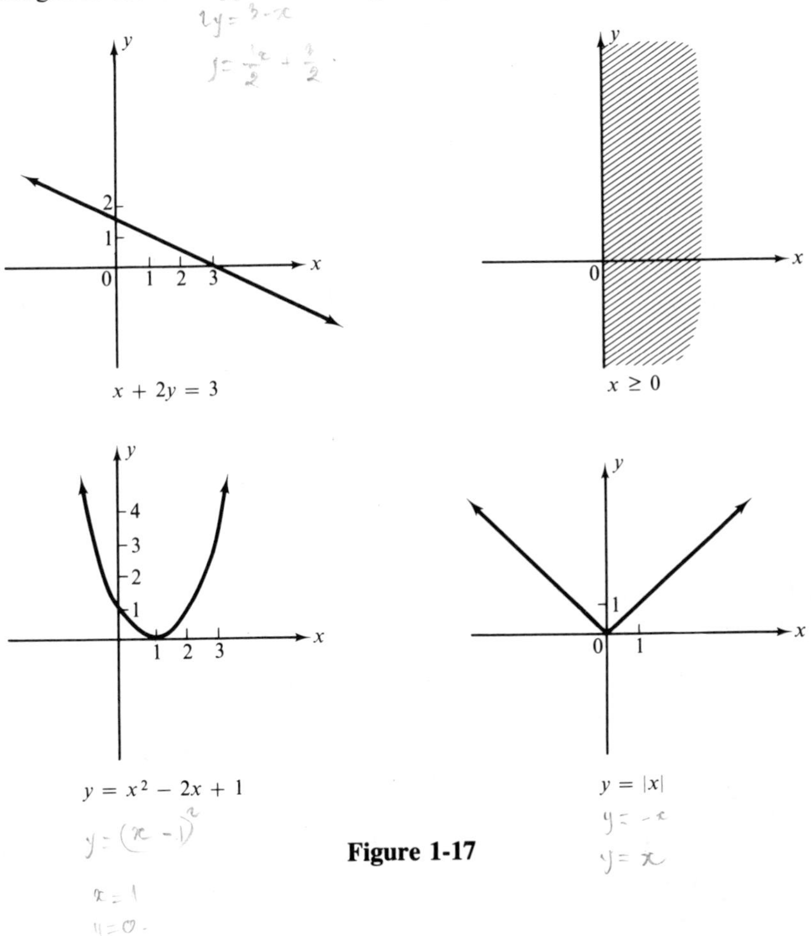

Figure 1-17

through B and A parallel to the x- and y-axes respectively, these lines intersect in a point C whose coordinates are (x_0, y_1) (Figure 1-18). ABC is a right triangle with its right angle at C. The distance from B to C is $|x_1 - x_0|$ and from A to C is $|y_1 - y_0|$, since these line segments are parallel to the x- and y-axes. By the Pythagorean Theorem, if d is the distance from A to B, then

$$d^2 = (|x_1 - x_0|)^2 + (|y_1 - y_0|)^2$$

But $(|x_1 - x_0|)^2 = (x_1 - x_0)^2$, hence

$$d = \sqrt{(x_1 - x_0)^2 + (y_1 - y_0)^2}$$

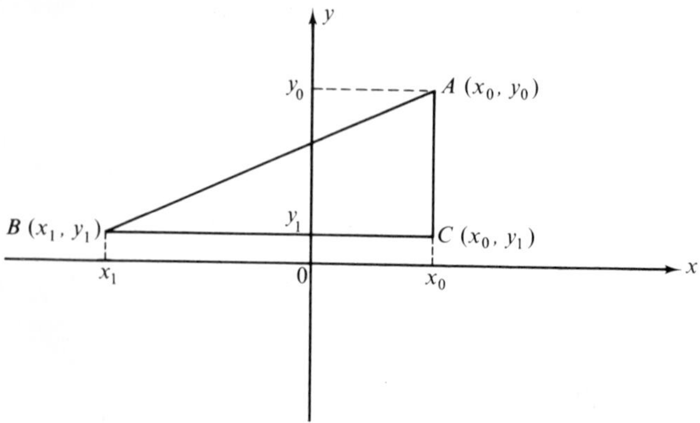

Figure 1-18

Example 1-15. Find the distance from $P_1(2, 1)$ to $P_2(-1, 3)$.

Solution: By the distance formula

$$\begin{aligned} d &= \sqrt{(2-(-1))^2 + (1-3)^2} \\ &= \sqrt{3^2 + (-2)^2} \\ &= \sqrt{13} \end{aligned}$$

Example 1-16. Show that the points $A(3, 2)$, $B(5, 7)$, and $C(-1, -8)$ are collinear.

Solution: If these points lie on one line and A is between B and C, as a sketch would suggest, then $BA + AC = BC$. Conversely, if $BA + AC = BC$, then the three points are collinear. Accordingly we find that $BA = \sqrt{29}$, $AC = \sqrt{116} = 2\sqrt{29}$, and $BC = \sqrt{261} = 3\sqrt{29}$.

The graph of a circle is the set of all points in the plane whose distance from some fixed point P (the center) is a positive constant r (the radius). Thus, the circle with center at $(-1, 2)$ and radius 2 can be described as the set of all points (x, y) whose distance from $(-1, 2)$ is equal to 2 units. To find an algebraic equation corresponding to this graph we use the distance formula.

$$2 = \sqrt{(x - (-1))^2 + (y - 2)^2}$$

or squaring both sides,

$$4 = (x + 1)^2 + (y - 2)^2$$

This is called the equation for the circle.

In general, the equation for a circle with center at (x_0, y_0) and radius r is

$$(x - x_0)^2 + (y - y_0)^2 = r^2$$

Example 1-16. The equation of a circle with center at $(0, 2)$ and radius 1 is

$$(x - 0)^2 + (y - 2)^2 = 1^2$$

or

$$x^2 + (y - 2)^2 = 1$$

If a point (x, y) is in the interior of a circle with center (x_0, y_0) and radius r, then its distance from the center is *less than* r units. These interior points must satisfy the inequality

$$(x - x_0)^2 + (y - y_0)^2 < r^2$$

In a similar manner, points in the exterior of the circle satisfy the inequality

$$(x - x_0)^2 + (y - y_0)^2 > r^2$$

Exercise 1-6

1. Locate each of the following points on a Cartesian coordinate system: $A(-1, 4)$; $B(0, 2)$; $C(-1, 0)$; $D(-1, -3)$; $E(4, 2)$; $F(0, -3)$; $G(3, -1)$; $H(2\frac{1}{2}, 0)$; $I(\sqrt{2}, 1)$; $J(\pi, 0)$; $K(0, 1 - \sqrt{2})$.
2. Give the rectangular coordinates of each of the points in Figure 1-19.
3. Give the coordinates of the vertices
   (a) of a square whose sides measure one unit, whose center is at the origin, and whose sides are parallel to the axes.

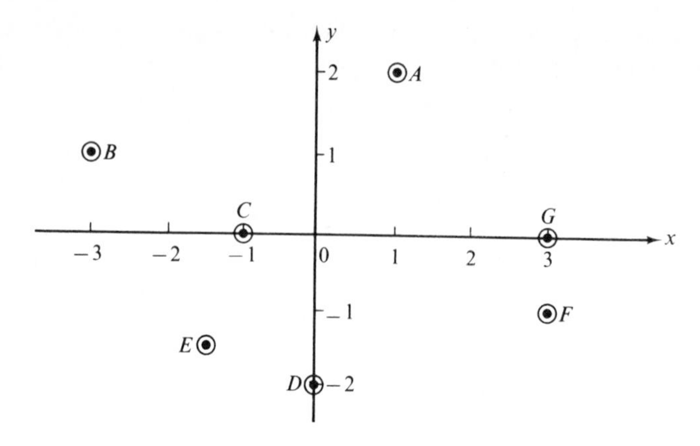

Figure 1-19

(b) of a square if two of the opposite vertices are (1, 1) and (2, 2).
(c) of a rectangle if two of the opposite vertices are (−2, 4) and (3, 8).

4. Find the distance between the following pairs of points:
   (a) $(2, -2)$ and $(0, -1)$ (b) $(-1, -3)$ and $(5, -1)$
   (c) $(\sqrt{2}, \sqrt{3})$ and $(0, 0)$ (d) (a, b) and $(a + c, b + d)$

5. Use the distance formula to determine whether or not the following sets of points are collinear:
   (a) $A(0, 1)$; $B(2, 5)$; $C(-1, -1)$
   (b) $A(1, 9)$; $B(-3, 13)$; $C(5, 5)$
   (c) $A(1, -3)$; $B(-1, -13)$; $C(3, 7)$

6. Use the distance formula and the Pythagorean Theorem to show that the triangle with vertices $(-3, 0)$, $(3, -6)$, and $(5, 8)$ is a right triangle.

7. Show that the triangle having vertices $(-1, 0)$, $(7, -8)$, and $(-2, -9)$ is isosceles.

8. Show that the points $A(1, 1)$, $B(4, 4)$, $C(2, 8)$, and $D(-1, 5)$ are vertices of a parallelogram.

9. (a) Determine x so that the point $(x, 0)$ will be 3 units from the point $(0, 1)$.
   (b) Determine y so that the point $(1, y)$ will be 2 units from the point $(2, 0)$.

10. Draw the graphs of each of the following algebraic sentences:
   (a) $x + y = 2$ (b) $2x - y = 4$
   (c) $x \geq 1$ (d) $y \leq -1$
   (e) $x^2 + y^2 = 4$ (f) $(x - 1)^2 + (y - 2)^2 = 1$
   (g) $-1 \leq x \leq 1$ (h) $0 \leq y \leq 2$

11. The *unit circle* is a circle with center at the origin and radius 1. Find an equation for the unit circle.

12. Give an equation for the circle with
    (a) center $(1, -1)$ and radius 2.
    (b) center $(-2, -4)$ and radius 3.
    (c) center $(0, -1/2)$ and radius 1.

13. If C is a circle with center at $(-3, 1)$ and radius 2, decide whether each of the following points is in the interior, the exterior, or is on the circle C: (a) $(4, 0)$; (b) $(-2, 1)$; (c) $(-1, 1)$; (d) $(2, -2)$.

Review Exercise

1. If $U = \{1, 2, 3, 4, 5, 6, 7\}$, $A = \{1, 3, 5, 7\}$, $B = \{2, 4, 6\}$, $C = \{1, 2\}$, find (a) $\overline{A}$; (b) $A \cup C$; (c) $A \cap B$; (d) $\overline{A} \cap \overline{C}$; (e) $\overline{A \cap B}$; (f) $\overline{B} \cup \overline{C}$; (g) $A \cup (B \cap C)$.

2. Copy the Venn diagram shown in Figure 1-20 and shade the region that represents the set $(A \cap \overline{B}) \cup (B \cap \overline{A})$.

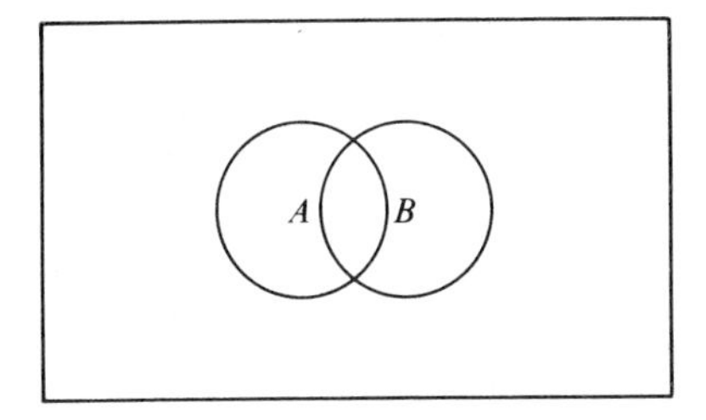

Figure 1-20

3. Describe in symbols the region shaded in the Venn diagram in Figure 1-21.

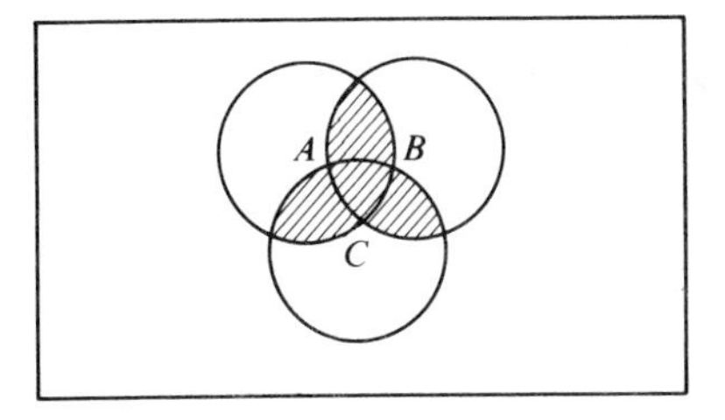

Figure 1-21

4. (a) Classify as rational or irrational the following: (i) $1 + \sqrt{2}$; (ii) $\sqrt{16}$; (iii) .201201201 . . . ; (iv) 0.

(b) Classify as real or nonreal the following: (i) $\sqrt{-7}$; (ii) i^2; (iii) 0; (iv) $1 - 2i$.

5. List the members of the set $\{x \mid x \text{ is an integer and } -3 < x \leq 5\}$. Describe the complement of this set if the universal set is taken to be the set of integers.

6. Find the solution set of each of the following inequalities. Sketch each solution set on the number line.
   (a) $\frac{x+1}{2} < \frac{2x+3}{4}$
   (b) $7 - 2x > -1$
   (c) $|x - 4| \leq 2$
   (d) $|2x + 1| < 3$
   (e) $|x| > 2$

7. Plot the points $(-1, 5)$ and $(-2, -1)$ on a rectangular coordinate system and find the distance between them.

8. Use the distance formula to show that the triangle whose vertices are $(1, -2)$, $(-4, 2)$, and $(1, 6)$ is isosceles.

9. Give an equation for the circle with center at $(1, -2)$ and radius 1.

Bibliography

Bell, E. T., *Men of Mathematics,* Ch. 3. New York: Simon & Schuster, Inc., 1937.

Davis, Philip J., "Number," in *Mathematics in the Modern World, Readings from Scientific American,* ed. Morris Kline. San Francisco and London: W. H. Freeman and Company, 1964.

Gamow, George, *One, Two, Three, . . . Infinity,* Ch. 1, 2. New York: Viking Press, 1947.

Vilenkin, N. Ya., *Stories About Sets,* Ch. 2. New York and London: Academic Press, 1968.

2

Functions

2-1 What Is an Elementary Function?

This book will be concerned with certain functions which are called *elementary functions*. These will include *polynomial functions* such as

$$y = x^2 + 2x + 3$$

exponential functions,

$$y = 2^x$$

logarithm functions,

$$y = \log_{10} x$$

trigonometric functions, of which

$$y = \sin x$$

and

$$y = \tan x$$

are two examples, and the *inverse trigonometric functions* such as

$$y = \arctan x$$

We might call these the *basic* elementary functions.

In addition to these we can build new elementary functions by combining any of these basic ones by the operations of addition, subtraction, multiplication, or division. Thus, for example

$$y = \frac{x^2 + 2^x \sin x}{\log_{10} x}$$

is an elementary function.

Another operation which when performed on elementary functions yields another elementary function is extraction of roots. Thus,

$$y = \sqrt{x}$$

and

$$y = \sqrt[5]{\sin x}$$

are also elementary functions.

Finally we can construct new elementary functions from old ones by substitution. Given a function such as

$$y_1 = x^2 + 2x + 3$$

we can construct another elementary function by substituting for the variable x a second elementary function, such as

$$y_2 = \sin x$$

This would yield the elementary function

$$y = (\sin x)^2 + 2(\sin x) + 3$$

In general an elementary function is one which can be constructed by applying to the basic elementary functions the operations of addition, subtraction, multiplication, division, extraction of roots, and substitution *a finite number of times*. In this book we will study chiefly the building blocks, the basic elementary functions.

Since the classification of elementary functions seems so all-inclusive, it might be interesting at this point to look at a function which is not elementary. Consider for example the ellipse in Figure 2-1. The area bounded by this ellipse depends on the two quantities a and b, half the lengths of the major and minor axes, respectively, and is given by the formula $A = \pi ab$. The circumference of the ellipse also depends on these two numbers, that is, it is a function of the two variables a and b. However, there exists no formula expressing this circumference in terms of a finite number of elementary functions. It is not an elementary function.

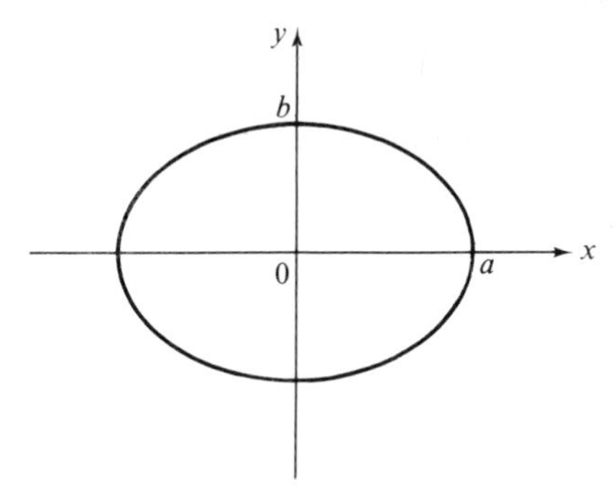

Figure 2-1
An ellipse

2-2 What Is a Function?

The idea of a function is one of the most important and basic concepts in mathematics. Before defining a function there are two other terms, *variable* and *constant*, which we must examine. The term variable is applied to a symbol, usually a letter such as x, which is allowed to assume different (usually numerical) values. In other words, its value *varies*. The set of values which it may take on is called *the range of the variable*. On the other hand, the term constant is applied to a symbol which represents a fixed object, usually a number. Its value is not allowed to change, at least during the course of the discussion or problem. Traditionally letters from the end of the alphabet such as w, x, y, z designate variables and letters from the beginning of the alphabet, a, b, c, and so on, represent constants. Of course numerals such as 2, -5, and π also are constants. In the expression $y = ax + b$, x and y are variables, while a and b are constants.

If two variables are related in such a way that by assigning a value to one variable *one and only one* value of the second variable is determined, then we say the second variable *is a function of* the first. Since the second variable depends on the first, it is called the *dependent variable.* The first variable is called the *independent variable.*

To make sure that these important concepts are clearly understood, let us look at three different ways of describing a function.

1. *By a formula*

This way of representing a function is probably most familiar to the student. When we write

$$y = 2x + 1$$

then every value we assign to x determines a unique value for y. Here x is the independent variable since we assign a value to x first, and y the dependent variable. On the other hand, if we solve for x in this equation and write

$$x = \frac{1}{2}y - \frac{1}{2}$$

then the roles have changed. Now y is the independent variable and x is the dependent variable. In general, in a formula if one variable is "solved for" then the implication is that we are thinking of this variable as the dependent one. In an equation such as $x + y = 1$, either variable can be taken to be independent.

The formula $x^2 + y^2 = 1$ does not represent a function. If we solve for y, we get

$$y^2 = 1 - x^2$$

or

$$y = \pm\sqrt{1 - x^2}$$

When $x = 0$, we get two values for y, $+1$ and -1, and it is not true that every value assigned to x determines one and only one value of y.

It is important to realize that the letters x and y are not reserved for the independent and dependent variables. Any pair of letters can be used. For example, in the formula for the area of a circle

$$A = \pi r^2$$

r is the independent variable and A the dependent.

2. *By a graph*

Since every point on a graph represents an ordered pair of numbers (x, y), a graph may represent a function. Given a value of x on the horizontal axis, we can find the value of y associated with it by moving vertically to the curve then horizontally to the y-axis. The graph will represent a function if for every x there corresponds at most one y. Geometrically this means that any vertical line will cut the graph at most once. (See Figure 2-2.)

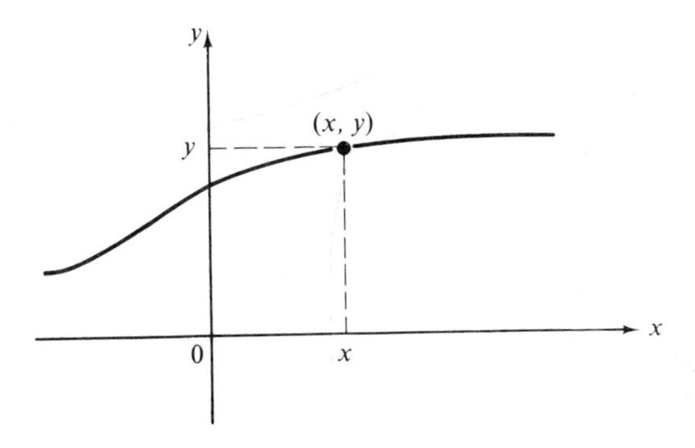

Figure 2-2

The graph of the circle in Figure 2-3 does not represent a function because we can draw a vertical line that cuts the graph in more than one point. The y-axis, for example, cuts the graph in two points. This means that there are two values of y, $+1$ and -1, corresponding to $x = 0$.

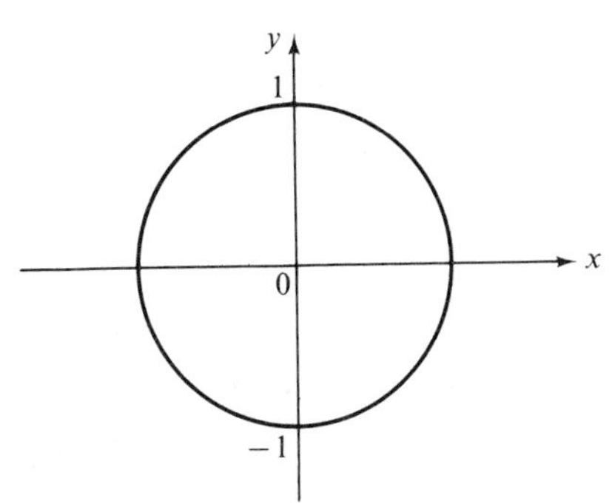

Figure 2-3

When a function is represented by a graph the independent variable is measured along the horizontal axis and the dependent variable along the vertical axis. It is not necessary that these axes be labeled x and y. The physicist observing a moving body may wish to graph its velocity v as a function of time t. In this case, t, the independent variable, will be measured along the horizontal axis, and v, the dependent variable, will be measured along the vertical axis. The economist may wish to plot cost C as a function of production P. The independent variable P is measured along the horizontal axis and the dependent variable C along the vertical axis. (See Figure 2-4.)

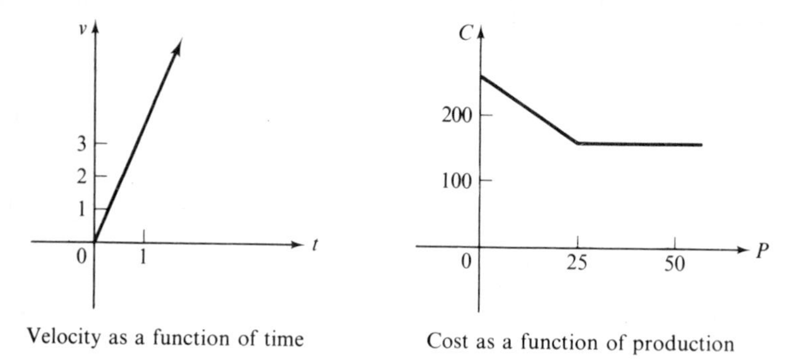

Figure 2-4

3. *By a table*

A two column table will represent a function provided each entry in the first column appears only once. Consider Table 2-1 which gives the population of the United States for the years 1890 through 1970.

To every number in the first column there corresponds exactly one number in the second column. The first column gives the set of values

Table 2-1

United States Population (Official Census), 1890–1970

Year	U.S. Population
1890	62,947,714
1900	75,994,575
1910	91,972,266
1920	105,710,620
1930	122,775,046
1940	131,669,275
1950	150,697,361
1960	179,323,175
1970	200,251,326

SOURCE: *The World Almanac and Book of Facts, 1971*, Luman H. Long, Editor, Newspaper Enterprise Assn., Inc., New York, N.Y., 1970.

the independent variable may assume, while the second column gives the corresponding values of the dependent variable. This example differs from the other two in that there are only a finite number of entries. In the other two examples, the independent variable could assume infinitely many different values.

Three different ways of describing a function have been given. The definition of a function should be broad enough to include all three types. These three descriptions have two things in common. First, in every case we have a collection of *pairs* of numbers. Moreover, these are *ordered* pairs, since the value of the independent and dependent variable cannot in general be interchanged. These pairs may be read directly as in the case of the table, or they may have to be computed from a formula. We use the symbol (a, b) for an ordered pair, where the letter a stands for the first element of the ordered pair and b stands for the second element of the ordered pair. Second, in all three cases, for every first element of the ordered pair there was one and only one corresponding second element. The first element of any ordered pair is a value assigned to the independent variable, while the second element is the corresponding value of the dependent variable.

Now we are ready to give our definition of a function:

Definition 2-1. A *function* is a set of ordered pairs having the property that two different second elements cannot correspond to the same first element.

This means that no function could contain the ordered pairs (a, b) and (a, c) unless $b = c$. A set of ordered pairs that does not meet this qualification is called a relation.

Definition 2-2. A *relation* is a set of ordered pairs.

Thus, a function is a special kind of relation. At one time relations which were not functions were called *multiple valued functions,* while what we have defined as a function was called a *single valued function.* Modern mathematicians prefer the terms function and relation as we have defined them, however. The formula

$$y = \pm\sqrt{1 - x^2}$$

represents a relation since for some choices of x there correspond two values of y.

Although technically a function is a certain set of ordered pairs which may be represented by a formula, it is common usage to refer to the formula itself as a function. Thus, we may say "the function $y = x + 1$" when we mean "the function represented by the formula $y = x + 1$."

2-3 Domain and Range of a Function

The set of all values that the independent variable may assume is called the *domain* of the function and the set of values that the dependent variable may take on is called the *range* of the function. Since we have defined a function as a certain set of ordered pairs, we can phrase these definitions as follows:

Definition 2-3. The set of all the first elements of the ordered pairs of a function is called the *domain* of the function.

Definition 2-4. The set of all the second elements of the ordered pairs of a function is called the *range* of the function.

The domain and range of a relation can be defined in the same way.

Do not confuse the phrase *range of a variable* with the *range of a function.* Recall that the range of a variable is simply the set of values that variable may assume. Using this terminology, the range of the independent variable is the domain of the function, and the range of the dependent variable is the range of the function. To keep confusion to a minimum we will not use the phrase "range of a variable" after this. When we use the word "range" it will refer to the range of a function.

The domain and range of the elementary functions we will be studying will be real numbers.

How can we find the domain and range of a given function? Sometimes this is very easy. If the function is described by a table, for instance, the domain is simply the set of elements in the first column, and the range is the set of elements in the second. In the example given in the previous section the domain is the set {1890, 1900, . . . , 1970} and the range is the set of numbers representing the population for these years.

Now suppose the function is defined by a graph. Since the domain is the set of all values the independent variable can assume and the independent variable is measured along the horizontal axis, we want to find the set of all values on the horizontal axis for which there is a point on the graph. One way of doing this is illustrated in Figure 2-5. From each point on the graph draw a perpendicular to the horizontal axis. The points on the horizontal axis shaded by this process will give the

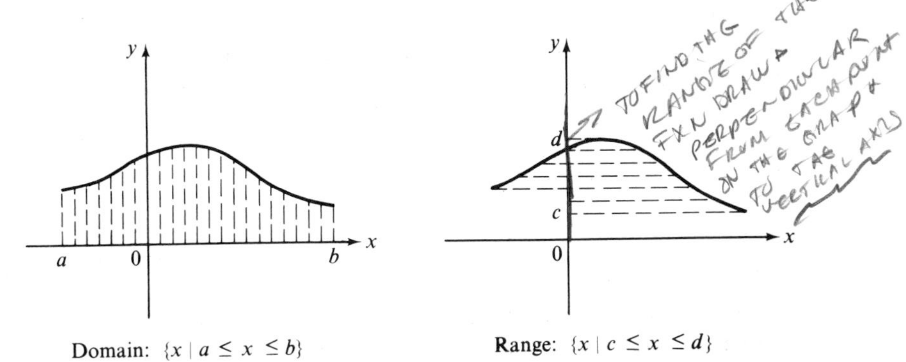

Domain: $\{x \mid a \leq x \leq b\}$

Range: $\{x \mid c \leq x \leq d\}$

Figure 2-5

domain of the function. Similarly to find the range of the function, draw a perpendicular from each point on the graph to the vertical axis (Figure 2-5). As an example of this consider the graph in Figure 2-6. The domain of this function is the set of all real numbers between -1 and $+1$ inclusive. In symbols we may write this set $\{x \mid -1 \leq x \leq 1\}$. The range of the function is the set of all real numbers between 0 and 1, inclusive, written $\{x \mid 0 \leq x \leq 1\}$.

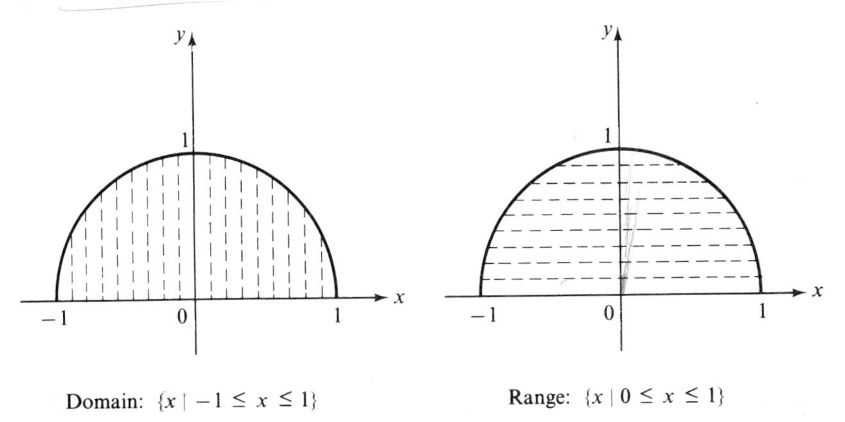

Domain: $\{x \mid -1 \leq x \leq 1\}$

Range: $\{x \mid 0 \leq x \leq 1\}$

Figure 2-6

If a function is described by a formula the domain and range are not so obvious. Sometimes the domain is specified as a part of the formula,

$$y = 2x + 1 \quad \text{if } x > 0$$

Here the domain is given to be $\{x \mid x > 0\}$. More often the domain is not given. To find the domain, we seek those values of the independent variable which yield acceptable values for the dependent variable. An acceptable value would be a real number, since the domain and range of the functions we will be studying are real numbers. In practice it is easier to find the values which are *not* in the domain. For example, if

$$y = \frac{1}{x}$$

the independent variable can be any real number except zero. Zero is not in the domain of the function because division by zero is not defined in the real number system. The domain of this function is the set of all nonzero real numbers, $\{x \mid x \neq 0\}$. If

$$y = \sqrt{x}$$

then no negative numbers can be in the domain of the function because the square root of a negative number is not a real number.

The domain of this function is the set of all nonnegative numbers, $\{x \mid x \geq 0\}$.

In short, those values of the independent variable are *excluded* which would result in either of the following:

(i) a zero in the denominator
(ii) a negative sign under a square root symbol

Although finding the range of a function given by a formula is not as easy as finding the domain, sometimes the set of values the dependent variable will assume is obvious. For example, in the function

$$y = x^2$$

clearly y is never negative. The range here is all nonnegative real numbers. The domain of this function is all real numbers since any choice of x will yield a real number value for y. If the range cannot be seen by inspection, we can graph the function and find the range in this way.

A function can be thought of as a "black box" with one input and one output. Each element from the domain goes into the box and is there transformed into an element of the range. (See Figure 2-7.)

Even though the domain and range of the functions studied in this text will be real numbers, a function could have as its domain and range *any* kind of object. For example, every living person has a unique blood type. The correspondence that assigns to each person his blood type is then a function. The domain of the function is the set of living people; the range the set of blood types. A table listing cities of the United States and their population at the last census represents a function. The domain is the set of cities of the United States and the range is a certain set of numbers. You can see from this that the idea of a

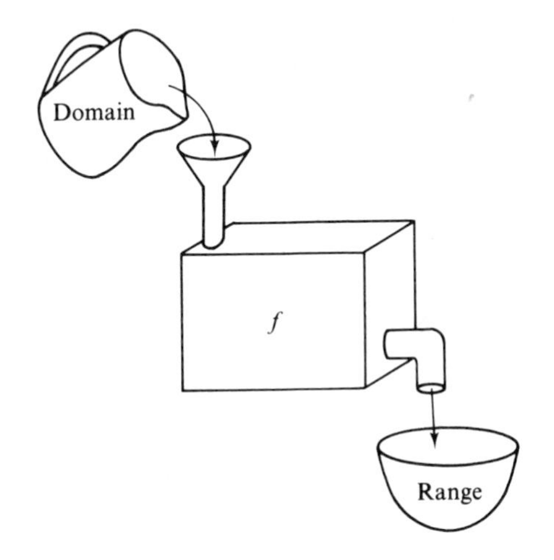

Figure 2-7

function is a very general concept that may apply in many situations which do not seem to be mathematical at all.

Exercise 2-3

1. Name the dependent and independent variable in each of the following formulas:
   (a) $s = 3t^2 - 2t + 1$ (b) $V = (4/3)\pi r^3$
   (c) $x = 3y - 1$ (d) $C = \pi d$

2. Rewrite the formula $2x + 3y = 1$ so that x is the independent variable and y the dependent variable; so that y is independent and x dependent.

3. Give the domain and range of each of the following. Which describe functions?

(a)

1	1
2	1
3	1
4	1

(b)

1	1
2	2
3	3
4	4

(c)

1	1
1	2
1	3
1	4

4. Give the domain and range of each graph in Figure 2-8. Which describe functions?

5. In the graphs shown in Figure 2-9 the arrows indicate that the curve continues indefinitely in the manner indicated. A solid dot indicates a point that is on the graph, a hollow dot one that is not. Give the domain and range of each of these. Which of the graphs represent functions?

6. Give the domain of each of the following functions:
   (a) $y = x^2 + 1$ (b) $y = \dfrac{1}{x^2 + 1}$

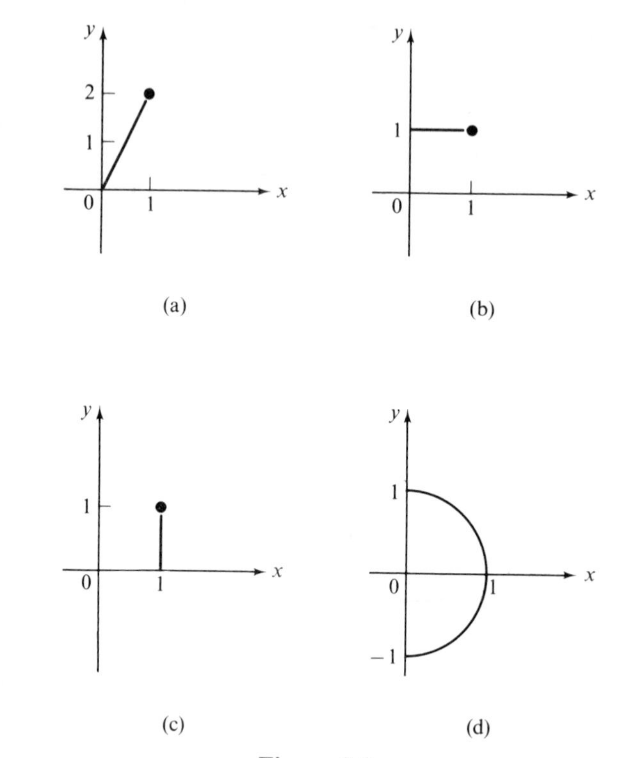

Figure 2-8

(c) $r = \dfrac{1}{s(s+2)}$ (d) $u = \sqrt{v-1}$

(e) $v = \dfrac{1}{t^2 + 3t + 2}$ (f) $z = \dfrac{1}{\sqrt{y-2}}$

(g) $v = \dfrac{1}{t^2 - 2}$ (h) $y = \dfrac{1}{\sqrt{x^2 - 1}}$

7. Boyles' Law states that if temperature is held constant then the product of the pressure p and the volume v of a gas is constant. We can write this $p \cdot v = a$, where the letter a stands for a nonzero constant. Express this as a function in which v is the independent variable and p the dependent variable and as a function in which p is independent and v dependent. What do you assume would be the domain and range of these functions?

8. Give a formula describing a function for which the independent variable is F, degrees Fahrenheit, and the dependent variable is C, degrees Celsius.

9. Does the mathematical sentence $y < x$, where x and y are real numbers, define a function? Why or why not?

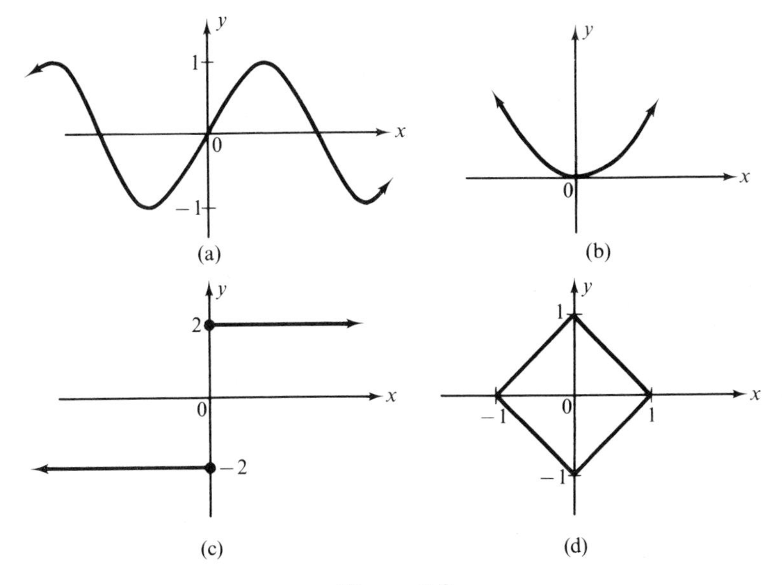

Figure 2-9

10. The operation of addition can be thought of as a function. What is the domain and range of this function?

11. A car travels at the rate of 60 mph for two hours, and at the rate of 50 mph for the succeeding three hours. Make a table expressing distance s as a function of time t. Graph this function.

12. Describe a function
    (a) whose domain is the set of books in the library and whose range is a set of numbers.
    (b) whose domain is the set of students at the university and whose range is a set of letters of the alphabet.
    (c) whose domain is a set of people and whose range is a set of colors.
    (d) whose domain is the set of students in this class and whose range is the letters A, B, C, D, and F.

2-4 Functional Notation

We will use letters such as f, g, h, or F, G, H to name functions, and we will often use the general formula $y = f(x)$ (read "y equals f of x") to represent a function. This form implies that x is the independent variable and y the dependent variable. Of course we could just as easily write $v = g(t)$ or $C = h(P)$, where g (or h) is the function, t(or P) the independent variable, and v (or C) the dependent variable.

If the number 2 is in the domain of the function $y = f(x)$, then $f(2)$ will mean the value of the dependent variable when the independent variable is equal to 2. For example, if

$$y = f(x) = 3x + 2$$

then

$$f(2) = 3 \cdot 2 + 2 = 8$$

That is, when $x = 2$, $y = 8$. In general, if a is any number in the domain of f, $f(a) = 3 \cdot a + 2$. In other words, to find $f(a)$ we *substitute* a for x in the formula $y = f(x)$. We can use this same notation if a stands for an algebraic expression instead of a number. Thus,

$$\begin{aligned} f(x^2 + 1) &= 3(x^2 + 1) + 2 \\ &= 3x^2 + 5 \end{aligned}$$

Here we have substituted $x^2 + 1$ for x in the formula for $f(x)$.

2-5 Some Examples of Functions

A function can be described by using more than one formula. In fact such functions are quite common. For example, the economist may have one formula to give the cost if production is less than 2,500 units and another different cost formula when production rises above this level. Such a function may be written

$$y = \begin{cases} \left(\frac{5}{3}\right)(12{,}500 - x) & \text{if } 0 \leq x \leq 2{,}500 \\ \frac{50{,}000}{3} & \text{if } 2{,}500 < x \end{cases}$$

We can graph such a function in two pieces. (See Figure 2-10.)

Another function which is described by more than one formula is the postage stamp function. Postage is 8¢ for letters weighing up to and including one ounce. Letters weighing more than one ounce but less than or equal to two ounces take 16¢ in postage and so on. We can describe such a function by the following formulas:

$$y = \begin{cases} .08 & \text{for } 0 < x \leq 1 \\ .16 & \text{for } 1 < x \leq 2 \\ \cdot \\ \cdot \\ \cdot \end{cases}$$

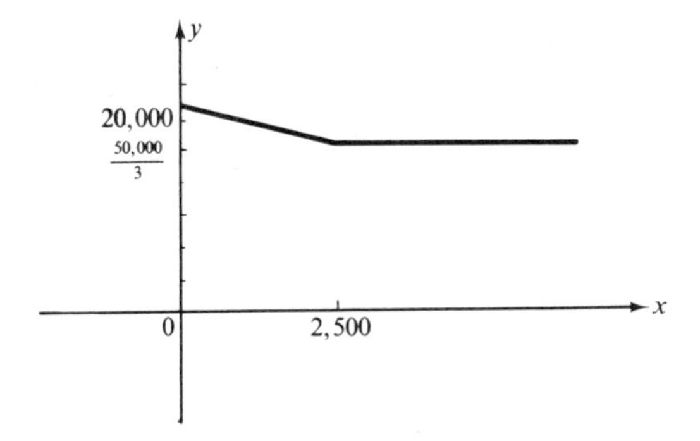

Figure 2-10

Part of the graph of this function is given in Figure 2-11. The domain of this function is the set of positive real numbers. The range is the set of positive multiples of .08. A function of this sort is called a *step function*.

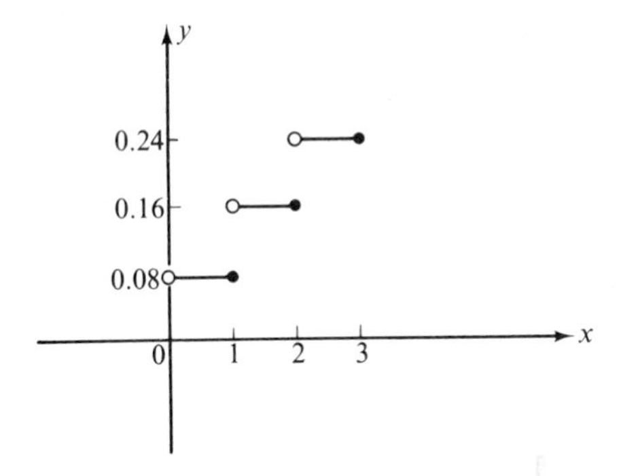

Figure 2-11

The postage stamp function

The *absolute value* function $y = |x|$ can also be described by two formulas, since

$$y = |x| = \begin{cases} x & \text{if } x \geq 0 \\ -x & \text{if } x < 0 \end{cases}$$

The domain of this function (Figure 2-12) is the set of all real numbers and its range is the set of nonnegative real numbers.

The *identity* function has the formula $y = x$. If we think of this function as a black box (described earlier in Figure 2-7), it is a box which does not change the input as it passes through the box. The output

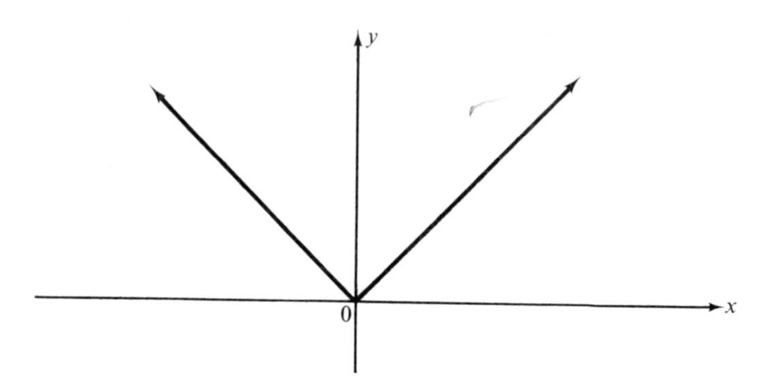

Figure 2-12

The absolute value function $y = |x|$

element is the same as the input element. This function is designated by the letter I and we write $I(x) = x$.

Exercise 2-5

1. If $f(x) = 3x - 1$ and $g(x) = x^2 - 2x + 1$, find (a) $f(0)$; (b) $g(-1)$; (c) $f(1) + g(0)$; (d) $f(-2)/g(1)$; (e) $f(x - 2)$; (f) $g(p)$; (g) $f(x^2)$; (h) $g(-x)$; (i) $f(1/x)$; (j) $f(g(x))$.

2. If

$$f(x) = \begin{cases} 3x + 1 & \text{if } x \leq -1 \\ 0 & \text{if } -1 < x \leq 2 \\ x^2 & \text{if } x > 2 \end{cases}$$

find $f(-3)$; $f(-1)$; $f(0)$; $f(3)$; $f(\mathrm{b})$.

Graph each of the following functions and give the domain and range of each:

3. $w = |z - 1|$

4. $y = \begin{cases} 2x & \text{if } x \leq 0 \\ x + 1 & \text{if } x > 0 \end{cases}$

5. $y = 6$

6. $y = \begin{cases} 1 & \text{if } x \text{ is an even integer} \\ 0 & \text{otherwise} \end{cases}$

7. $y = x + |x|$

8. Let

$$f(x) = \begin{cases} 1 & \text{if } x \text{ is rational} \\ 0 & \text{if } x \text{ is irrational} \end{cases}$$

Is this a function? What is the domain and the range? What is $f(1/3)$? $f(\sqrt{2})$? $f(\pi)$? $f(\sqrt{4})$?

9. Find a function f such that $f(a+b)=f(a)+f(b)$ for all choices of a and b in the domain. Find another function for which this equation does not hold.

10. Let

$$f(x)=\begin{cases}1 & \text{if } x \text{ is divisible by } 2\\ 2 & \text{if } x \text{ is divisible by } 3\end{cases}$$

where the domain of f is the set of integers. Is this a function? Why or why not?

11. Graph the identity function I. What is its domain? What is its range? What is $I(0)$? $I(2)$? $I(-1)$?

12. A constant function assigns to every element in the domain the same number. An example of a constant function is $f(x)=2$. Graph this function. What is its domain? What is its range?

13. We define the *signum* function, abbreviated *sgn*, as follows:

$$sgn(x)=\begin{cases}-1 & \text{if } x<0\\ 0 & \text{if } x=0\\ 1 & \text{if } x>0\end{cases}$$

The name comes from the Latin word for "sign." Graph the *signum* function. Give its domain and range. What is $sgn(-5)$? $sgn(14)$? $sgn(0)$?

14. Suppose we define $f(x)$ to be the number of prime factors of x, where x is a positive integer. Thus, $f(12)=f(2\cdot 2\cdot 3)=3$; $f(7)=1$, etc. Is this a function? What is its range? Is zero in the range? What is $f(2)$? $f(100)$? $f(1)$?

15. Let $d(x)$ be defined to be the denominator of x, where x is a rational number. Thus, $f(1/2)=2$; $f(3/4)=4$. Is this a function? What is its range?

16. Let us define the *bracket function* $y=[x]$ as follows: The symbol $[x]$ is defined to be the largest integer less than or equal to x, where x is a real number. Thus, $[1]=1$, $[2.5]=2$, $[-1.2]=-2$. Graph the bracket function. What is its domain? What is its range?

17. Let the functions f and g be defined by the tables below. What is $f(a)$? $g(b)$? $f(b)\cdot g(c)$? $g(d)\div f(d)$? $f(c)+g(c)$?

	f
a	1
b	2
c	3
d	4

	g
a	2
b	4
c	6
d	8

2-6* Implicit and Parametric Forms

Not all formulas representing functions are in the form $y = f(x)$. Sometimes the two variables are related by an equation which would be difficult or even impossible to put in this form. Nevertheless, it may be true that y is a function of x or x is a function of y, or both.

Example 2-1. $x^3 + x + y^5 + y = 0$

Equations such as this are called *implicit functions* because the relationship between the two variables is implied rather than explicitly stated.

Another form the formula might have is called *parametric form.* In this form both x and y are expressed as functions of a third auxiliary variable called the *parameter.*

Example 2-2. $x = 3t + 1$

$$y = t^2$$

Here t is the parameter. Indirectly this form expresses y as a function of x, since for every value of x a unique value of t is determined, which in turn gives a unique value of y. This form is common in applications where the parameter t often stands for time, and the equations give the position (x, y) of a particle at any time t.

To graph a function given in parametric form, we simply assign values to t and compute the corresponding values for x and y. (See Figure 2-13.)

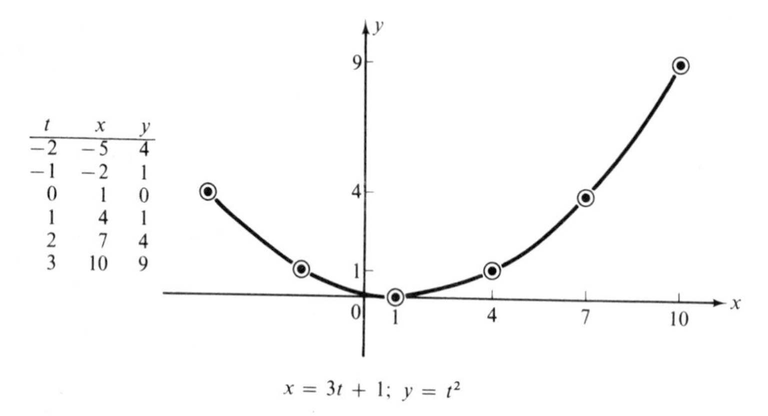

t	x	y
−2	−5	4
−1	−2	1
0	1	0
1	4	1
2	7	4
3	10	9

$x = 3t + 1;\ y = t^2$

Figure 2-13

In some cases it is possible to change from parametric form to the form $y = f(x)$ or to implicit form by eliminating the parameter. In Example 2-2 we might solve the first equation for t and substitute this in the second equation, getting $y = ((x - 1)/3)^2$.

To change $y = x^2 + 2x$ to parametric form we might set $x = t$, in which case $y = t^2 + 2t$. Or we could rewrite the equation as $y = x^2 + 2x + 1 - 1 = (x + 1)^2 - 1$ and set $t = x + 1$. Then $x = t - 1$; $y = t^2 - 1$ is another parametric form of this equation. Clearly there are many ways to write an equation in parametric form.

Exercise 2-6

Sketch the graph of each of the following. Change from parametric form to the form $y = f(x)$ or to implicit form by eliminating the parameter.

1. $x = 2 - t$
   $y = t + 1$
2. $x = t + 1$
   $y = t^2 + 1$
3. $x = t$
   $y = \sqrt{1 - t^2}$
4. $x = t^2$
   $y = t^3$
5. $x = t^2 + 1$
   $y = 1/(t^2 + 1)$

Change to parametric form by making the indicated substitution.

6. $xy^2 + x^3 + y^4 = 0$; $y = tx$
7. $y = x^2 + 4x - 7$; $x = t - 2$
8. $x + 2y = 1$; $x = t - 1$
9. $x + 2y = 1$; $x = t + 1$
10. $x^2 + y^2 = x$; $y = tx$

2-7 Symmetry—Even and Odd Functions

Working with a function can frequently be simplified if we are aware of any symmetries the function may have.

Definition 2-5. A curve is said to be *symmetric with respect to a line* L (called the line of symmetry) if for every point p on the curve and not on L there corresponds a point q, also on the curve, such that the line of symmetry is the perpendicular bisector of the line segment joining the points p and q. (See Figure 2-14.)

This means that if the curve were drawn on a piece of paper and the paper folded along the line of symmetry the two portions of the curve would coincide point for point. Each side is a mirror image of the other.

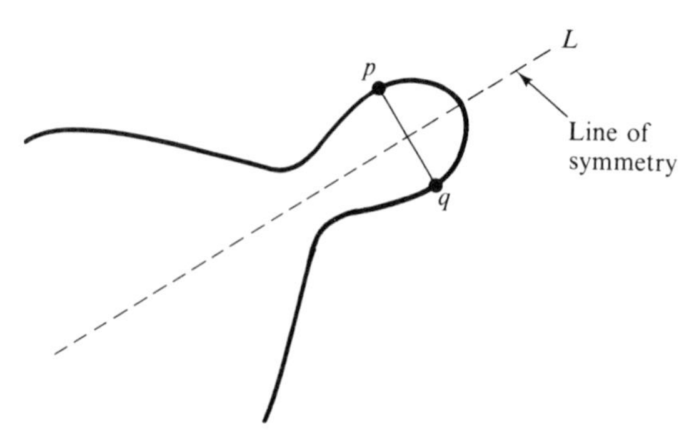

Figure 2-14

Some symmetries can be detected by looking at the equation which represents the function. If the graph is symmetric with respect to the y-axis, for instance, this means that to each point (x, y) on the graph there corresponds a point $(-x, y)$ also on the graph. Thus, if $y = f(x)$ represents such a function, $f(x) = f(-x)$. In other words, if we replace x by $-x$ in the equation, the equation will be unchanged.

Example 2-3. The graph of $f(x) = x^4 - x^2$ is symmetric with respect to the y-axis since $f(-x) = (-x)^4 - (-x)^2 = x^4 - x^2 = f(x)$. Its graph is shown in Figure 2-15.

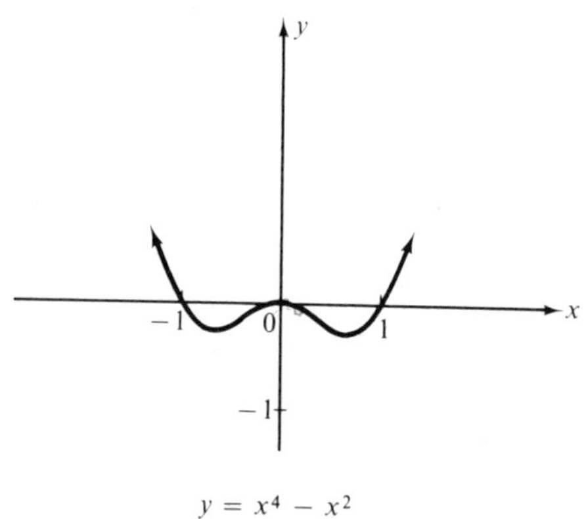

Figure 2-15

The graph of a relation (not a function) may be symmetric with respect to the x-axis. If this is the case, then for every point (x, y) on the graph the point $(x, -y)$ will also be on the graph.

Example 2-4. The graph of $y^2 - x^3 = 0$ is symmetric with respect to the x-axis since if y is replaced by $-y$ the equation will not be changed. (See Figure 2-16.) The graph is not symmetric with respect to the y-axis however.

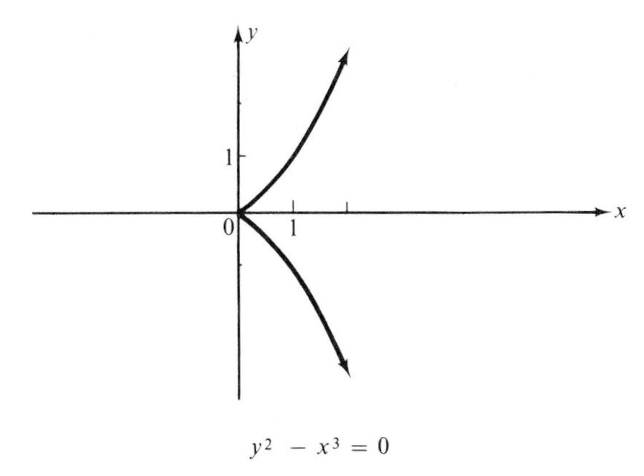

Figure 2-16

Example 2-5. The graph of $x^2 + y^2 = 1$ is symmetric with respect to *both* the x- and the y-axes. (See Figure 2-17.)

Definition 2-6. A curve is said to be *symmetric with respect to a point* O if for every point p on the curve $(p \neq O)$ there corresponds a point

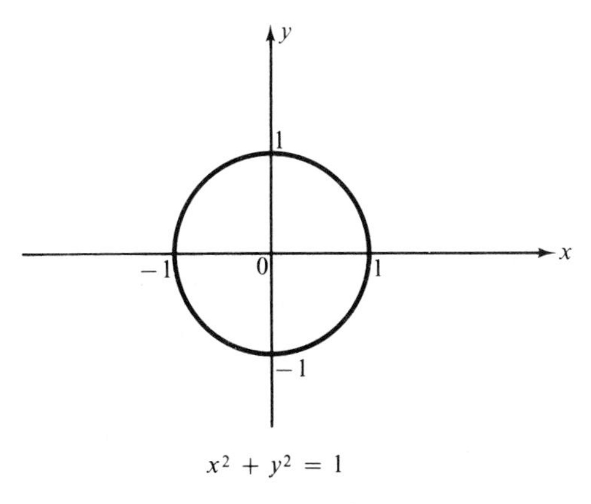

Figure 2-17

q also on the curve such that O is the midpoint of the line segment joining the points p and q. (See Figure 2-18.)

Note that if Figure 2-18 is rotated 180° about the point O it will look the same as before.

Symmetry with respect to the origin can be detected by inspecting the equation representing the function (or relation). The graph will be symmetric with respect to the origin if for any point (x, y) on the graph, the point $(-x, -y)$ is also on the graph. In other words, if substituting $-x$ for x and $-y$ for y at the same time gives an equivalent equation, then the graph is symmetric with respect to the origin.

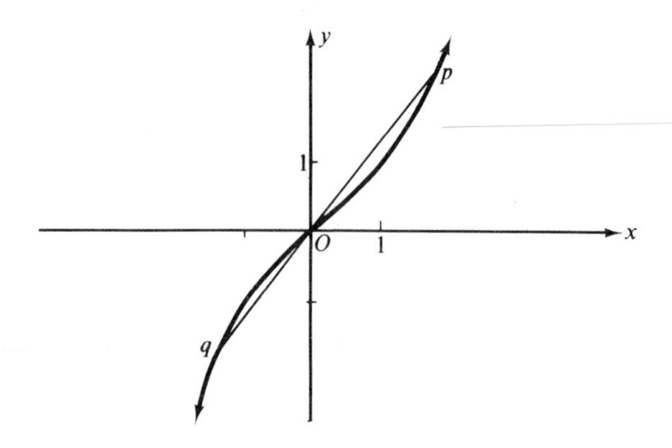

Figure 2-18

Symmetry with respect to the origin

Example 2-6. The graph of $y = x^3$ is symmetric with respect to the origin, because if we substitute $-x$ for x and $-y$ for y we get $(-y) = (-x)^3$ or $-y = -x^3$. This is equivalent to the original equation, that is, they represent the same set of ordered pairs.

Definition 2-7. A function is called an *even* function if for all x in the domain $f(-x) = f(x)$. A function is called an *odd* function if for all x in the domain $f(-x) = -f(x)$.

Thus, an even function is one whose graph is symmetric with respect to the y-axis, and the graph of an odd function is symmetric with respect to the origin. Any polynomial having only even powers of the variable will be an even function, while a polynomial having only odd powers and no constant term will be odd.

Example 2-7. The function $f(x) = x^6 + 7x^4 - 2x^2 + 4$ is an even function since $f(-x) = (-x)^6 + 7(-x)^4 - 2(-x)^2 + 4 = x^6 + 7x^4 - 2x^2 + 4 =$

$f(x)$. The function $f(x) = 2x^5 - 3x^3 + x$ is an odd function since $f(-x) = 2(-x)^5 - 3(-x)^3 + (-x) = -2x^5 + 3x^3 - x = -(2x^5 - 3x^3 + x) = -f(x)$.

Not every function can be classified as either even or odd, since there are many functions which are neither. However, every function whose domain is all the reals or some interval $\{x \mid -a \leq x \leq a\}$ centered about the origin can be written as the sum of two functions one of which is even and the other odd. (See Exercise 2-7, problem 8.)

Exercise 2-7

Determine whether the graph of each of the following is symmetric with respect to the x-axis, the y-axis, the origin.

1. $y = x^2$
2. $y = 1/x$
3. $x^3 + y^3 = x$
4. $xy^2 = 4$
5. $x + y = 0$
6. $x^2 + xy^2 = 0$
7. $y = x^4 + 3x^2 + 1$
8. Every function whose domain is all the reals or some interval $\{x \mid -a \leq x \leq a\}$ centered about the origin can be written as the sum of two functions one of which is even and the other odd.

   Proof: Let $f(x)$ be any such function. Then $f(x) = f(x) + (1/2)f(-x) - (1/2)f(-x) = [(1/2)f(x) + (1/2)f(-x)] + [(1/2)f(x) - (1/2)f(-x)]$. The first function in brackets is even, the second is odd. ■

   Write each of the following as the sum of an even function and an odd function.

   (a) $y = 3x^4 + 7x^3 - 8x^2 + 3x - 2$
   (b) $y = x^3 + 2x^2 + 3$
   (c) $y = \sqrt{x + 1}; -1 \leq x \leq 1$
9. Is a constant function even, odd, or neither?
10. Is $y = |x|$ even, odd, or neither?
11. Consider the function

$$y = \begin{cases} 1 & \text{if } x \text{ is rational} \\ 0 & \text{if } x \text{ is irrational} \end{cases}$$

    Is this function even, odd, or neither?

2-8 The Inverse of a Function

If we interchange the first and second elements of the ordered pairs of a function f, we get a relation which may or may not be a function. If it is a function, then it is called the *inverse* of f and is designated by the symbol f^{-1} (read "f inverse"). Do not interpret the -1 in this symbol as an exponent; f^{-1} does *not* mean $1/f$. The inverse of the function $f(x) = x + 1$ for example is the function $f^{-1}(x) = x - 1$ and this is not at all the same as the *reciprocal* of f which is $1/(x + 1)$.

It is easy to see that f will have an inverse if for every element of the range there corresponds one and only one element of the domain. Since f is a function, it is also true that to every element in the domain there corresponds one and only one element of the range. If both of these conditions hold then we say that f is a one-to-one function.

Definition 2-8. A function f is said to be *one-to-one* if two different first elements cannot correspond to the same second element.

Definition 2-9. The function obtained by interchanging the first and second elements of the ordered pairs of a one-to-one function f is called the *inverse* of f and is denoted by the symbol f^{-1}.

Clearly the domain of f is the range of f^{-1}, and the range of f is the domain of f^{-1}.

If a function is described by a table, it will be one-to-one provided no entry appears in the second column more than once. The inverse is easily found by interchanging the first and second columns.

f	
1	0
2	-1
3	4

f^{-1}	
0	1
-1	2
4	3

A function which is represented by a graph will have an inverse if every horizontal line cuts the graph at most once. We can find the graph of the inverse function by rotating the sheet of paper on which the graph is drawn about the line $y = x$ so that the positive y-axis falls horizontally to the right and the positive x-axis rises vertically. (See Figure 2-19.)

It is not easy to tell by looking at the formula whether a function has an inverse or not. For example, $y = x^3$ has an inverse, but $y = x^4$ does not. If the function does have an inverse, a formula for the inverse

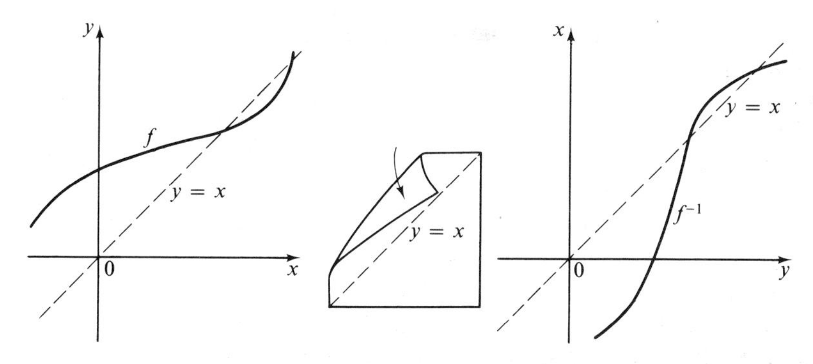

Figure 2-19

function can sometimes be found by solving for the independent variable. To find a formula for the inverse of $y = 3x - 1$ we solve for x, getting $x = (1/3)(y + 1)$. Thus, if $f(x) = 3x - 1$, then $f^{-1}(x) = (1/3)(x + 1)$.

It is not always possible to solve for the independent variable, however. The function $y = x^5 + x$ does have an inverse, however, it is impossible to solve for x in terms of y. The inverse is not an elementary function. This does not mean that the inverse does not exist. We can even graph it by first graphing the function, then rotating the graph as described above. (See Figure 2-20.)

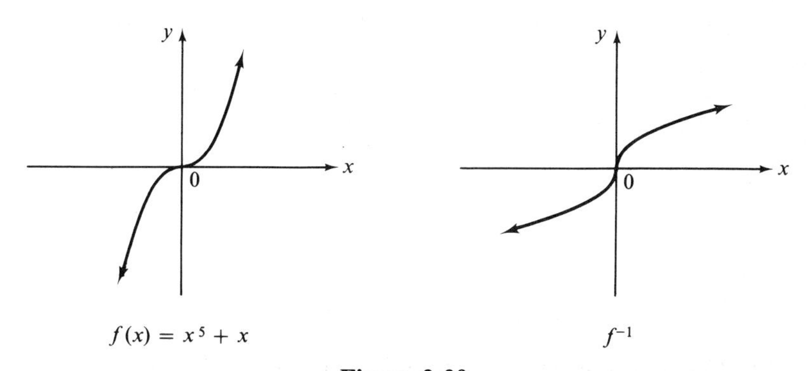

Figure 2-20

If a function does not have an inverse, sometimes it can be broken up into two or more pieces, each of which will have an inverse. For example, $y = x^2$ does not have an inverse. This can be seen from the graph in Figure 2-21, since a horizontal line can be drawn which cuts the graph in more than one point. However, if we take only that portion of the graph for which x is positive or zero, then this graph will represent a function which has an inverse. This inverse is given by the

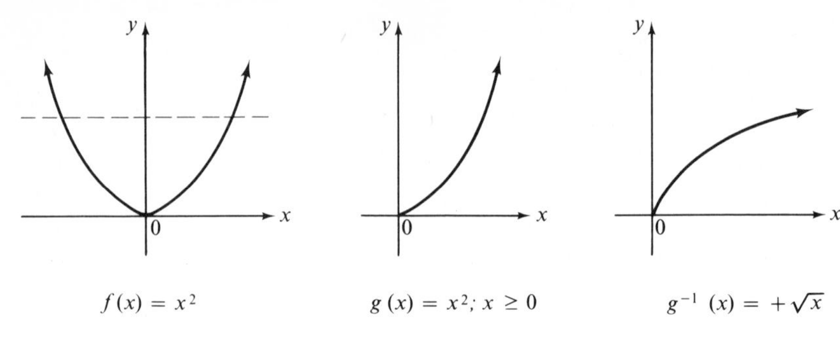

Figure 2-21

formula $x = +\sqrt{y}$. As an alternative we could take that portion of the graph for which x is zero or negative, and this portion would represent a function which has an inverse also.

The two formulas

$$y = x^2$$

and

$$y = x^2 \qquad x \geq 0$$

are very similar, but clearly they are not the same set of ordered pairs. (See Figure 2-21.) They do not represent the same function because they do not have the same domain. The domain of the first is the set of all real numbers, while the domain of the second is the set of non-negative real numbers.

This raises the question, when are two functions equal? Since a function is a set of ordered pairs, two functions f and g will be equal when they contain exactly the same ordered pairs. This will happen when they have the same domain and when $f(x) = g(x)$ for every x in that common domain.

Definition 2-10. Two functions f and g are *equal* if they have the same domain and if for every element x of the domain $f(x) = g(x)$.

Example 2-8. The formulas $f(x) = x$ and $g(x) = x^2/x$ do not represent the same function. The domain of f is the set of all real numbers, while the domain of g does not contain the number zero.

Exercise 2-8

1. Find a formula for the inverse of each of the following functions:
   (a) $y = x + 1$ (b) $y = 3x$

(c) $y = 2x + 3$ (d) $y = -3x + 7$
(e) $y = 1/x$ (f) $y = 1/(x - 1)$

2. Graph $y = x^3$ and its inverse on the same set of axes.

3. Draw the graph of the inverse of each of the functions shown in Figure 2-22 by tracing the coordinate axes and the curve on thin paper and then rotating the paper about the line $y = x$.

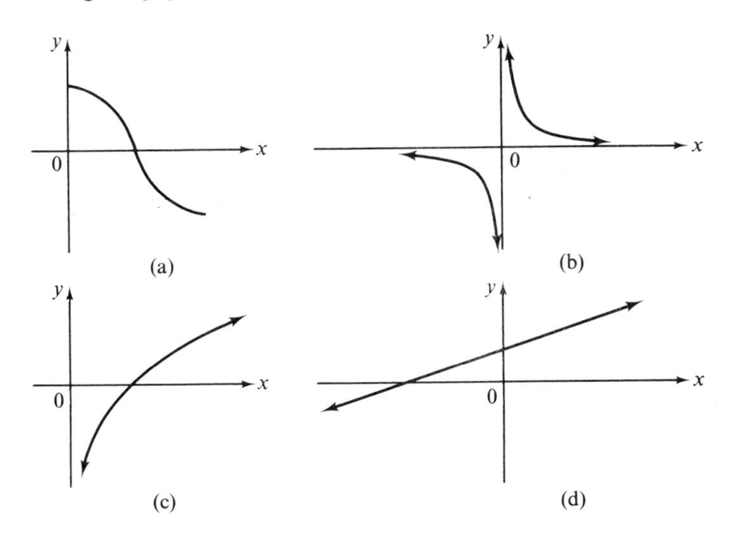

Figure 2-22

4. Which of the functions in Figure 2-23 has an inverse? Sketch the inverse function if it exists.

5. Which of the functions described below has an inverse? Give the inverse if it exists.

(a)

f	
1	1
2	0
3	1

(b)

g	
1	4
2	5
3	6

(c)

h	
1	1
2	2
3	3

6. Does the postage stamp function have an inverse? Does $y = |x|$ have an inverse? Why or why not?

7. If f is a function with inverse f^{-1}, then what is the inverse of f^{-1}? That is, what is $(f^{-1})^{-1}$?

8. Are the two functions $f(x) = x - 2$ and $g(x) = x(x - 2)/x$ equal? Why or why not? Are the two functions $F(x) = 1/x$ and $G(x) = x/x^2$ equal? Why or why not?

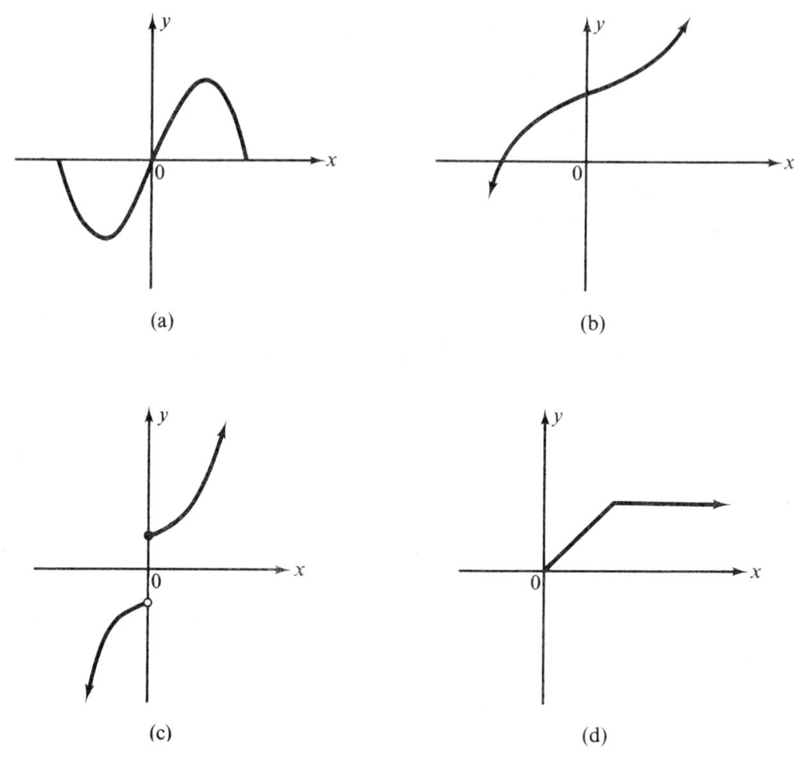

Figure 2-23

9. Show that the three functions $f(x) = |x|$, $g(x) = \sqrt{x^2}$, and $h(x) = x \cdot sgn(x)$ are equal.

10. If $f(x) = x^2 + 1$, $x \geq 0$, find f^{-1}. What is the domain of f^{-1}?

11. If $g(x) = -\sqrt{x+1}$, $x \geq -1$, find g^{-1}. What is the domain of g^{-1}?

12. Consider the function that assigns to each person in this class his age. Does this function have an inverse?

2-9 Operations on Functions

Functions whose range is the set of real numbers may be added, subtracted, multiplied, and divided just as real numbers, provided they have the same domain or share some part of their domains. The result of each of these operations will be a new function which is valid on some subset of that part of the domains which the two functions have in common.

The *sum* of two functions f and g is denoted by $f+g$ and is defined

$$(f+g)(x)=f(x)+g(x) \quad \text{for } x \in D_f \cap D_g$$

where D_f and D_g stand for the domains of f and g respectively.

In other words the function $f+g$ assigns to every element x in the domain of both f and g the sum of the two y values, $f(x)$ and $g(x)$. If f and g are represented by tables, for example,

f	
1	4
2	5
3	−1

g	
1	−6
2	8
4	−5

$f+g$	
1	−2
2	13

In this case $D_f=\{1, 2, 3\}$, $D_g=\{1, 2, 4\}$, and $D_f \cap D_g=\{1, 2\}$.

If f and g are described by graphs, we can construct the graph of $f+g$ by drawing f and g on the same set of axes. (See Figure 2-24.) For every value of x in the common domain we can find the corresponding y value by adding algebraically the distances from each curve to the x-axis. If both curves are above (or below) the x-axis at a given point x, then the distances are added. If one curve is above and the other below, then the distances are subtracted.

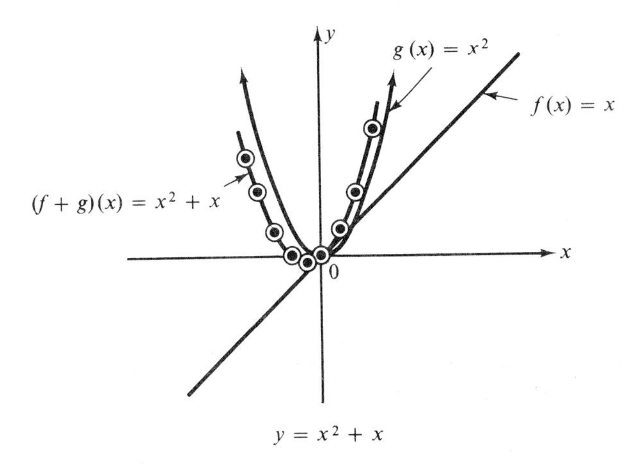

Figure 2-24

If the two functions are described by formulas, then $f+g$ is simply written as the sum of the two formulas.

Example 2-9. $f(x)=\sqrt{x}$; $D_f=\{x \mid x \geq 0\}$
$g(x)=1/x$; $D_g=\{x \mid x \neq 0\}$
$(f+g)(x)=\sqrt{x}+1/x$; $D_{f+g}=\{x \mid x > 0\}$

The difference, product, and quotient of two functions can be defined in the same way, except that in the case of the quotient function, the domain does not include any values for which the denominator function is zero.

Definition 2-11. Let f and g be two functions with domains D_f and D_g respectively whose ranges are subsets of real numbers. Then

$$(f+g)(x) = f(x) + g(x)$$
$$(f-g)(x) = f(x) - g(x)$$
$$(f \cdot g)(x) = f(x) \cdot g(x)$$
$$\left(\frac{f}{g}\right)(x) = \frac{f(x)}{g(x)} \qquad g(x) \neq 0$$

for all $x \in D_f \cap D_g$.

Another operation on functions is called *composition* of functions. To picture how this operation works, let us go back to our description of a function as a black box with one input and one output. Suppose we have two of these black boxes, f and g, and we hook up the *output* of f to the *input* of g. (See Figure 2-25.)

Now clearly for this kind of set up to work, the elements that come out of f must be elements that g will accept. In other words, *the range of f must be contained in the domain of g.* If this is the case, the result of this method of combining two functions will be a function whose

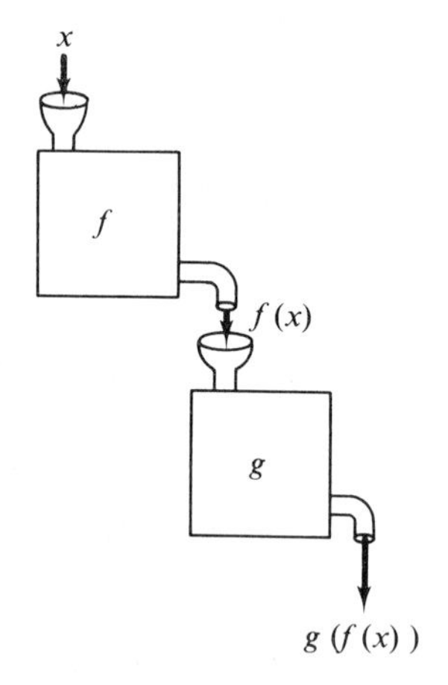

Figure 2-25

The function $(g \circ f)(x) = g(f(x))$

domain is the domain of f and whose range is a subset of the range of g. This function is called the *composite* of the two functions and is denoted by the symbol $g \circ f$. This symbol needs some explanation. It is read from right to left and is interpreted as f first, then g. We can define the function $g \circ f$ by the formula

$$(g \circ f)(x) = g(f(x))$$

This means, for any x in the domain of f, first find $f(x)$. Since this is some element in the domain of g, we can find the value of the function g corresponding to it or $g(f(x))$.

Definition 2-12. If f and g are two functions such that the range of f is a subset of the domain of g, then the *composite* of f and g, written $g \circ f$, will be a function which is defined by the formula

$$(g \circ f)(x) = g(f(x)) \quad \text{for all } x \in D_f$$

Example 2-10. Let $f(x) = x^2$, $g(x) = x + 1$. Then $(g \circ f)(2) = g(f(2)) = g(4) = 5$.

We can find a formula for $g \circ f$ if we recall that the symbol $g(a)$ means *substitute a* for the independent variable in the formula for g (Section 2-4). Thus, since $g(x) = x + 1$, then $(g \circ f)(x) = g(f(x)) = g(x^2) = x^2 + 1$.

The operation of composition is not commutative, that is, $f \circ g$ is not in general the same function as $g \circ f$. In Example 2-10, $(f \circ g)(x) = f(g(x)) = f(x + 1) = (x + 1)^2 = x^2 + 2x + 1$. In this case both $f \circ g$ and $g \circ f$ are defined since the domain of both f and g is the set of all real numbers.

However, consider the two functions $F(x) = \sqrt{x - 1}$ and $G(x) = 1/x$. Now in order for $G \circ F$ to be defined, the range of F must be contained in the domain of G. The domain of G is the set of all real numbers except zero, and unfortunately zero is in the range of F, since $F(1) = 0$. Therefore, as the functions are defined, $G \circ F$ would make no sense. We can get around this difficulty by restricting the domain of F to exclude the number 1. The function $G \circ F$ then will be defined so long as $x \neq 1$.

$$(G \circ F)(x) = G(F(x)) = G(\sqrt{x - 1}) = 1/\sqrt{x - 1} \qquad x \neq 1$$

To define $F \circ G$ the range of G must be contained in the domain of F. Since the domain of F is the set of all real numbers greater than or equal to one, we must restrict the domain of G so that only these numbers are in its range. The function $G(x) = 1/x$ will be greater than or equal to one when x is less than or equal to one. Thus, $(F \circ G)(x) = F(G(x)) = F(1/x) = \sqrt{(1/x) - 1}$ will be defined so long as $x \leq 1$.

Can you picture what will be the result if we take the composite of a function f with its inverse f^{-1}? Since $f(x) = y$ and $f^{-1}(y) = x$, then for every x in the domain of f, $(f^{-1} \circ f)(x) = f^{-1}(f(x)) = f^{-1}(y) = x$. Thus, the result of the composition of f with f^{-1} will be the identity function I for x in the domain of f. Similarly $f \circ f^{-1} = I$ for x in the domain of f^{-1}.

Example 2-11. If $f(x) = 2x$, then $f^{-1}(x) = (1/2)x$ and $(f^{-1} \circ f)(x) = f^{-1}(2x) = 2x/2 = x$. Similarly $(f \circ f^{-1})(x) = f(f^{-1}(x)) = f(x/2) = 2(x/2) = x$.

Exercise 2-9

1. If $f(x) = \sqrt{x}$ and $g(x) = 1/x$, give the formulas for $f + g$, $f - g$, $f \cdot g$, f/g, and g/f. Give the domain of each.

2. If $f(x) = x$, $g(x) = 3$, and $h(x) = -2x$, find the following:
   (a) $f \cdot g + h$ (b) $f \cdot f \cdot g + h \cdot g - f \cdot h$
   (c) $f \cdot h \cdot g$ (d) $g \cdot f \cdot f + h \cdot f + g$

3. If f and g are given by the tables below, find the tables for $f + g$, $f - g$, $f \cdot g$, and f/g. What is the domain of each?

f	
1	−1
2	1
3	−4
4	2

g	
0	1
1	0
2	−2
3	3

4. Find the graph of $f + g$ by tracing the graphs in Figure 2-26 and adding distances algebraically as illustrated in Figure 2-24.

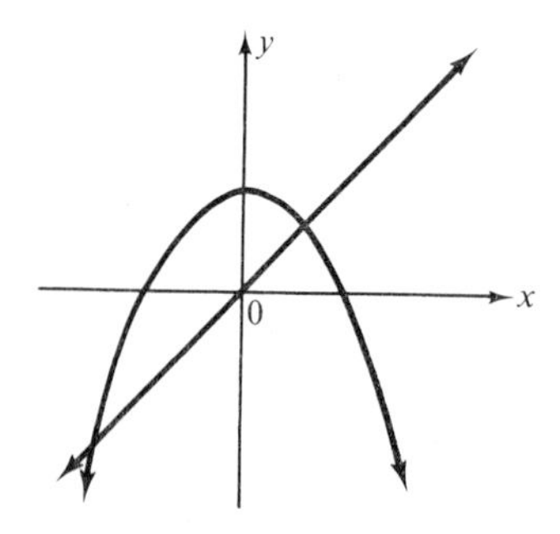

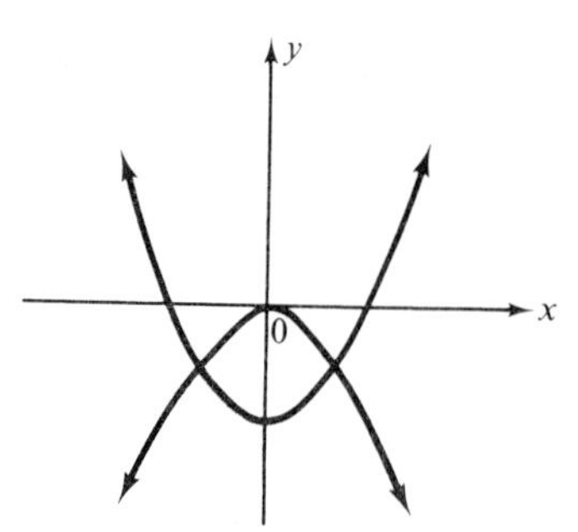

Figure 2-26

5. If $f(x) = 2x + 3$ and $g(x) = 1/x$, find $(g \circ f)(0)$, $(g \circ f)(-1)$, $(f \circ g)(2)$, $(f \circ g)(-1/3)$.

For each of the following pairs of functions, find a formula for $g \circ f$ and $f \circ g$. Give the domain of each.

6. $f(x) = x^2 + 1$; $g(x) = 3x - 1$
7. $f(x) = \sqrt{x+2}$; $g(x) = x - 1$
8. $f(x) = 2x$; $g(x) = \sqrt{x-1}$
9. $f(x) = 1/x$; $g(x) = 2x + 1$
10. $f(x) = 1/(x-1)$; $g(x) = x^2 + 1$
11. $f(x) = x^2 + 2x - 7$; $g(x) = 3x - 1$
12. $f(x) = 2x - 1$; $g(x) = \sqrt{x-1}$
13. $f(x) = 1/(x^2 + 1)$; $g(x) = x^3$
14. $f(x) = 1/(\sqrt{x})$; $g(x) = x + 1$
15. $f(x) = (x+1)/(x-1)$; $g(x) = \sqrt{x}$
16. $f(x) = 2$; $g(x) = x^2$
17. $f(x) = 1/(x+1)$; $g(x) = 1/(x+1)$
18. $f(x) = 1/(x+1)$; $g(x) = x + 1$
19. $f(x) = 1/(2x)$; $g(x) = 2x$
20. $f(x) = 1/(\sqrt{x})$; $g(x) = \sqrt{x}$
21. $f(x) = 1/(2x)$; $g(x) = 1/(2x)$
22. Show that the composite of a constant function $f(x) = k$ and any function g whose domain contains the constant k is always a constant function.
23. Recall that I is the symbol for the identity function, $I(x) = x$. If $f(x) = x^2 + x - 1$, find $I \circ f$, $f \circ I$. If g is any function of x whatsoever will $I \circ g$ and $g \circ I$ always be defined? If these composite functions are defined, what will they be?
24. If $f(x) = 2x - 1$, find $f^{-1}(x)$. Verify that $(f \circ f^{-1})(x) = x$ and that $(f^{-1} \circ f)(x) = x$ for all x.
25. Let $f(x) = ax + b$; $a \neq 0$. Find f^{-1} and verify that $(f \circ f^{-1})(x) = (f^{-1} \circ f)(x) = x$ for all x.
26. Can you find a function f (other than the identity function) for which $(f \circ f)(x) = x$?
27. Let f be any function and let us define the two functions

$$f^{+} = \frac{(f + |f|)}{2}$$

and

$$f^- = \frac{(f - |f|)}{2}$$

If $f(x) = 2x + 1$, graph f^+ and f^-. If $f(x) = x^2$, graph f^+ and f^-. What is $f^+ + f^-$?

Review Exercise

1. Which of the following are functions?

   (a) $y = 2x - 3$
   (b) $y = \begin{cases} 1 & \text{if } x \geq 0 \\ -1 & \text{if } x \leq 0 \end{cases}$
   (c) $x^2 + y^2 = 25$
   (d) $s = r^2 + 1$
   (e) $x < y$

2. Give the domain of each of the following functions:
   (a) $y = 1/(x^2 - 1)$
   (b) $y = \sqrt{x - 2}$
   (c) $y = (x + 1)/x$
   (d) $y = \sqrt{x} + (1/x)$

3. Give the domain and range of each of the following. Which describe a function? Give the inverse of those which are one-to-one.

   (a)

1	1
2	3

   (b)

1	1
0	1

   (c)

1	1
1	2

   (d)

0	4
4	0

4. Graph
   $$y = \begin{cases} 2x + 1 & \text{if } x \geq 0 \\ 1 & \text{if } x < 0 \end{cases}$$
   Give the domain and range of this function.

5. A rectangle has length twice its width. State a function whose independent variable is w (width) and whose dependent variable is A (area).

6. A wheel 2 ft in diameter revolves at a rate of 5 revolutions per second. State a function whose independent variable is t (time) and whose dependent variable is d (distance traveled).

7. Give the domain and range of (a) through (d) in Figure 2-27. Which of these describe functions?

*8. Change the following to implicit form or to the form $y = f(x)$ by eliminating the parameter:

   (a) $x = t^2$
   $y = 2 + 3t$

   (b) $x = t^3$
   $y = 1/t$

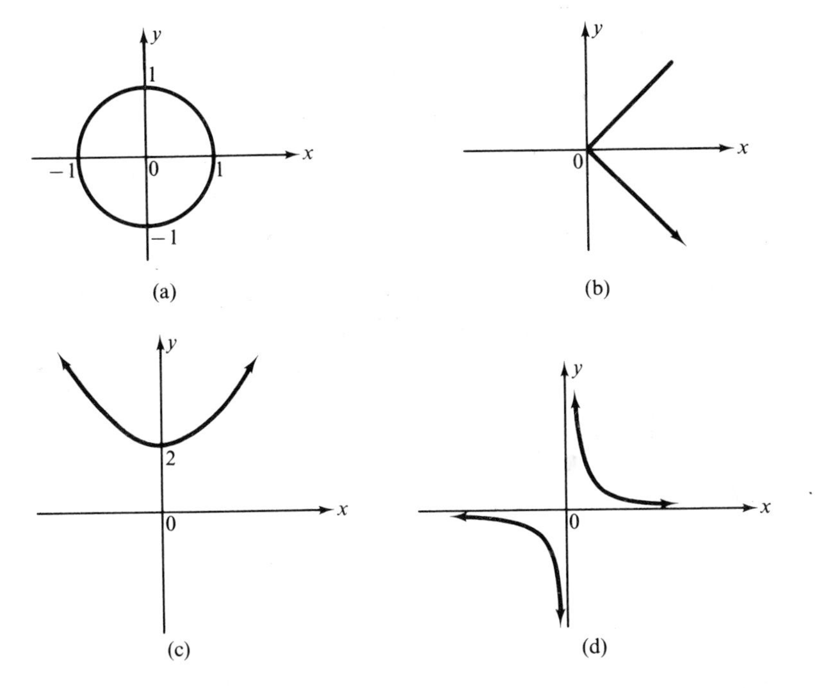

Figure 2-27

9. Determine whether the graph of each of the following is symmetric with respect to the x-axis, the y-axis, the origin.
   (a) $y = 3x^2$ (b) $x^2y^3 = 1$
   (c) $x^5 + x^3 = y$ (d) $x + y = 1$

10. If $f(x) = \sqrt{x}$ and $g(x) = 1/x$, find a formula for $f + g, f - g, f \cdot g, f/g$, and g/f. Give the domain of each.

11. If f and g are given by the tables below, find the tables for $f + g$, $f - g$, $f \cdot g$, and f/g. What is the domain of each?

f	
1	−1
2	1
3	−4
4	2

g	
0	1
1	0
2	−2
3	3

12. If $f(x) = \sqrt{x - 1}$, and $g(x) = 2x - 1$, find a formula for $g \circ f$, and for $f \circ g$. Are these two functions the same? What is the domain of each?

13. Are the two functions $f(x) = x$ and $g(x) = [x(x - 2)]/(x - 2)$ equal? Explain.

14. If $f(x) = 3x + 2$, find $f^{-1}(x)$. Verify that $(f \circ f^{-1})(x) = x$ and that $(f^{-1} \circ f)(x) = x$ for all x.

15. Graph $y = x^3 + x$ and its inverse on the same set of axes.

Bibliography

Courant, Richard, and Hubert Robbins. *What is Mathematics?*, pp. 272–286. London, New York, Toronto: Oxford University Press, 1941.

Eves, Howard, *An Introduction to the History of Mathematics,* pp. 371–372. New York: Holt, Rinehart and Winston, 1964.

Weyl, Hermann, "Symmetry," in *The World of Mathematics,* ed. James R. Newman, pp. 671–724. New York: Simon and Schuster, 1956.

3

Polynomial Functions

3-1 Linear Functions

Definition 3-1. A linear function is a function which can be represented by a formula of the form

$$y = ax + b$$

where a and b are real number constants.

The number a is called the coefficient of x and b is called the constant term. An example is the function $y = 2x - 1$. Here $a = 2$ and $b = -1$. If $a = 0$, then the function is also called a *constant* function. An example is the function $y = 4$. These functions are called linear because their graph is a straight line. (See Figure 3-1.)

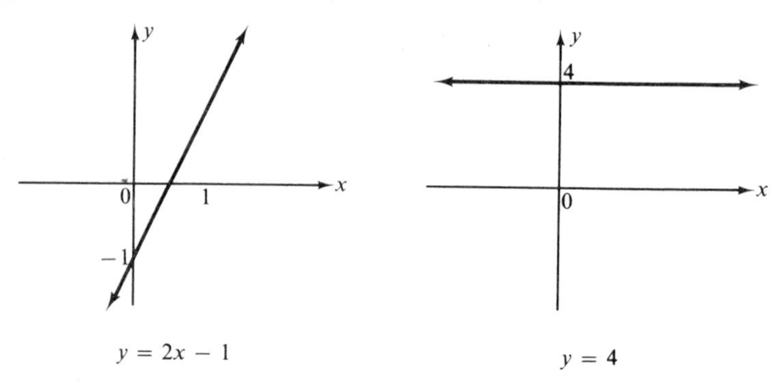

Figure 3-1

The domain of any linear function is the set of all real numbers, while the range is either the set of all real numbers if $a \neq 0$ or a single real number in the case of a constant function. If a is not zero, then the linear function is one-to-one and will have an inverse. A formula for this inverse can easily be found by solving for x.

$$x = \frac{(y - b)}{a} \qquad a \neq 0$$

The value of x for which $ax + b = 0$ is called the *root* or a *solution* to this equation. If $a \neq 0$, it is easy to see that this value is $-b/a$. On the other hand if $a = 0$, then either there is no root (if $b \neq 0$) or *any* value of x will be a root (when $b = 0$). Geometrically we can interpret the root of $ax + b = 0$ as the value of x at the point where the graph of the function $y = ax + b$ crosses the x-axis. If $a \neq 0$, there will be exactly one such point, while if $a = 0$ the graph will either be parallel to the x-axis, in which case there is no point of intersection, or it will coincide with the x-axis ($y = 0$), and every point will be a point of intersection.

3-2 Quadratic Functions

Definition 3-2. A *quadratic function* is a function which can be represented by a formula of the form

$$y = ax^2 + bx + c$$

where a, b, and c are real number constants and a is not zero.

The numbers a and b are called the coefficients of x^2 and x respectively and c is called the constant term. Examples of quadratic functions are $y = x^2$ and $y = -2x^2 + 4x - 1$. In the first example, $a = 1$ and b and c are both zero. In the second, $a = -2$, $b = 4$, and $c = -1$.

The graph of any quadratic function is a *parabola*. (See Figure 3-2.) If a is positive, the parabola will open upward; if a is negative, it will open downward.

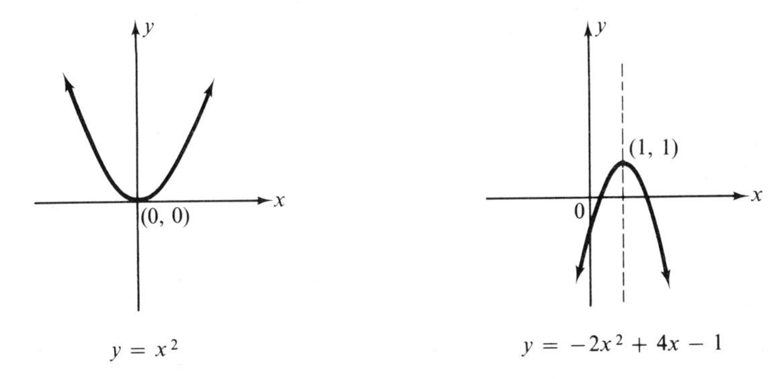

Figure 3-2

If a is positive, the graph will have a lowest point, and if a is negative, the graph will have a highest point. This point is called the *vertex* of the parabola, and its x-coordinate is given by the formula $x = -b/(2a)$.

We can see why this is so by rewriting $y = ax^2 + bx + c$ in the form $y = a[x^2 + (b/a)x + (c/a)]$. If we add and subtract $(b/2a)^2$ in the parentheses, we have

$$y = a\left[x + \frac{b}{a}x + \left(\frac{b}{2a}\right)^2 + \frac{c}{a} - \left(\frac{b}{2a}\right)^2\right]$$
$$= a\left[\left(x + \frac{b}{2a}\right)^2 + \left(\frac{c}{a} - \frac{b^2}{4a^2}\right)\right]$$

Since the first parentheses is squared, it is never negative; therefore, y will be smallest (if a is positive) when $[x + (b/2a)]^2 = 0$, i.e., when $x = -b/(2a)$. If a is negative, then y will be largest when $x = -b/(2a)$.

The vertex of $y = x^2$ for example has coordinates $x = -0/(2 \cdot 1) = 0$; $y = 0$. The vertex of $y = -2x^2 + 4x - 1$ is $x = -4/[2(-2)] = 1$; $y = -2(1)^2 + 4(1) - 1 = 1$.

The graph of any quadratic function is symmetric with respect to a vertical line through the vertex. This means that if we fold the paper (on which the graph is drawn) along this line of symmetry, the two sides of the parabola will coincide. This makes it very simple to draw the graph of a quadratic function. All we have to do is find the vertex and a few points on the graph on one side of this line of symmetry.

Connect these points with a smooth curve, then use its symmetry to sketch the other side.

The domain of any quadratic function is the set of all real numbers. The range is the set of all real numbers greater than or equal to the y value at the vertex if the parabola opens upward, or the set of all real numbers less than or equal to the y value at the vertex if the parabola opens downward. The range of the function $y = x^2$ is $\{y \mid y \geq 0\}$, while the range of $y = -2x^2 + 4x - 1$ is $\{y \mid y \leq 1\}$.

Quadratic functions are of some importance in physics. The position s of a falling object at any time t (if we neglect air resistance) is given by the formula

$$s = -16t^2 + v_0t + s_0$$

where v_0 and s_0 are constants standing for the velocity and position of the object at time $t = 0$. If $y = 0$, then the resulting equation is called a quadratic equation.

The *roots* of a quadratic equation are the values of x for which $ax^2 + bx + c = 0$. If $ax^2 + bx + c$ can be factored, then the roots are easy to find.

Example 3-1. Find the roots of $x^2 + 5x - 6 = 0$.

Solution: Factoring, we have $(x + 6)(x - 1) = 0$. If this product is to be zero, then one of the factors must be zero, that is, either $x + 6 = 0$ or $x - 1 = 0$. Thus, the roots are -6 and 1.

If the quadratic cannot be factored, then we can use a formula to find the roots. To find a formula for these roots, first subtract c from both sides of the equation, then multiply both sides by $4a$.

$$ax^2 + bx + c = 0$$

$$ax^2 + bx = -c$$

$$4a^2x^2 + 4abx = -4ac$$

Adding b^2 to both sides of the equation will make the left-hand side a perfect square.

$$4a^2x^2 + 4abx + b^2 = b^2 - 4ac$$
$$(2ax + b)^2 = b^2 - 4ac$$

Taking the square root of both sides, then solving for x, we have

$$2ax + b = \pm \sqrt{b^2 - 4ac}$$

$$x = \frac{-b \pm \sqrt{b^2 - 4ac}}{2a}$$

This is called the *quadratic formula.*

Example 3-2. Find the roots of $-2x^2 + 4x - 1 = 0$.

Solution: By the quadratic formula $x = (-4 \pm \sqrt{16 - 4(-2)(-1)})/2(-2) = (-4 \pm \sqrt{8})/-4 = (-4 \pm 2\sqrt{2})/-4 = (-2 \pm \sqrt{2})/-2$.

There is evidence that by 2,000 B.C. the Babylonians were able to solve quadratic equations, however, all the problems they solved had numerical coefficients. The idea of using letters so that a general solution might be found came much later. This idea is due to Francois Vieta (1540–1603), a French lawyer whose hobby was mathematics. The Greeks used geometric methods to solve quadratic equations. Propositions 28 and 29 of Euclid's *Elements* are construction problems which turn out to be methods of constructing line segments whose lengths are the solutions to the quadratic equations $x^2 - px \pm q^2 = 0$. (For a geometric method of solving the quadratic, see Exercise 3-2, problem 19.)

The number $b^2 - 4ac$ is called the *discriminant* of the quadratic equation. The sign of this number will determine the nature of the roots of the equation. If $b^2 - 4ac$ is positive, then the equation will have two different real roots, $x = (-b + \sqrt{b^2 - 4ac})/(2a)$ and $x = (-b - \sqrt{b^2 - 4ac})/(2a)$. If $b^2 - 4ac = 0$, then $x = -b/(2a)$, and $ax^2 + bx + c$ will be a perfect square. Since in this case the equation becomes $[x + (b/2a)]^2 = 0$, with the factor $[x + (b/2a)]$ appearing twice, $-b/(2a)$ is called a *double root* or a repeated root. If we agree to count a double root twice then we can make the statement that every quadratic equation has exactly two roots.

If $b^2 - 4ac$ is negative, then the formula contains the square root of a negative number. But no real number has a square which is negative. If we set $i = \sqrt{-1}$, then the roots will be of the form $p \pm qi$, where p and q are real numbers. Such numbers are called *complex* numbers. In this case we have two complex roots, $p + qi$ and $p - qi$. Pairs of complex numbers like these are called *complex conjugates.*

In summary,

if $b^2 - 4ac > 0$, the two roots are real and unequal.
if $b^2 - 4ac = 0$, the two roots are real and equal.
if $b^2 - 4ac < 0$, the two roots are complex.

Geometrically, these three types of roots correspond to the cases in which the graph of the function $y = ax^2 + bx + c$ cuts the x-axis in two different points, touches the x-axis at exactly one point (the vertex), or does not cross the x-axis at all. (See Figure 3-3.)

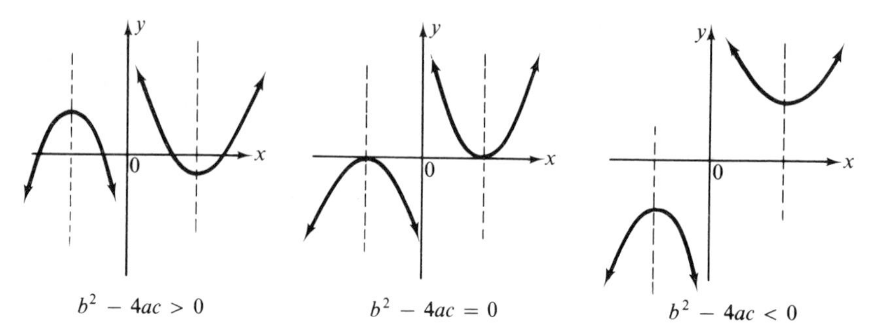

Figure 3-3

Example 3-3. In the equation $x^2 - 2x - 3 = 0$, $b^2 - 4ac = (-2)^2 - 4(1)(-3) = 16$ and the equation has two different, real roots. From the formula we find $x = -1$ and $x = 3$ to be roots. These roots can also be found by factoring the left-hand side, since $x^2 - 2x - 3 = (x + 1)(x - 3)$.

Example 3-4. If $x^2 + 2x + 3 = 0$, then $b^2 - 4ac = (2)^2 - 4(1)(3) = -8$. Since the discriminant is negative, the roots must be complex numbers. By the quadratic formula, we get $x = (-2 \pm \sqrt{-8})/2$. Writing $\sqrt{-8} = \sqrt{-1} \cdot \sqrt{8} = i\sqrt{8} = i2\sqrt{2}$, we have $x = (-2 \pm i(2\sqrt{2}))/2$, or $x = -1 \pm i\sqrt{2}$.

Example 3-5. When $x^2 + 4x + 4 = 0$, $b^2 - 4ac = (4)^2 - 4(1)(4) = 0$, and the equation has a double root. Factoring the left-hand side (or using the quadratic formula) we find that this root is -2.

Exercise 3-2

1. Every linear function can be represented by the graph of a line. Is the converse of this statement true? That is, does every line represent a linear function? Explain.

2. Choose two linear functions f and g and use them to illustrate each of the following statements:
   (a) If f and g are linear functions, then $f + g$ is a linear function.
   (b) If f and g are linear functions, then $f - g$ is a linear function.
   (c) If f and g are linear functions, then $g \circ f$ and $f \circ g$ are both linear functions.

3. Sketch on one set of axes the functions $f(x) = mx + 1$; $m = -3, -2, -1, 0, 1, 2$. The coefficient of x in this linear function is called the *slope* of the line. Why do you think it is called this?

4. Sketch on one set of axes the functions $f(x) = 2x + b$; $b = -2, -1, 0, 1, 2, 3$. The constant term b in this linear function is called the *y-intercept* of the line. Why do you think it is called this? All of these functions have the same slope. What do you conclude about the graphs of two linear functions that have the same slope?

5. Sketch $y = x$ and $y = -x$ on the same set of axes. Graph $y = 2x$ and $y = -(1/2)x$ on the same set of axes. What conclusion can you draw about the graphs of the functions $y = mx$ and $y = -(1/m)x$, $m \neq 0$?

6. Which of the following are linear functions? Which are quadratic functions? Which are neither?

   (a) $y = 3$ (b) $y = (x - 1)(2x + 3)$
   (c) $2x + 3y = 1$ (d) $xy = 2$
   (e) $f(x) = 2x^2 + 7$ (f) $y = x(x + 1)/x$

7. Find the vertex of each of the following and tell whether the graph opens up or down. Graph.

   (a) $y = x^2 - 2x + 1$ (b) $y = -x^2 + 3x + 7$
   (c) $y = -2x^2 + 5$ (d) $y = x^2 - 2$

8. Give the range of each of the quadratic functions in problem 7.

9. Can a quadratic function ever have an inverse? Explain.

10. Find the discriminant, $b^2 - 4ac$, of each of the following and use this to classify the roots as real and unequal, real and equal, or complex:

    (a) $2x^2 - 3x + 7 = 0$ (b) $x^2 + 5x + 1 = 0$
    (c) $-3x^2 + x - 1 = 0$ (d) $x^2 + 4x + 4 = 0$
    (e) $2x^2 - 3 = 0$ (f) $x^2 + 7x = 0$

11. Find the roots of the equations in problem 10.

12. Find all values for the constant k so that the roots of $x^2 - 2x + k = 0$ are real and unequal; real and equal; complex.

13. Find all values for the constant k so that the roots of $x^2 + kx + 1 = 0$ are real and unequal; real and equal; complex.

14. Find all values for the constant k so that the roots of $kx^2 + 3x - 2 = 0$ are real and unequal; real and equal; complex.

15. Graph each of the following and use the graph to estimate the roots. Check by using the quadratic formula.

    (a) $f(x) = x^2 - 5x + 6$ (b) $f(x) = 2x^2 - 3$
    (c) $f(x) = x^2 - 1$ (d) $f(x) = x^2 - 2x + 1$

16. Choose quadratic functions f and g to illustrate each of the following statements:
    (a) If f and g are quadratic functions then $f+g$ may or may not be a quadratic function.
    (b) If f and g are quadratic functions then neither $f \circ g$ nor $g \circ f$ will be a quadratic function.

17. If f is a linear function and g a quadratic function, give an example to show that $f \circ g$ and $g \circ f$ are both quadratic functions. Give an example to show that $f+g$ is a quadratic function.

18. A Babylonian tablet of about 1800 B.C. gives the following problem: "An area A, consisting of the sum of two squares, is 1,000. The side of one square is 10 less than 2/3 of the side of the other square. What are the sides of the squares?" Solve this problem.

19. The following geometric method of solving a quadratic equation is called Carlyle's method. Given the quadratic equation $ax^2+bx+c=0$, on the Cartesian coordinate plane plot the points $A(0, 1)$ and $B(-b/a, c/a)$. Draw the circle whose diameter is AB. If this circle cuts the x-axis in the points P and Q, then the directed distances OP and OQ (where O represents the origin) will represent the real roots of the quadratic equation. Use this method to find the roots of $x^2+x-2=0$; of $x^2+6x+9=0$. Check your answers by the quadratic formula. In what position would you expect to find the circle if the roots of the quadratic equation were complex?

3-3 The General Polynomial Function

Definition 3-3. A *polynomial function* is a function which can be represented by a formula of the form

$$y = a_n x^n + a_{n-1}x^{n-1} + \ldots + a_2 x^2 + a_1 x + a_0$$

where $a_0, a_1, a_2, \ldots, a_n$ are real number constants and n is a whole number, that is, a number from the set $\{0, 1, 2, 3, \ldots\}$.

The numbers $a_1, a_2, \ldots, a_n$ are called the coefficients of $x, x^2, \ldots, x^n$ respectively, and a_0 is called the constant term. Before going any further, let us explain the symbols we are using here. In describing the general quadratic function we used the letters a, b, and c for coefficients. Now we are speaking of a function which may have a great many terms (there will be a finite number, however). If it had more than 26 terms, we would run out of letters to use for coefficients. Even if we agreed to use capital and lower case letters both, we would not be able to describe a polynomial function with more than 52 terms. What we need is an inexhaustible supply of symbols to use for coefficients. A very common inexhaustible set is the set of counting numbers, 1, 2, 3, ... We will not use these numbers as the coefficients themselves, but we will use them to index the coefficients. We pick some letter, say a, then tag it with a numerical subscript. Thus, a_1 (read "a sub one") and a_2 stand for two different coefficients. So that the subscript and the exponent on x will be the same, we start indexing with zero and let a_0 stand for the constant term.

Thus, for example in the polynomial function $y = 3x^4 + 7x^2 + x - 1$, $a_4 = 3$; $a_3 = 0$ (since the x^3 term is missing); $a_2 = 7$; $a_1 = 1$; and $a_0 = -1$.

The *degree* of a polynomial function is the largest exponent n for which $a_n \neq 0$. The degree of a quadratic function is two. The degree of the linear function $y = ax + b$ is one if $a \neq 0$, and the degree of the constant function $y = b$ is zero provided $b \neq 0$. The function $y = 0$ is not assigned a degree.

A polynomial function of degree three is called a *cubic*. Some examples of cubic functions are $y = x^3$ and $y = x^3 - 2x^2 - x + 2$. The graphs of these two functions are typical and are shown in Figure 3-4.

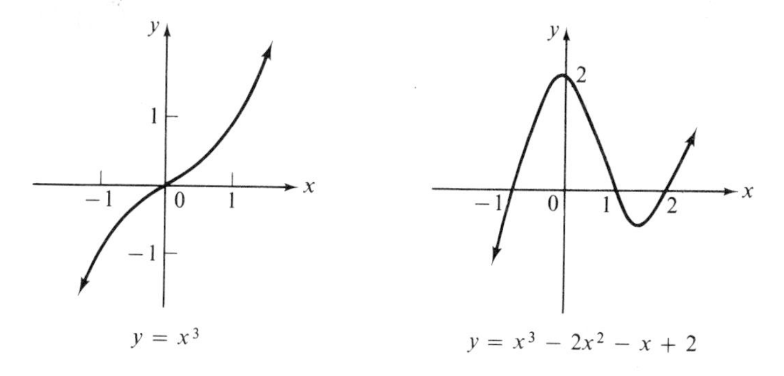

Figure 3-4

For large values of x the value of the cubic function is dominated by the x^3 term. By this we mean that for large x the x^3 term is so much larger than the other terms that the addition or subtraction of these terms makes very little difference in the value of y. (Compute $x^3 - 2x^2 - x + 2$ for $x = 10$, for instance.) Thus, for large positive values of x the graph of this function rises, while for negative values whose absolute value is large it falls, since the cube of a negative number is negative. If the sign of the coefficient of the x^3 term is negative, the situation is reversed. The curve drops as we move to the right and rises as we move to the left. From these facts it is easy to see that the graph must cross the x-axis at least once, and it may cross it three times (Figure 3-4). Clearly the graph cannot cross the x-axis exactly twice. Since points where the graph crosses the x-axis correspond to roots of the cubic equation, this means that a cubic can have one or three real roots, but it cannot have exactly two real roots.

A polynomial function of the fourth degree is called a *quartic*. The graphs of some typical quartic functions are given in Figure 3-5. For large values of x the x^4 term dominates, and since the exponent is an even number, x^4 will always be nonnegative. Thus, if the coefficient of x^4 is positive, the curve will rise for either positive or negative x as x gets larger in absolute value. If the coefficient of x^4 is negative, the opposite is true and the curve falls.

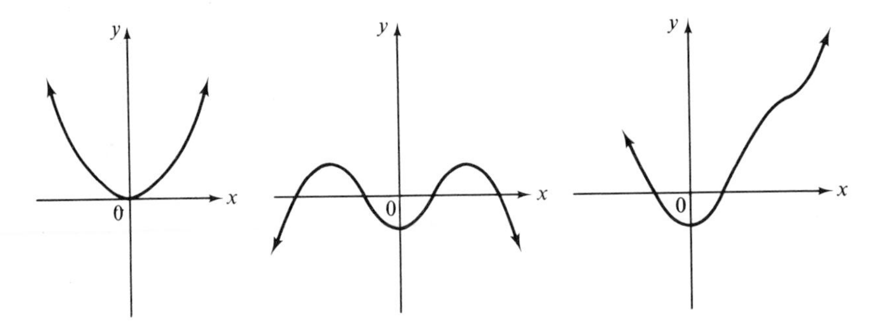

Figure 3-5

Graphs of typical quartic functions

Since this is the case, the curve can cross the x-axis as many as four times, two times, or not at all (Figure 3-5). It would be impossible for the curve to cross the x-axis an odd number of times. (The graph of $y = x^4$ does not *cross* the x-axis. It is tangent to it at $x = 0$, indicating a repeated root. See Figure 3-5(a).) The algebraic interpretation of these geometric facts is that a quartic equation must always have an even number of real roots. (Provided that we count multiple roots as many times as they occur.)

There are formulas for finding the roots of a cubic and a quartic equation; however, these are quite complicated and we will not study them here. The interested reader is referred to Dickson's *New First Course in the Theory of Equations*, Chapter V. In the sections which follow, we will solve polynomial equations of higher degree by other methods.

The solution of the cubic has an interesting history. The most common method of solution is called *Cardan's* method. Cardan was a unique combination of mathematician and rogue. Born in 1501 in Italy he was a physician, astrologer, gambler, liar, and cheat. He wrote a famous book *On Games of Chance* in which he discussed the probabilities in gambling and even gave tips on how to cheat.

Another famous mathematician of this time was Nicolo of Brescia, better known as Tartaglia (the stammerer) because of his cleft palate. Tartaglia discovered a method of solving cubic equations which he confided to Cardan but asked him to keep secret. At this time it was common for a mathematician who had made some discovery to keep it secret and challenge other mathematicians to problem solving contests in which this discovery gave him an advantage. Cardan however was writing a book on algebra, called *Ars Magna,* and breaking his promise he published Tartaglia's method of solution of the cubic. Even though Cardan gave credit to Tartaglia in his book, this led to a bitter and undignified dispute which, unfortunately, is not unique in the history of mathematics. Cardan's book was published in 1545 and it also contained a method of solving quartic equations which had been discovered by Ferrari, one of Cardan's pupils.

After these discoveries, mathematicians turned their attention to equations of the fifth, sixth, and higher degrees. If a formula could be found for the roots of first, second, third, and fourth degree equations, they reasoned, surely they could do the same for equations of higher degree. Three hundred years of work by many mathematicians brought no success. Then finally in 1832 a French mathematician Evariste Galois showed that it is impossible to find a general algebraic formula for the roots of equations of degree higher than four. The roots of such equations cannot be found by applying the operations of addition, subtraction, multiplication, division, and extracting of roots to the coefficients a finite number of times.

Galois' life was brief and tragic. Killed in a duel at the age of twenty, he spent the night before feverishly committing to paper some of the mathematical discoveries which filled his mind. In one of the most tragic scenes in the history of mathematics, we see this unfortunate young man, forseeing his death, frantically writing against time, breaking off from time to time to scribble in the margin, "I have not time."

Exercise 3-3

1. Give the degree of each of the following polynomial functions:
   (a) $f(x) = 2x^3 + 3x^2 - 4$ (b) $f(x) = x^8 - x^5 - x^3 + 1$
   (c) $f(x) = 2x^2 - x^4 + x + 7$ (d) $f(x) = 5x^3 - 8x + 3x^5 + 25$

(e) $f(x) = x + 2$ (f) $f(x) = 3x$
(g) $f(x) = 10$ (h) $f(x) = 0$

2. What is the minimum number of real roots a fifth degree polynomial may have? Can a fifth degree polynomial have exactly two real roots? Exactly three? Explain.

3. What is the minimum number of real roots a sixth degree polynomial may have? Can a sixth degree polynomial have exactly one real root? Exactly two? Explain.

4. What is the domain of the general polynomial function? Describe the range of the general polynomial function, classifying these functions as being of even or odd degree.

5. Any polynomial function can be generated by combining the identity function $I(x) = x$ and the constant functions $f(x) = k$ by the operations of addition and multiplication. For example, if $f(x) = x^2 + 2$, $f = I \cdot I + 2$. Write each of the following functions as combinations of the identity function and constant functions:
(a) $f(x) = -3x^2 + 7x - 1$ (b) $f(x) = x^3 + 2x^2 - x + 2$
(c) $f(x) = x^5 + x^3 - x$ (d) $f(x) = x^4 - 1$

6. The graph of $y = x^4$ is similar to that of $y = x^2$ only flatter near the origin and steeper to the right of $x = 1$ and to the left of $x = -1$. In the graphs of $y = x^{2n}$, $n = 3, 4, \ldots$ these features become more and more pronounced. As n gets larger the flatness and steepness becomes more and more accentuated until the curve looks very much like the heavy line in Figure 3-6. Graph $y = x^6$ choosing $x = 0, \pm.1, \pm.5, \pm.8, \pm1, \pm1.2, \pm1.4$.

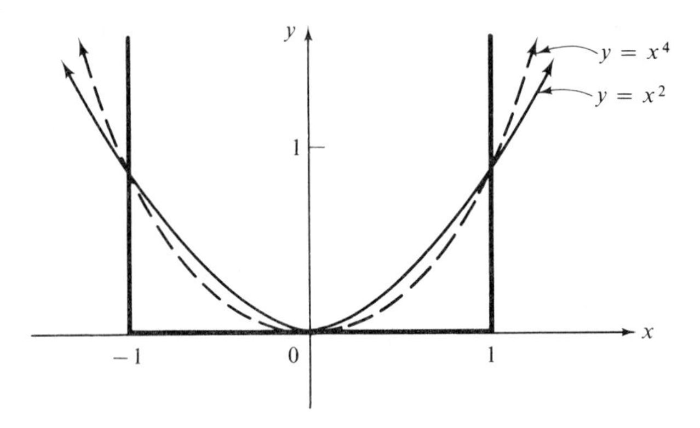

Figure 3-6

7. The graphs of $y = x^{2n-1}$ for $n = 1$ and $n = 2$ are given in Figure 3-7. As n gets larger and larger the curve will become closer and closer to the heavy

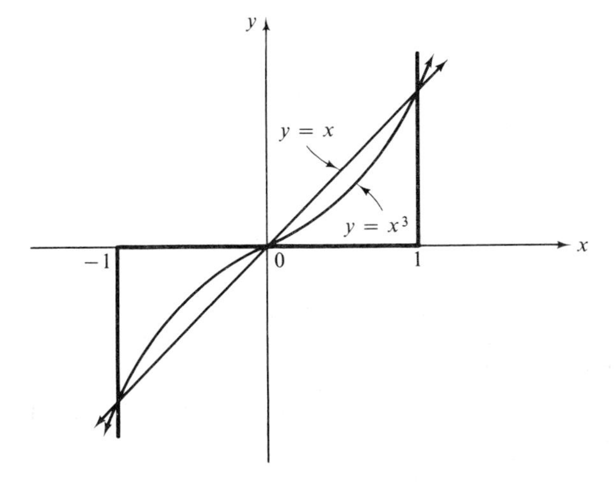

Figure 3-7

lines in the figure. Graph $y = x^5$, choosing $x = 0, \pm.1, \pm.5, \pm.8, \pm1, \pm1.2, \pm1.4$.

8. The discriminant of the cubic equation, $x^3 + ax^2 + bx + c = 0$ is given by the formula $\Delta = a^2b^2 + 18abc - 4b^3 - 4a^3c - 27c^2$. If $\Delta > 0$, then the equation has three real, distinct roots; if $\Delta < 0$, it has one real and two complex roots; if $\Delta = 0$, it has three real roots and at least two of them are equal. Classify the roots of the following equations:
   (a) $x^3 + 1 = 0$ (b) $x^3 + x^2 - 2x - 2 = 0$
   (c) $x^3 - 3x^2 + 3x - 1 = 0$ (d) $x^3 - x^2 - x - 1 = 0$

9. Let f and g be polynomial functions, and let $\deg(f)$ and $\deg(g)$ stand for the degree of f and the degree of g respectively. Tell whether each of the following is true or false, and give an example to illustrate the truth or falsity of each. (Assume that all of the functions named have a degree, that is, none is the function $y = 0$.)
   (a) $\deg(f + g) = \deg(f) + \deg(g)$
   (b) $\deg(f + g) = \max(\deg(f), \deg(g))$*
   (c) $\deg(f + g) \leq \max(\deg(f), \deg(g))$
   (d) $\deg(f \cdot g) = \deg(f) \cdot \deg(g)$
   (e) $\deg(f \cdot g) = \deg(f) + \deg(g)$
   (f) $\deg(f \circ g) = \deg(g \circ f)$
   (g) $\deg(g \circ f) = \deg(f) \cdot \deg(g)$
   (h) $\deg(g \circ f) = \deg(f) + \deg(g)$

---

* If a and b are two numbers, then $\max(a, b)$ is the larger of the two if they are different, or their common value if they are equal. Thus, $\max(2, 4) = 4$; $\max(2, 2) = 2$.

3-4 The Remainder and Factor Theorems

Since we cannot hope to find formulas for the roots of polynomial equations of higher degree, we must turn to other methods of finding the roots. These methods are a part of a branch of mathematics called Theory of Equations.

Before we proceed, let us briefly review the notation we will be using. A *polynomial equation* is an equation of the form

$$a_nx^n + a_{n-1}x^{n-1} + \ldots + a_2x^2 + a_1x + a_0 = 0$$

If we use the symbol $P(x)$ for the expression on the left-hand side of this equation, then we can write this equation $P(x) = 0$. The expression $P(x)$ itself is called a *polynomial.* The *polynomial function* would be written $y = P(x)$. A number a is called a *root* of this equation if it satisfies the equation. That is, if we substitute a for x in the left-hand side and perform the indicated operations, the result will be zero. Symbolically, we can write this $P(a) = 0$.

By way of introduction to our first theorem, the Remainder Theorem, let us recall a few facts from elementary arithmetic. If a number P is divided by a divisor D, we get a quotient Q and a remainder R. To check our work we take the product of the divisor D and the quotient Q, add the remainder R, and, if we did the work correctly, we get the number P. In symbols, we have the equation $P = D \cdot Q + R$. This relationship is called the *division algorithm.*

The same relationship holds for polynomials. If $P(x)$ is a polynomial and we divide it by the polynomial $x - a$, the quotient $Q(x)$ will be a polynomial of degree one less than $P(x)$, and the remainder R will be a constant.

$$P(x) = (x - a)Q(x) + R$$

We will use this to prove our first theorem.

Theorem 3-1. *The Remainder Theorem.* If a polynomial $P(x)$ is divided by $x - a$ until a constant remainder R is obtained, then $R = P(a)$.

Proof: By the division algorithm

$$P(x) = (x - a)Q(x) + R$$

This relation is an identity in x, that is, it is true for any value assigned to x, and in particular it is true for $x = a$. Substituting a for x, we have

$$P(a) = (a - a)Q(a) + R$$

or

$$P(a) = R \blacksquare$$

Theorem 3-1 says that the number we get by substituting a for x in the polynomial $P(x)$ is exactly the same number as the remainder obtained when $P(x)$ is divided by $x - a$.

Example 3-6. Let $P(x) = x^4 - 3x^3 + 2x - 7$, and $a = 1$. Then $P(1) = 1^4 - 3 \cdot 1^3 + 2 \cdot 1 - 7 = -7$. If we divide $P(x)$ by $x - 1$, the remainder is also -7.

$$\begin{array}{r|l}
 & x^3 - 2x^2 - 2x \\
x - 1 & x^4 - 3x^3 \qquad\qquad + 2x - 7 \\
 & \underline{x^4 - x^3} \\
 & \quad - 2x^3 \\
 & \quad \underline{- 2x^3 + 2x^2} \\
 & \qquad\qquad - 2x^2 + 2x \\
 & \qquad\qquad \underline{- 2x^2 + 2x} \\
 & \qquad\qquad\qquad\qquad - 7
\end{array}$$

The following theorem follows almost immediately from the Remainder Theorem:

Theorem 3-2. *The Factor Theorem.* A number a is a root of a polynomial equation $P(x) = 0$ if and only if $x - a$ is a factor of $P(x)$.

Comment: An "if and only if" theorem is really two theorems in one. This statement implies two things: if a is a root, then $x - a$ is a factor, and conversely, if $x - a$ is a factor, then a is a root.

Proof: Part 1. If a is a root of $P(x) = 0$, then $P(a) = 0$ and by the Remainder Theorem, $R = 0$. Thus, the division algorithm becomes

$$P(x) = (x - a)Q(x)$$

which is to say $x - a$ is a factor of $P(x)$.

Part 2. If $x - a$ is a factor of $P(x)$, then $P(x)$ can be written in factored form

$$P(x) = (x - a)Q(x)$$

Substituting a for x, we have $P(a) = (a - a)Q(a) = 0$. Since $P(a) = 0$, a is a root of $P(x) = 0$. $\blacksquare$

By the Factor Theorem, if 2 is a root of $P(x) = 0$, then $x - 2$ is a factor of $P(x)$; if zero is a root, then $x - 0$, or x, is a factor; and if -1 is a root, then $x - (-1) = x + 1$ is a factor.

The Factor Theorem can be very useful in finding all of the roots of an equation. If by some means we can find one root a, then $x - a$ is a factor of $P(x)$ and we can find the other factor $Q(x)$ by dividing $x - a$ into $P(x)$. The other roots of the equation must be roots of the factor $Q(x)$. Since the degree of $Q(x)$ is one less than the degree of $P(x)$, it will be easier to work with than $P(x)$. In particular, if $P(x)$ is a cubic, then $Q(x)$ is a quadratic, and we can find the roots of any quadratic by using the quadratic formula.

Example 3-7. Find all of the roots of $x^3 - 2x^2 + 3x - 2 = 0$.

Solution: We notice that $x = 1$ makes the left-hand side zero; hence, $x - 1$ must be a factor. By division we find that

$$x^3 - 2x^2 + 3x - 2 = (x - 1)(x^2 - x + 2) = 0$$

The roots of $x^2 - x + 2 = 0$ (called the *depressed equation*) by the quadratic formula are $x = (1 \pm i\sqrt{7})/2$; hence, the three roots of this cubic equation are 1, $(1 + i\sqrt{7})/2$, and $(1 - i\sqrt{7})/2$.

By the Factor Theorem if an equation is written as a product of linear factors, then we can simply read off the roots.

Example 3-8. The equation $2x(x + 1)(x - 3)(x + 1/2)^2 = 0$ has the root 0, from the factor x; the root -1, from the factor $x + 1$; the root 3, from the factor $x - 3$; and $-1/2$ is a double root, since the factor $x + (1/2)$ appears twice. This equation then has five roots, $0, -1, 3, -1/2, -1/2$.

Example 3-9. Find an equation whose only roots are 2, -3, and 0.

Solution: Since 2, -3, and 0 are roots, then $x - 2$, $x + 3$, and x must be factors of the polynomial equation. Therefore, an equation having these roots would be

$$(x - 2)(x + 3)x = 0$$

or

$$x^3 + x^2 - 6x = 0$$

This answer is not unique. The equation $2(x - 2)(x + 3)x = 0$ or $2x^3 + 2x^2 - 12x = 0$ also has these same three roots.

Exercise 3-4

1. Without dividing find the remainder when
   (a) $x^3 - 2x^2 + 3x - 8$ is divided by $x - 2$.
   (b) $x^4 + 3x - 1$ is divided by $x + 1$.
   (c) $x^6 + 4x - 2$ is divided by $x + 2$.
   (d) $x^{10} + 1$ is divided by $x - 1$.
   (e) $17x$ is divided by $x + 21$.
   (f) 21 is divided by $x - (1/2)$.

2. Find a polynomial equation whose roots are given below.
   (a) $2, 0$ (b) $-1, -2, 0$
   (c) $\sqrt{2}, -\sqrt{2}$ (d) $i, -i, 1, -3$

3. Construct two different polynomials whose roots are $0, 1, -1$.

4. The only roots of a polynomial $P(x)$ are $1, -2$, and 3. Find $P(x)$ if $P(0) = 14$.

5. Find all of the roots of each of the following equations. How many roots does each equation have?
   (a) $(x + 1)^2(x - 3)(x + 4) = 0$
   (b) $2[x - (1/2)][x + (3/2)]^2(x + 1)^3 = 0$
   (c) $x^3(x + \sqrt{3})(x - \sqrt{3}) = 0$
   (d) $(x + i)^2(x - i)^2(x + 1)^2 = 0$

6. If one root of $x^3 + 2x^2 - 6x - 12 = 0$ is -2, find the other two roots.

7. If one root of $x^3 - 2x^2 + 7x - 30 = 0$ is 3, find the other two roots.

8. Find the remainder when
   (a) $x^5 + a^5$ is divided by $x + a$; by $x - a$.
   (b) $x^6 + a^6$ is divided by $x + a$; by $x - a$.
   (c) $x^{10} - a^{10}$ is divided by $x + a$; by $x - a$.
   (d) $x^{17} - a^{17}$ is divided by $x + a$; by $x - a$.

9. Use the Factor Theorem to show that $x - a$ is a factor of $x^n - a^n$ for any positive integer n. Show that $x + a$ is a factor of $x^n - a^n$ when n is a positive even integer, but not when n is odd. Show that $x + a$ is a factor of $x^n + a^n$ when n is a positive odd integer, but not when n is even.

10. Find the value of k for which $2x^3 + 7x^2 - 8x + k$ is exactly divisible by $x + 2$.

11. Find the value of k for which $x^4 + 2x^2 - 3x + k$ has remainder -2 when divided by $x - 3$.

12. Use the Factor Theorem to show that $x^4 + x^2 + 1$ does not have the factor $x - a$ for any real number a.

13. Use long division to find the quotient and remainder when $x^3 + 2x^2 - 7x + 1$ is divided by $2x - 1$; by $x - (1/2)$. What differences do you observe?

14. Let $f(x) = 0$ and $g(x) = 0$ be two polynomial equations both having the root r. Prove that
    (a) r is a root of $(f + g)(x) = 0$.
    (b) r is a double root of $(f \cdot g)(x) = 0$.

15. Graph:
    (a) $y = x(x - 1)(x + 1)$
    (b) $y = x^2(x - 1)$

3-5 The Fundamental Theorem of Algebra

The Fundamental Theorem of Algebra states that every polynomial equation whose degree is greater than or equal to one and whose coefficients are real or complex numbers has at least one root, which may be real or complex. Thus, we are assured that the equation $x^3 + 3x^2 + 7 = 0$ has at least one root. On the other hand, the equation $2 + \sqrt{x} = 0$ has no root, either real or complex, since $\sqrt{x}$ is never negative. (See Section 4-1.) This does not contradict the Fundamental Theorem, however, because this is not a polynomial equation.

This important theorem was proved by Carl Friedrich Gauss in 1799 when he was only 22. Gauss is universally considered to be one of the greatest mathematicians who ever lived. The proof of this theorem is quite difficult, and we will not give it here. The interested reader is referred to the appendix of Dickson's *New First Course in the Theory of Equations.*

Using the Factor Theorem and the Fundamental Theorem of Algebra, it is not hard to prove that *every polynomial equation of degree n has exactly n roots.*

We will prove this interesting and useful theorem for $n = 3$. Suppose $P(x) = 0$ is a polynomial of degree 3. By the Fundamental Theorem, $P(x) = 0$ has at least one root, call it a_1. By the Factor Theorem we can factor $P(x)$ into two factors,

$$P(x) = (x - a_1)Q_1(x)$$

where $Q_1(x)$ is a polynomial of degree 2. Again by the Fundamental Theorem $Q_1(x)$ has a root, call it a_2. By the Factor Theorem, $Q_1(x)$ can be factored into $(x - a_2)Q_2(x)$, and we have

$$P(x) = (x - a_1)(x - a_2)Q_2(x)$$

where $Q_2(x)$ is a polynomial of degree one. Now $Q_2(x)$ has a root a_3 and can be factored into $(x - a_3)Q_3(x)$, where $Q_3(x)$ is a polynomial of degree zero, that is, it is a nonzero constant c. Thus, we have

$$P(x) = c(x - a_1)(x - a_2)(x - a_3)$$

We have shown that $P(x) = 0$ has *at least* three roots, a_1, a_2, and a_3. Could $P(x) = 0$ have *more* than three roots? Suppose a_4 is any number different from a_1, a_2, and a_3. Then clearly

$$P(a_4) = c(a_4 - a_1)(a_4 - a_2)(a_4 - a_3)$$

is not zero, since none of these factors is zero. Thus, a_4 is not a root, and $P(x) = 0$ has *exactly* three roots, no more, no less. Of course the three roots a_1, a_2, and a_3 do not have to be different. If $x - a$ appears as a factor k times, a is called a *root of multiplicity* k, and we must count it as a root k times in order for the theorem to hold.

Example 3-10. The polynomial equation

$$2x^4(x - 2)^2(x + 3)(x + 1)^3 = 0$$

has ten roots. Zero is a root of multiplicity four $[x^4 = (x - 0)^4]$; 2 is a double root; -3 a single root; and -1 is a root of multiplicity three. If you were to multiply this out, you would get a polynomial of degree ten.

We have already seen that if a quadratic equation having real coefficients has a negative discriminant, then the two roots are complex numbers of the form $a + bi$ and $a - bi$, where a and b are real numbers and $i = \sqrt{-1}$. We called these complex numbers *conjugates*. This is a special case of the following useful theorem:

If a polynomial equation $P(x) = 0$ has real number coefficients, then any complex roots it has will occur in conjugate pairs. In other words, if $a + bi$ is a root of such an equation, then so is $a - bi$.

Example 3-11. The equation $x^3 - 3x^2 - 6x - 20 = 0$ has the complex root $-1 + i\sqrt{3}$. Since its coefficients are real, $-1 - i\sqrt{3}$ must also be a root. By the Factor Theorem $[x - (-1 + i\sqrt{3})] = (x + 1 - i\sqrt{3})$ and $[x - (-1 - i\sqrt{3})] = (x + 1 + i\sqrt{3})$ must both be factors. Since

$$\begin{aligned}(x + 1 - i\sqrt{3})(x + 1 + i\sqrt{3}) &= (x + 1)^2 - (i\sqrt{3})^2 \\ &= (x + 1)^2 + 3 \\ &= x^2 + 2x + 4\end{aligned}$$

must be a factor of $x^3 - 3x^2 - 6x - 20$, we can find the remaining factor,

$x - 5$, by division. Thus, the three roots of this equation are $-1 + i\sqrt{3}$, $-1 - i\sqrt{3}$, and 5.

Example 3-12. The two roots of $x^2 + ix = 0$ are 0 and $-i$. This does not contradict the theorem however, as the coefficients of this equation are not real numbers.

It follows from this theorem that if a polynomial has real coefficients, then it must have an even number of complex roots, provided it has any at all. We conclude that a polynomial of odd degree whose coefficients are real numbers must have at least one real root. We had already deduced this fact for the cubic by examining its graph (Section 3-3).

There is a similar theorem about equations whose coefficients are rational numbers. If $P(x) = 0$ is a polynomial equation with rational coefficients, and $a + \sqrt{b}$ is a root, where a and b are rational numbers, and $\sqrt{b}$ is irrational, then $a - \sqrt{b}$ is also a root.

Example 3-13: Suppose that we know that $x^4 - 4x^3 + 4x - 1 = 0$ has the root $2 + \sqrt{3}$. Since its coefficients are rational, we conclude that it also has the root $2 - \sqrt{3}$. By the Factor Theorem, two factors are $[x - (2 + \sqrt{3})] = (x - 2 - \sqrt{3})$ and $[x - (2 - \sqrt{3})] = (x - 2 + \sqrt{3})$. Thus, $(x - 2 - \sqrt{3})(x - 2 + \sqrt{3}) = (x - 2)^2 - (\sqrt{3})^2 = x^2 - 4x + 1$ is a factor. By division we find that the other factor is $x^2 - 1$, thus the four roots are $2 + \sqrt{3}$, $2 - \sqrt{3}$, 1, and -1.

Let us summarize the theorems of this section.

> **Theorem 3-3.** *The Fundamental Theorem of Algebra.* Every polynomial equation whose degree is greater than or equal to one and whose coefficients are real or complex numbers has at least one root.

> **Theorem 3-4.** Every polynomial equation of degree n has exactly n roots, provided we count a root of multiplicity k exactly k times.

> **Theorem 3-5.** If a polynomial equation with real coefficients has the complex root $a + bi$, then it also has the root $a - bi$.

> **Theorem 3-6.** If a polynomial equation with rational coefficients has the root $a + \sqrt{b}$, where a and b are rational numbers and $\sqrt{b}$ is irrational, then it also has the root $a - \sqrt{b}$.

Exercise 3-5

1. Show that if a polynomial equation has the complex roots $a + bi$ and $a - bi$ then it has the quadratic factor $x^2 - 2ax + a^2 + b^2$, where a and b are real numbers. Thus, every polynomial with real coefficients can be factored into linear and quadratic factors having real coefficients.

2. A polynomial equation with real coefficients has the roots $2 + 3i$, $-i$, and 7. Write this polynomial as a product of linear and quadratic factors having real coefficients.

3. Show that if a polynomial equation has the roots $a + \sqrt{b}$ and $a - \sqrt{b}$ then it has the quadratic factor $x^2 - 2ax + a^2 - b$, where a and b are rational.

4. A polynomial equation with rational coefficients has the roots $3 - \sqrt{2}$, $\sqrt{3}$, and -1. Write this polynomial as a product of linear and quadratic factors having rational coefficients.

5. Find all of the roots of $x^3 + x^2 - 4x + 6 = 0$ if $1 + i$ is one of the roots.

6. Find all of the roots of $x^5 - 2x^4 - 2x^3 + 8x^2 - 3x + 10 = 0$ if i and $2 - i$ are two roots of this equation.

7. Given that $x^3 - x^2 - 4x - 2 = 0$ has the root $1 - \sqrt{3}$, find the other two roots.

8. Find all of the roots of $2x^5 + x^4 - 2x^3 - x^2 - 4x - 2 = 0$ if $\sqrt{2}$ and i are two of the roots.

9. (a) The equation $x^4 - ix - 2 = 0$ has the root i. Does it follow from Theorem 3-5 that it must have the root $-i$? Explain.
   (b) The equation $x^3 + (4 - \sqrt{2})x^2 + (13 - 4\sqrt{2})x - 13\sqrt{2} = 0$ has the root $-2 + 3i$. Does it follow from Theorem 3-5 that it must have the root $-2 - 3i$? Explain.

10. (a) Find an equation of lowest degree with real coefficients having the roots -2, $1 - 3i$.
    (b) Find an equation of lowest degree, whose coefficients may be real or complex, having the roots -2, $1 - 3i$.

11. (a) Find an equation of lowest degree with rational coefficients having the roots 1, $1 + \sqrt{2}$.
    (b) Find an equation of lowest degree, whose coefficients are real numbers, having the roots 1, $1 + \sqrt{2}$.

12. The polynomial $f(x) = 5$ does not have a root, that is, there is no number a such that $f(a) = 0$. Explain why this does not contradict the Fundamental Theorem of Algebra.

13. Find a polynomial of degree 3 whose only root is 1. Find a polynomial of degree 4 whose only roots are i and $\sqrt{2}$.

3-6 Synthetic Division

The Remainder Theorem says that we can find $P(a)$ by dividing $P(x)$ by $x - a$ and looking at the remainder. You may think that this is a rather useless bit of information. The process of division of polynomials is so cumbersome, surely it is easier to find $P(a)$ by substitution. In this section we will develop an abbreviated version of this division called *synthetic division* that is actually shorter and easier than substitution in most cases.

Consider the usual division algorithm:

$$
\begin{array}{rrrrrr}
 & 2x^3 & +\;x^2 & +\;2x & +\;5 & \\ \cline{2-6}
x-2\,) & 2x^4 & -\;3x^3 & & +\;x & +\;1 \\
 & 2x^4 & -\;4x^3 & & & \\ \cline{2-3}
 & & x^3 & & & \\
 & & x^3 & -\;2x^2 & & \\ \cline{3-4}
 & & & 2x^2 & +\;x & \\
 & & & 2x^2 & -\;4x & \\ \cline{4-5}
 & & & & 5x & +\;1 \\
 & & & & 5x & -\;10 \\ \cline{5-6}
 & & & & & 11
\end{array}
$$

The various powers of x are not really essential to the division process since they are mainly functioning as place holders. Our first step in simplifying the algorithm will be to leave out all these powers of x. Accordingly, we rewrite the example using only the coefficients. If we agree always to write the polynomials in order of descending powers of x, inserting zeros when any terms are missing, then our example will look like this:

$$
\begin{array}{lrrrrrr}
(1) & & 2 & 1 & 2 & 5 & \\ \cline{3-7}
(2) & 1-2\,) & 2 & -3 & 0 & 1 & 1 \\
(3) & & 2 & -4 & & & \\ \cline{3-4}
(4) & & & 1 & 0 & & \\
(5) & & & 1 & -2 & & \\ \cline{4-5}
(6) & & & & 2 & 1 & \\
(7) & & & & 2 & -4 & \\ \cline{5-6}
(8) & & & & & 5 & 1 \\
(9) & & & & & 5 & -10 \\ \cline{6-7}
(10) & & & & & & 11
\end{array}
$$

Since the first number in lines (3), (5), (7), and (9) needlessly duplicates the number directly above it, we lose nothing by leaving these numbers out. Moreover, the last number in lines (4), (6), and (8) repeats numbers found in the dividend. Deleting these, our example becomes

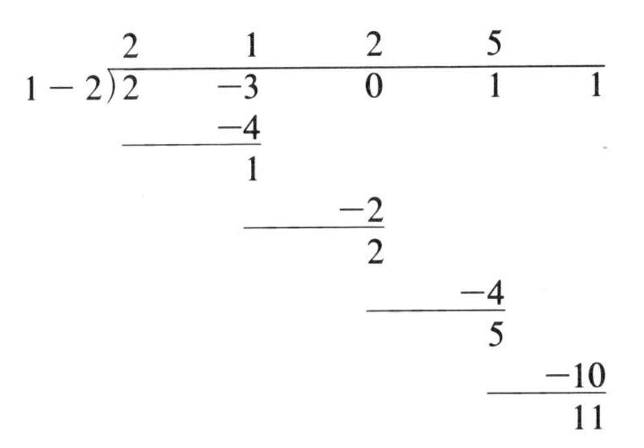

Since the divisor will always be of the form $x - a$, the first 1 in the divisor is unnecessary. Now adding signed numbers is simpler than subtracting them, so it would be an improvement if we could add at each step instead of subtracting. If we change the sign of the 2 in the divisor, we can do this. Now we have

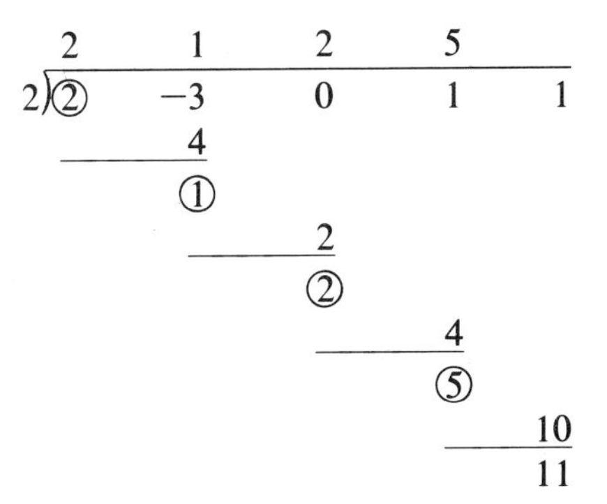

Finally note that the circled numbers repeat the quotient, thus we really do not need the quotient either! Now we can write the entire example on just three lines.

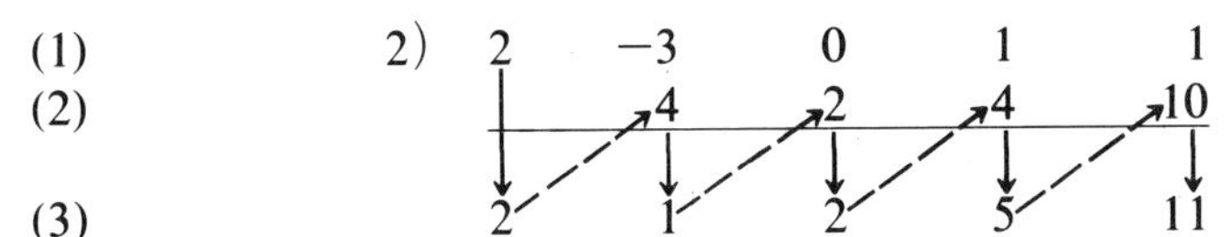

Our method of simplified division is now clear.

(a) Write the coefficients of the polynomial in order of descending exponents in row (1), *inserting zeros for missing terms*. Draw a line between rows (2) and (3).

(b) If the divisor is $x - a$, write the number a at one side in row (1).
(c) Write the first coefficient in the first column of row (3).
(d) Multiply this number by a and write the product in the next column of row (2).
(e) Add the next column and write the sum in row (3).
(f) Repeat steps (d) and (e) until the last column is added.

This algorithm is shown diagrammatically above. The solid lines indicate addition and the broken lines multiplication by 2. The last number in row (3) is the remaining $P(2)$, while the other numbers are the coefficients of the quotient in order of descending exponents.

Example 3-14. Find the quotient and remainder when $x^4 + 7x^3 - 32x - 8$ is divided by $x + 3$.

Solution:

$$\begin{array}{r|rrrrr} -3) & 1 & 7 & 0 & -32 & -8 \\ & & -3 & -12 & 36 & -12 \\ \hline & 1 & 4 & -12 & 4 & -20 \end{array}$$

In this case a is -3 since $x + 3 = x - (-3)$. The remainder is -20. This is also the number you would get if you substituted -3 for x in the polynomial. (Compute $P(-3)$ by substitution and compare the labor involved.) The quotient is $x^3 + 4x^2 - 12x + 4$.

Synthetic division has many uses. In graphing a polynomial function, we need to compute $P(x)$ for many different choices of x. This can be done quickly and accurately by synthetic division. Another use of synthetic division is in finding roots. To test if a number a is a root of $P(x) = 0$, divide synthetically by $x - a$. If the remainder is zero, then a is a root; if the remainder is not zero, then a is not a root.

Example 3-15. Use synthetic division to determine whether or not -2 is a root of $x^4 + 3x^2 + 7x - 14 = 0$.

Solution:

$$\begin{array}{r|rrrrr} -2) & 1 & 0 & 3 & 7 & -14 \\ & & -2 & 4 & -14 & 14 \\ \hline & 1 & -2 & 7 & -7 & 0 \end{array}$$

Since the remainder is zero, -2 is a root. By the Factor Theorem $x - (-2) = x + 2$ is one factor and the other factor is the quotient $x^3 - 2x^2 + 7x - 7$.

Although synthetic division is usually easier than substitution, this is not always the case. For example, if $P(x) = x^{16} + 2$, and you want to find $P(1)$, clearly substitution is shorter. It is necessary to exercise some judgment in deciding which method to use.

Exercise 3-6

1. Use synthetic division to find $P(-2)$, $P(2)$, $P(3)$, $P(-1)$, if $P(x) = 2x^3 - 6x^2 + 7x - 20$.

2. Use synthetic division to test 1, −1, 3, 2, and 1/2 to see if any of these numbers are roots of $2x^4 - 3x^3 - 7x^2 - 8x + 6 = 0$.

3. Use synthetic division to find the quotient and the remainder in each of the following division problems:
   (a) $3x^4 - 7x^2 + 10x - 1 \div x - 3$
   (b) $3x^4 - 7x^2 + 10x - 1 \div x + 2$
   (c) $6x^5 + x^4 - 4x^3 + x^2 + 7 \div x - 1/2$
   (d) $x^5 + 32 \div x - 2$
   (e) $2x^2 - 3x - 4 \div x + \sqrt{2}$
   (f) $-x^7 + 2x^4 + 7x^3 - x \div x - 2$
   (g) $2x^3 + 3x^2 + 100 \div x + 5$
   (h) $x^6 - x^3 + 1 \div x + 1$

4. If $y = 2x^3 - 3x^2 - 12x + 7$, use synthetic division to complete the following table:

x	0	1	2	3	−1	−2	−3
y							

   Use this information to graph the function.

5. If $y = x^4 - 2x^2 + 10$, use synthetic division to complete the following table:

x	0	1	2	3	−1	−2	−3
y							

   Use this information to graph the function.

6. The polynomial $x^3 - a^3$ clearly has the root $x = a$, and $x^3 + a^3$ has the root $x = -a$. Use synthetic division to factor $x^3 - a^3$ and $x^3 + a^3$ into the product of linear and quadratic factors.

7. Use synthetic division to find the value of k in each of the following polynomials if the number given is to be a root.
   (a) $2x^3 - 7x^2 + 3x + k$; −1
   (b) $3x^5 - 2x^3 + 9x + k$; 0
   (c) $x^2 + kx - 10$; 5
   (d) $kx^2 - x - 12$; −3

8. If a polynomial $P(x)$ is divided by $2x - 1$, do you get the same quotient and remainder as when you divide by $x - (1/2)$? What is the difference? Devise a rule for using synthetic division to find the quotient and remainder when $P(x)$ is divided by $ax + b$, $a \neq 0$.

3-7 The Rational Root Theorem

If we can find one root a of a polynomial equation $P(x) = 0$, then $x - a$ is a factor of $P(x)$ and the other factor $Q(x)$ is a polynomial whose degree is one less than the degree of $P(x)$. The remaining roots of $P(x)$ will be the roots of $Q(x)$ and these may be easier to find since the degree of $Q(x)$ is less. In particular, if $P(x)$ is a cubic equation, then $Q(x)$ is a quadratic and we can always find these roots by the quadratic formula. The problem is, How can we find that first root? The following theorem tells us where to look:

Theorem 3-7. *The Rational Root Theorem.* Let

$$a_n x^n + a_{n-1} x^{n-1} + \ldots + a_1 x + a_0 = 0$$

be a polynomial equation whose coefficients are integers. If this equation has a rational root p/q, where p and q have no factors in common (other than ± 1), then p divides a_0 and q divides a_n.

Comment: We say that an integer a *divides* another integer b provided there exists some integer k such that $a \cdot k = b$. Thus, for example, 2 divides -4, since $2(-2) = -4$, but 3 does not divide 5.

Proof: If p/q is a root, then it satisfies the equation, i.e.,

$$a_n(p/q)^n + a_{n-1}(p/q)^{n-1} + \ldots + a_1(p/q) + a_0 = 0$$

Multiplying both sides of the equation by q^n, we have

$$a_n p^n + a_{n-1} p^{n-1} q + \ldots + a_1 p q^{n-1} + a_0 q^n = 0 \qquad \textbf{(3-1)}$$

Adding $-a_0 q^n$ to both sides and factoring p out of the left-hand side, we get

$$p(a_n p^{n-1} + a_{n-1} p^{n-2} q + \ldots + a_1 q^{n-1}) = -a_0 q^n$$

Since p is a factor of the left-hand side, it must be a factor of the right-hand side as well. But we assumed that p and q have no factors in common, therefore, p cannot divide q^n (unless $p = \pm 1$). We conclude that p must divide a_0. Adding $-a_n p^n$ to both sides of Equation (3-1) and factoring out q, by the same reasoning we conclude that q divides a_n. ■

Theorem 3-7 does not guarantee that a polynomial equation will have a rational root. It merely states that if $P(x)$ does have a rational root, then this root is to be found in a particular finite set of numbers which we can list.

Example 3-16. If the equation $3x^3 - 2x^2 + 9x - 6 = 0$ has a rational root p/q, then p is a divisor of 6 and q is a divisor of 3. We list all possible choices for p/q.

Possible numerators: $\pm 6, \pm 1, \pm 2, \pm 3$

Possible denominators: $\pm 3, \pm 1$

Possible rational roots: $\pm 2, \pm 6, \pm 1/3, \pm 1, \pm 2/3, \pm 3$

Using synthetic division to test these possibilities, we find that 2/3 is a root and $3x^3 - 2x^2 + 9x - 6 = [x - (2/3)](3x^2 + 9) = 0$. Setting $3x^2 + 9$ equal to zero we find that the other two roots are $\pm i\sqrt{3}$.

Since the denominator must be a factor of the leading coefficient, it follows that if this coefficient is 1 any rational roots must be integers.

Corollary 3-1. If

$$x^n + a_{n-1}x^{n-1} + \ldots + a_1 x + a_0 = 0$$

is a polynomial equation with integral coefficients, then any rational root must be an integer.

Example 3-17. If $x^3 + x^2 + x + 1$ has any rational roots, they must be integers and factors of 1. The only possibilities are 1 and -1. Testing we find that -1 is a root and $x^3 + x^2 + x + 1 = (x + 1)(x^2 + 1) = 0$. The three roots are $-1, \pm i$.

Example 3-18. Find all of the roots of $2x^4 + 9x^3 + 15x^2 + 6x - 8 = 0$.

Solution: The possible rational roots are $\pm 8, \pm 1, \pm 2, \pm 4$, and $\pm 1/2$. Starting with the smallest positive integer, we test $+1, +2, +4$, and $+8$ and find that none of these are roots. Testing the negative integers, -1 and -2, we find that -1 is not a root, but -2 is.

$-2)$	2	9	15	6	-8
		-4	-10	-10	8
	2	5	5	-4	0

The remaining roots must be roots of the quotient $2x^3 + 5x^2 + 5x - 4 = 0$. The possible rational roots of this equation (called the depressed equation) are $\pm 1, \pm 2, \pm 4$, and $\pm 1/2$. Of these possibilities we have already eliminated 1, 2, 4, and -1, since if they are not roots of the original equation they cannot be roots of the quotient. Consequently we test -2 (again), -4, and $\pm 1/2$, and find that $1/2$ is a root.

$$\begin{array}{r|rrrr} 1/2) & 2 & 5 & 5 & -4 \\ & & 1 & 3 & 4 \\ \hline & 2 & 6 & 8 & 0 \end{array}$$

We have found that -2 and $1/2$ are roots and the new depressed equation is $2x^2 + 6x + 8 = 0$. The roots of this quadratic, $(-3 \pm i\sqrt{7})/2$, are the remaining two roots of the quartic equation.

In this search for roots there are several points to keep in mind.

1. After a root is found we continue the search for roots with the depressed equation, not the original equation.
2. If a number a is *not* a root of the original equation, then it is not a root of any of the depressed equations. In other words, when we eliminate a possible root, it is eliminated permanently.
3. On the other hand, if a number a is found to be a root of the original equation, it should *not* be eliminated as a possible root of the depressed equation, since it may very well be a repeated root.

Exercise 3-7

1. List all possible rational roots of the following:
   (a) $3x^3 - 7x^2 + 10x - 21 = 0$ (b) $x^4 + 7x^2 - 12 = 0$
   (c) $4x^3 - 3x^2 + x - 9 = 0$ (d) $2x^3 + x^2 - 7x + 24 = 0$

2. Use the Rational Root Theorem to find all of the roots of the following:
   (a) $x^3 - 2x^2 - 7x + 2 = 0$ (b) $2x^3 + 7x^2 - x - 21 = 0$
   (c) $2x^3 + 13x^2 + 24x + 9 = 0$ (d) $4x^3 + 3x^2 - 5x + 1 = 0$
   (e) $x^4 - 20x - 21 = 0$ (f) $x^4 + 4x^2 - 5 = 0$
   (g) $x^4 - x^3 + 7x^2 - 15x + 8 = 0$ (h) $x^4 - 4x^3 + x^2 - 6x + 36 = 0$

3. Prove that $\sqrt[3]{2}$ is irrational by showing that the equation $x^3 - 2 = 0$ has no rational roots.

4. Prove that $\sqrt[4]{2}$ is irrational.

5. Prove that $\sqrt[3]{10}$ is irrational.

6. (a) The equation $x^4 - (1/2)x^3 + 4x - 2 = 0$ has a rational root. Is it necessarily an integer? Explain.
   (b) The equation $x^3 + (2\sqrt{2} - 1)x^2 + (2 - 2\sqrt{2})x - 2 = 0$ has a rational root. Will it necessarily be found in the set of numbers ± 2, ± 1? Explain.

7. If a polynomial equation does not have integral coefficients, then the

Rational Root Theorem does not apply. However, if the coefficients are rational numbers an equation with integral coefficients having the same roots can be found by multiplying both sides of the equation by the lowest common denominator of all of the fractional coefficients. Find the roots of the following equations by first finding an equation with integral coefficients having the same roots.
(a) $(1/6)x^3 - (1/3)x^2 + (1/2)x - 1 = 0$
(b) $(1/2)x^3 + (1/4)x^2 + 4x + 2 = 0$

3-8 Upper and Lower Bounds for the Roots

We have seen in the last section that by using the Rational Root Theorem we can list the possible rational roots of a polynomial equation, then proceed to test them by synthetic division. If the number of factors is large, this could involve a great deal of work! A procedure which would eliminate some of these possibilities, without involving any extra labor would certainly be worthwhile. Such a procedure is finding upper and lower bounds for the roots.

Definition 3-4. A real number U is called an *upper bound* for the real roots of a polynomial equation if every real root is less than or equal to U. A real number L is called a *lower bound* for the real roots of a polynomial equation if every real root is greater than or equal to L.

Clearly there can be many upper bounds and many lower bounds. There are many ways of finding upper and lower bounds for the roots, but the simplest method uses synthetic division.

Test for an upper bound for the real roots: Let $P(x) = 0$ be a polynomial equation and U a nonnegative real number. Use synthetic division to divide $P(x)$ by $x - U$. If the numbers in the bottom row are all nonnegative, then U is an upper bound for the real roots.

To use this test, the leading coefficient of $P(x) = 0$ must be positive. If it is not, simply multiply both sides of the equation by -1.

Example 3-19. Find the smallest nonnegative integer which the test of this section shows is an upper bound for the real roots of $2x^3 + 3x^2 - x - 12 = 0$.

Solution: If all of the coefficients were positive, then zero would be an upper bound, i.e., the equation would have no positive roots. Since this is not the case here, we start with the smallest positive integer, 1.

$$\begin{array}{r|rrrr} 1) & 2 & 3 & -1 & -12 \\ & & 2 & 5 & 4 \\ \hline & 2 & 5 & 4 & -8 \end{array}$$

The test does not show that 1 is an upper bound. Next we try 2.

$$\begin{array}{r|rrrr} 2) & 2 & 3 & -1 & -12 \\ & & 4 & 14 & 26 \\ \hline & 2 & 7 & 13 & 14 \end{array}$$

Since all of the numbers in the last line are positive, 2 is an upper bound for the roots.

It is not hard to see why this test works. If in the last example we divided synthetically by any number larger than 2, we would get a number larger than 4 in the second row. This in turn would give a number larger than 7 when we added, which in turn would give a number larger than 14 when multiplied, and so on. Thus, the remainder would be larger than 14, and the number we are dividing by cannot be a root.

Test for a lower bound for the real roots: Let $P(x) = 0$ be a polynomial equation and L a nonpositive real number. Use synthetic division to divide $P(x)$ by $x - L$. If the numbers in the bottom row are alternately positive and negative, then L is a lower bound for the real roots.

In this test, zero can be counted as either positive or negative, but it cannot be skipped. Thus, if the last row is

$$2 \quad 0 \quad 3 \quad -4 \quad 1$$

then we count the zero as negative and we have a $+ - + - +$ pattern, so the number being tested is a lower bound. But if the last row is

$$2 \quad 0 \quad -3 \quad 4 \quad -1$$

then no matter what sign we give zero, we do not have a $+ - + - + -$ pattern, and the test does not tell us we have found a lower bound.

Example 3-20. Find the largest nonpositive integer which the test shows to be a lower bound for the real roots of $2x^3 + 3x^2 - x - 12 = 0$.

Solution: If the coefficients of $P(x) = 0$ are alternately positive and negative, then zero is a lower bound for the roots, i.e., there are no negative roots. The missing terms must be considered here. Their coefficients are zero, which can be considered either positive or negative. Since this is not the case in this example, we first test -1.

$-1)$	2	3	-1	-12
		-2	-1	2
	2	1	-2	-10

Since we do not have a $+ - + -$ pattern, this test does not show -1 to be a lower bound. Next we try -2.

$-2)$	2	3	-1	-12
		-4	2	-2
	2	-1	1	-14

We have found that -2 is a lower bound for the roots.

The reasoning behind this test is easy to see. Suppose we divide by a number smaller than -2. Since we are dividing by a negative number, the first number in the second row will be a negative number smaller than -4. The sum then will be a negative number smaller than -1. Multiplying by a negative number will give us a positive number greater than 2 in the next position, which in turn will yield a sum larger than 1, and so on. Thus, we will still get a $+ - + -$ pattern in the bottom row, and all of the positive numbers will be larger and all of the negative numbers will be smaller than the corresponding ones for the divisor -2. Thus, we can never get a zero in the last position and no number smaller than -2 can be a root.

Now let us see how we can use these new ideas in finding roots.

Example 3-21. Find all of the roots of $6x^4 + 13x^3 - 3x^2 + 26x - 30 = 0$.

Solution: The possible rational roots are ± 30, ± 1, ± 15, ± 2, ± 10, ± 3, ± 5, ± 6, $\pm 1/6$, $\pm 5/2$, $\pm 1/3$, $\pm 5/3$, $\pm 1/2$, $\pm 5/6$, $\pm 15/2$, $\pm 3/2$, $\pm 2/3$, $\pm 10/3$; thirty-six possibilities in all! We will try to eliminate some of these by finding an upper and lower bound for the roots. We will look for the upper bound among the positive integers that are possible roots, starting with the smallest, 1.

$1)$	6	13	-3	26	-30
		6	19	16	42
	6	19	16	42	12

We find that 1 is an upper bound (and incidentally that 1 is not a root). This eliminates all of the possible positive roots except 1/6, 1/3, 1/2, 5/6, and 2/3. Testing these, we find that 5/6 is a root.

5/6)	6	13	−3	26	−30
		5	15	10	30
	6	18	12	36	0

The depressed equation is $6x^3 + 18x^2 + 12x + 36 = 0$. Since all of its coefficients are positive, we know that there are no more positive roots.

We now proceed to look for a lower bound for the negative roots, using the depressed equation and starting with the largest negative integer, −1.

−1)	6	18	12	36
		−6	−12	0
	6	12	0	36

−2)	6	18	12	36
		−12	−12	0
	6	6	0	36

−3)	6	18	12	36
		−18	0	−36
	6	0	12	0

We find that −3 is a lower bound, since zero can be considered either positive or negative, and it is also a root. The new depressed equation is $6x^2 + 12 = 0$, hence the other two roots of our quartic are $\pm i\sqrt{2}$.

Note that in looking for an upper bound for the roots we start with the smallest positive integer 1. If this is not an upper bound (or a root), we proceed to the next largest positive integer *which is a possible root.* Thus, if 1 and 3 are possible roots, but 2 is not, we test 1, then 3. The idea here is to have every synthetic division we perform serve two purposes: as a test for a root and as a test for an upper bound.

In looking for a lower bound we follow the same procedure, starting with −1, and if this is not a lower bound, proceeding to the next smallest negative integer which is a possible root.

If used in this way, the test for upper and lower bounds will involve no extra labor and may very well save a tremendous amount of work.

Exercise 3-8

Find the smallest nonnegative integer and the largest nonpositive integer which the tests of this section show to be an upper and a lower bound, respectively, for the real roots of the following:

1. $3x^3 + 7x^2 + 8x + 1 = 0$

2. $x^4 - 2x^3 + 7x^2 - x + 1 = 0$

3. $2x^5 - x^3 + x + 1 = 0$

4. $x^3 + 3x^2 - 7x - 2 = 0$

5. $x^4 - 1 = 0$

6. $x^5 - 1 = 0$

Using the methods of this section, find all of the roots of the following:

7. $6x^4 - x^3 + 34x^2 - 6x - 12 = 0$

8. $2x^4 + x^3 - 18x^2 - 6x + 36 = 0$

9. $6x^4 - 21x^3 - 10x^2 - 7x - 4 = 0$

10. $5x^5 + 8x^4 + 21x^3 + 30x^2 + 4x - 8 = 0$

3-9* The Nature of the Roots and Descartes' Rule of Signs

The roots of a polynomial equation $P(x) = 0$ may be classified as either real or complex. We have already seen that if the coefficients of $P(x)$ are real numbers, then complex roots occur in pairs. Thus, the number of complex roots will be an even number.

The real roots of $P(x) = 0$ may in turn be classified as positive, negative, or zero. Now zero roots are easy to detect. Zero is a root of $P(x) = 0$ if and only if the constant term is zero. In fact, we can say that zero is a root of multiplicity k if and only if the constant term and the coefficient of every term whose exponent is less than k is zero, but the coefficient of x^k is not zero. In other words, zero is a root of multiplicity k if we can factor $P(x)$ into $x^kQ(x)$, and zero is not a root of $Q(x)$.

Example 3-22. Zero is a root of multiplicity three of

$$3x^6 - 7x^5 + 10x^4 - x^3 = 0$$

There is a theorem called *Descartes' Rule of Signs* that tells us something about the number of positive and negative roots of a polynomial equation. Suppose the terms of $P(x)$ are arranged in order of descending exponents. We will ignore any missing terms. Then we will say we have a *variation in sign* if two consecutive terms have different signs. For example, in $x^5 - 2x^3 - 7x + 20$, we find two variations in sign, since the first two terms and the last two have different signs. The equation $x^5 - 2x^3 + 7x - 20$ has three variations in sign.

Descartes' Rule of Signs. Let $P(x) = 0$ be a polynomial equation with real coefficients, and assume that the terms of $P(x)$ are arranged in order of descending exponents. Suppose that when so arranged $P(x)$ has s variations in sign. Then the number of positive real roots of $P(x) = 0$ is either equal to s or is less than s by a positive even integer.

A root of multiplicity k here is counted k times. For a proof of this rule see Dickson's *New First Course in Theory of Equations,* pp. 76–80.

Example 3-23. The equation $x^5 - 2x^3 - 7x + 20 = 0$ has two variations of sign, hence it must have either two positive roots or none.

Example 3-24. The equation $x^5 - 2x^3 + 7x - 20 = 0$ has three variations of sign, so either it has three positive roots, or it has one.

Note that in general, Descartes' Rule of Signs does not tell us *exactly* how many positive roots $P(x) = 0$ has, it merely helps us to list several possibilities. However, if the coefficients of $P(x)$ are all positive or all negative, then there are *no* variations of sign and we can say the equation has no positive roots. Moreover, if $P(x)$ has only one variation in sign, then we can say that $P(x) = 0$ has exactly one positive root.

To investigate the number of negative roots, first consider what happens if we substitute $-x$ for x in $P(x)$. We will get a new polynomial $P(-x)$ in which the signs of the coefficients of *odd* powers of x will be changed, while the signs of the constant term and the coefficients of *even* powers of x will remain the same.

Example 3-25.

$$\begin{aligned} P(x) &= x^4 - 3x^3 + 7x^2 + 8x - 2 \\ P(-x) &= (-x)^4 - 3(-x)^3 + 7(-x)^2 + 8(-x) - 2 \\ &= x^4 + 3x^3 + 7x^2 - 8x - 2 \end{aligned}$$

Now it is not hard to see that if a is a root of $P(x) = 0$, then $-a$ will be a root of $P(-x) = 0$. In other words, the positive roots of $P(x) = 0$ are the negative roots of $P(-x) = 0$, and the negative roots of $P(x) = 0$ are the positive roots of $P(-x) = 0$. Geometrically, substituting $-x$ for x in the polynomial function $y = P(x)$ has the effect of rotating the graph about the y-axis. See Figure 3-8.

Accordingly if we want to count the negative roots of $P(x) = 0$, we can instead count the positive roots of $P(-x) = 0$.

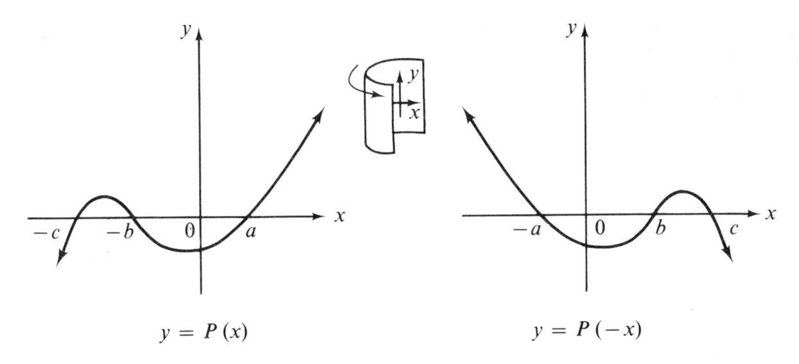

Figure 3-8

Corollary to Descartes' Rule of Signs. Let t be the number of variations in sign of $P(-x) = 0$. Then the number of negative roots of $P(x) = 0$ is either equal to t or is less than t by a positive even integer.

Example 3-26. If $P(x) = x^5 - 2x^3 - 7x + 20$, then $P(-x) = -x^5 + 2x^3 + 7x + 20$. Since $P(-x)$ has one variation in sign, $P(x) = 0$ has exactly one negative root.

Example 3-27. If $P(x) = x^5 - 2x^3 + 7x - 20$, then $P(-x) = -x^5 + 2x^3 - 7x - 20$. Since $P(-x)$ has two variations in sign, $P(x)$ has either two or no negative roots.

If we combine the information we have obtained about zero, positive, and negative roots with the fact that a polynomial equation of degree n has exactly n roots, then we can list all of the possible classifications of roots of $P(x) = 0$.

Example 3-28. The equation $P(x) = x^5 - 2x^3 - 7x + 20 = 0$ has no zero roots, two or no positive roots, and one negative root. Since $P(x) = 0$ has exactly five roots and any roots that are neither zero, positive, nor negative must be complex, we have two possible cases.

Zero	Positive	Negative	Complex
0	2	1	2
0	0	1	4

Example 3-29. Classify the roots of $P(x) = x^6 + 3x^5 - 7x^4 + x^3 + 2x^2 = 0$. $P(x)$ has two variations in sign, hence $P(x) = 0$ has either two or no positive roots. The polynomial $P(-x) = x^6 - 3x^5 - 7x^4 - x^3 + 2x^2$

has two variations of sign, hence $P(x) = 0$ has either two or no negative roots. There are two zero roots.

Zero	Positive	Negative	Complex
2	2	2	0
2	0	2	2
2	2	0	2
2	0	0	4

Note that we must list all possible combinations of positive and negative roots.

Exercise 3-9

1. Show that $x^4 + 3x^2 + 2x - 1 = 0$ has exactly two complex roots.
2. (a) Show that if n is even and $a > 0$, then $x^n - a$ has exactly two real roots.
   (b) Show that if n is even and $a > 0$, then $x^n + a$ has no real roots.
3. (a) Show that if n is odd and $a > 0$, then $x^n + a$ has exactly one real root.
   (b) Show that if n is odd and $a > 0$, then $x^n - a$ has exactly one real root.
4. Show that $x^4 + 3x^2 + 7 = 0$ has no real roots.

Classify the roots of the following:

5. $x^4 - 6x^3 + x^2 + 3x - 1 = 0$
6. $x^5 - 2x^3 + 2x^2 + 3x = 0$
7. $x^5 + x^3 - 2x^2 + 7x - 5 = 0$
8. $x^5 + 2x^3 - x^2 + 3x - 3 = 0$
9. $x^7 + 7x^4 - 3x^3 - x^2 = 0$
10. $x^9 - x^2 + 2x = 0$

3-10* Testing for Rational Roots Two at a Time

In this section we will investigate a method of testing the possible rational roots of a polynomial equation *two at a time,* thus cutting the work in half. This technique involves dividing the polynomial by a quadratic of the form $x^2 - a^2$. This can be done by a process very similar to synthetic division. First let us look at an example.

Example 3-30. Divide $3x^4 + 7x^3 - x + 10$ by $x^2 - 4$.

Solution: We proceed at first as in synthetic division, writing down the coefficients in order of descending exponents, and inserting zeros for any missing terms. The number 4 from the divisor $x^2 - 4$ is written on the same line.

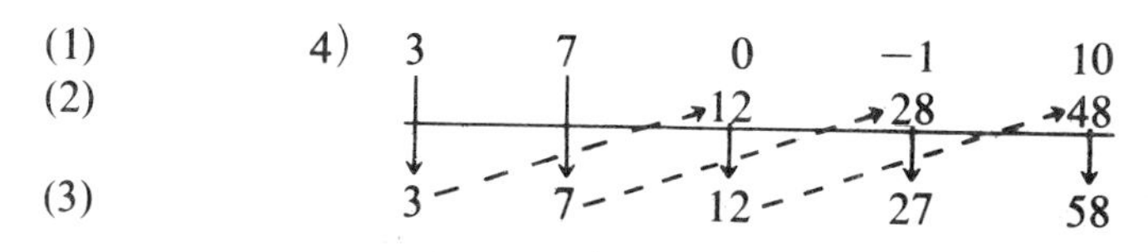

First the leading coefficient 3 is written in the first column in line (3). Multiply this 3 by the 4 from the divisor and write the product, not in the next column as in ordinary synthetic division, but in the *second column over,* column 3. Next bring down the coefficient in column 2. Multiply this number, 7, by 4 and write the product in the second column over, which will be column 4. Now add column 3. Multiply this sum by 4 and write the product in the second column over, column 5. Now add the last two columns.

Note that the only difference between this and ordinary synthetic division is that the products are written in the second column over, instead of the next column. This is shown diagrammatically above. The solid lines indicate addition and the dotted lines indicate multiplication by 4.

Since the divisor is a quadratic, the remainder will be of the form $mx + n$, and the degree of the quotient will be two less than the degree of the original polynomial. In Example 3-30, the last two numbers in row (3) are the coefficients of the remainder and the remaining numbers are the coefficients of the quotient. The quotient here is $3x^2 + 7x + 12$, and the remainder is $27x + 58$.

In summary, the algorithm for dividing a polynomial $P(x)$ by $x^2 - a^2$ is as follows:

(a) Write the coefficients of the polynomial in order of descending exponents in row (1), inserting zeros for any missing terms. Draw a line between rows (2) and (3).
(b) If the divisor is $x^2 - a^2$, write the number a^2 at one side in row (1).
(c) Write the first coefficient in the first column of row (3).
(d) Multiply this number by a^2 and write the product in the second column over in row (2). (That is, if the number multiplied is in column k, write the product in column $k + 2$.)
(e) Add the next column (column $k + 1$) and write the sum in row (3).

(f) Repeat steps (d) and (e) until a number is written in the last column and the next to last column is added.
(g) Add the last column and write the sum in row (3).

Example 3-31. Divide $5x^5 - 2x^4 + x^2 - 5x + 2$ by $x^2 - 1$.

Solution: 1)

5	−2	0	1	−5	2
		5	−2	5	−1
5	−2	5	−1	0	1

The quotient is $5x^3 - 2x^2 + 5x - 1$ and the remainder is $0x + 1$, or 1.

Using the Rational Root Theorem, we list possible rational roots in pairs, $\pm a$, $\pm b$, and so on. Now let us see how we can test the two possible roots $\pm a$ in one operation.

Theorem 3-8. Let $P(x)$ be a polynomial whose degree is greater than or equal to two. Let $mx + n$ be the remainder obtained when $P(x)$ is divided by $x^2 - a^2$. Then a is a root of $P(x) = 0$ if and only if $a = -n/m$. Similarly, $-a$ is a root if and only if $-a = -n/m$.

Proof: By the division algorithm,

$$P(x) = Q(x)(x^2 - a^2) + (mx + n)$$

where $Q(x)$ is the quotient polynomial. Suppose a is a root of $P(x) = 0$. Then

$$0 = P(a) = Q(a)(a^2 - a^2) + (ma + n)$$

Thus, $ma + n = 0$, and $a = -n/m$.

Conversely, suppose $a = -n/m$. Then

$$P(a) = Q(a)(a^2 - a^2) + m(-n/m) + n$$
$$P(a) = 0 + 0 = 0$$

and a is a root.

The proof is similar for $-a$. ■

Using Theorem 3-8 and the algorithm for dividing $P(x)$ by $x^2 - a^2$, we find the remainder $mx + n$. If either a or $-a$ is equal to $-n/m$, then this number is a root of $P(x) = 0$. On the other hand, if neither a nor $-a$ is equal to $-n/m$, then neither is a root, and we have eliminated two possible roots in one operation. Of course, if the remainder is zero ($m = n = 0$), then $x^2 - a^2$ is a factor of $P(x)$ and both a and $-a$ are roots.

Example 3-32. Find all of the roots of $3x^3 - 2x^2 - 5x - 6 = 0$.

Solution: The possible rational roots of this equation are $\pm 1, \pm 2, \pm 3, \pm 6, \pm 1/3, \pm 2/3$. To test the pair ± 1, we divide by $x^2 - 1$.

1)	3	−2	−5	−6
			3	−2
	3	−2	−2	−8

Here $m = -2$, $n = -8$, $-n/m = -(-8/-2) = -4$. Since this is equal to neither 1 nor −1, we conclude that neither is a root. Next we test the pair ± 2 by dividing by $x^2 - 4$.

4)	3	−2	−5	−6
			12	−8
	3	−2	7	−14

Here $-n/m = -(-14/7) = 2$, hence 2 is a root.

Now we divide synthetically by $x - 2$ to find the depressed equation.

2)	3	−2	−5	−6
		6	8	6
	3	4	3	0

The depressed equation is $3x^2 + 4x + 3 = 0$, and since this is a quadratic, we can use the quadratic formula to find the remaining two roots $(-2 \pm i\sqrt{5})/3$.

Exercise 3-10

1. Use the algorithm of this section to find the quotient and remainder in the following division problems:
   (a) $4x^3 - 7x + 3 \div x^2 - 1$
   (b) $x^4 + 2x^3 - x + 24 \div x^2 - 4$
   (c) $3x^4 - 2x^3 - 25x^2 - 9 \div x^2 - 9$
   (d) $x^5 - 7x^4 + 10x^2 - 11 \div x^2 - 4$

Use the method of this section to find all of the roots of the following:

2. $x^4 + x^3 - 4x^2 + 2x - 12 = 0$

3. $x^3 - 9x^2 + 24x - 20 = 0$

4. $x^4 - 3x^3 - 12x^2 + 52x - 48 = 0$

5. $x^3 - 11x^2 + 37x - 35 = 0$

6. $x^4 - 8x^3 + 25x^2 - 36x + 20 = 0$

7. $x^3 - 2x^2 - 15x + 36 = 0$

3-11* Irrational Roots

In the last few sections we have outlined methods for finding the rational roots of a polynomial equation, roots which are of the form p/q, where p and q are integers and $q \neq 0$. What can we do if the equation has roots which are irrational—roots such as $\sqrt{2}$, for example? It is not always possible to find such roots *exactly*. Usually the best we can hope for is to find a rational number which approximates the irrational root to any desired degree of accuracy. For example, if $\sqrt{2}$ is a root, we might settle for the rational approximation 1.4.

One way of finding an approximation to the irrational roots would be to graph the function $y = P(x)$ very carefully. Since roots of $P(x) = 0$ correspond to the values of x at those points where the graph touches the x-axis, we could estimate these values from the graph. This would not be particularly accurate, however. In this section we will describe a method of finding an approximation to an irrational root to any desired degree of accuracy.

The first step in finding an approximation to an irrational root will be to locate the root between two consecutive integers. To do this we use the Location Principle.

The Location Principle. If $P(x)$ is a polynomial and if $P(a)$ and $P(b)$ have different signs, then $P(x) = 0$ has a root between a and b.

Geometrically this means that if the graph of $y = P(x)$ is above the x-axis for $x = a$ ($P(a) > 0$) and below the x-axis for $x = b$ ($P(b) < 0$), then the graph must cross the x-axis at some point between a and b. Since every polynomial function is a smooth unbroken curve, this is intuitively obvious.

Example 3-33. The equation $P(x) = x^4 + x^3 - 4x^2 - 5x - 5 = 0$ has no rational roots. By Descartes' Rule of Signs, it has exactly one positive root. We compute $P(0) = -5$; $P(1) = -12$; $P(2) = -7$; $P(3) = 52$. We conclude by the Location Principle that $P(x) = 0$ has a root between 2 and 3.

Now to find a better approximation than this, we will make an assumption. We will assume that the graph of $y = P(x)$ between the two successive integers is very close to that of a straight line. It is not hard to find the point $(x, 0)$ where a straight line connecting the points $(a, P(a))$ and $(b, P(b))$ will cross the x-axis. The graph of $y = P(x)$ will cut the x-axis either to the right or to the left of this point. In Figure 3-9(a) two possibilities for $y = P(x)$ are represented by dotted lines.

If we assume that the graph is a straight line between the two points $(a, P(a))$ and $(b, P(b))$, then a pair of similar triangles is formed, [Figure 3-7(b)] and two of the corresponding sides have lengths $|P(a)|$ and $|P(b)|$. If we designate the distance from a to x by the letter d, then the corresponding side of the other triangle will have length $(b - a) - d$, since the distance from a to b is $b - a$, $a < b$. Since corresponding sides of similar triangles are proportional, we have

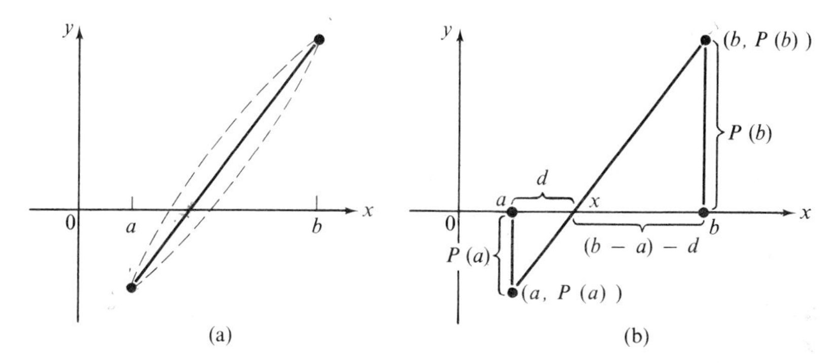

Figure 3-9

$$\frac{|P(a)|}{d} = \frac{|P(b)|}{b - a - d}$$

Solving for d, we get

$$d = \frac{|P(a)|(b - a)}{|P(a)| + |P(b)|} \tag{3-2}$$

Example 3-34. In Example 3-33, we found that $P(2) = -7$ and $P(3) = 52$. Since $a = 2$ and $b = 3$, $b - a = 1$ and we find

$$d = \frac{7}{52 + 7} \simeq .1$$

Since d is the distance from 2 to x, the point where the line crosses the x-axis, we find that $x \simeq 2.1$.

To find the root to a higher degree of accuracy, we can repeat the process. First we find $P(2.1) = -4.5$. Since this is negative and the curve is rising, the root must be larger than 2.1. Next we find $P(2.2) =$

-1.5. Since $P(2.2)$ is negative, the graph is still below the x-axis at $x = 2.2$, and our first approximation, 2.1, is too small. We find $P(2.3) = 2.6$ and we conclude that the root is between 2.2 and 2.3. Using Equation (3-2) again, where this time $b - a = .1$, we find

$$d = \frac{1.5(.1)}{1.5 + 2.6} \simeq .04$$

A second approximation to our root then is $2.2 + .04 = 2.24$. The last digit 4 is not significant, since the root may be closer to 2.25 or 2.23. The digit 2, however, is significant. To get three digit accuracy, in general, it would be necessary to carry out the process to four digits and then round off.

This method can be summarized as follows:

1. Use the Location Principle to locate a root between consecutive integers.
2. Sketch the graph in the neighborhood of this root.
3. Either estimate from your graph or use Equation (3-2) to find a first approximation to the nearest tenth.
4. Locate the root between successive tenths.
5. Use Equation (3-2) to find a second approximation to the nearest hundredth.
6. This process may be continued until the root is found to any desired degree of accuracy.

This method will be successful if the root is not a multiple root and if it is not too close to another root. In these cases other methods must be used.

Exercise 3-11

1. Find the positive root of $x^3 - 2x^2 - 5x + 4 = 0$ to three significant digits.
2. Find the positive root of $2x^3 + 3x^2 - 5x - 5 = 0$ to three significant digits.
3. Find the negative root of $x^3 - 3x^2 - 2x + 4 = 0$ to three significant digits.
4. Find the negative root of $x^4 + 2x^3 - 2x^2 - 10x - 15 = 0$ to three significant digits.
5. Use the methods of this section to find the cube root of 3 to four significant digits. (Hint: You want to find the positive root of $x^3 - 3 = 0$.)
6. Find the fifth root of 4 to three significant digits.

3-12 Rational Functions

A *rational function* is one which can be written as the ratio of two polynomials. If $P(x)$ and $Q(x)$ are polynomials and $Q(x)$ is not the zero polynomial, then $R(x) = P(x)/Q(x)$ is a rational function.

Example 3-35. $R(x) = x/(x^2 - 1)$

The domain of a polynomial is the set of all real numbers, but this is not the case for a rational function. Since division by zero is undefined, those numbers which make the denominator zero cannot be in the domain of the function. In Example 3-35, the domain of R is the set of all real numbers excluding 1 and -1, since these are the zeros of the denominator.

Example 3-36. $f(x) = [(x + 2)(x - 2)]/(x - 2)$

The domain of this function is the set of all real numbers excluding 2. Notice that the function $g(x) = x + 2$ is not the same function, although $f(x)$ and $g(x)$ have the same values for all $x \neq 2$. They do not have the same domain however, since 2 is in the domain of $g(x)$. Recall that two functions f and g are *equal* if they have the same domain and if $f(x) = g(x)$ for all x in that common domain. (See Section 2-8.) Cancelling common factors in a rational function may result in a different function. The graph of $f(x)$ is a line with a hole in it, since the point corresponding to $x = 2$ is missing. (See Figure 3-10.)

If $x - a$ is a factor of the denominator of a rational function, but $x - a$ is not a factor of the numerator, then the behavior of the function near $x = a$ is interesting.

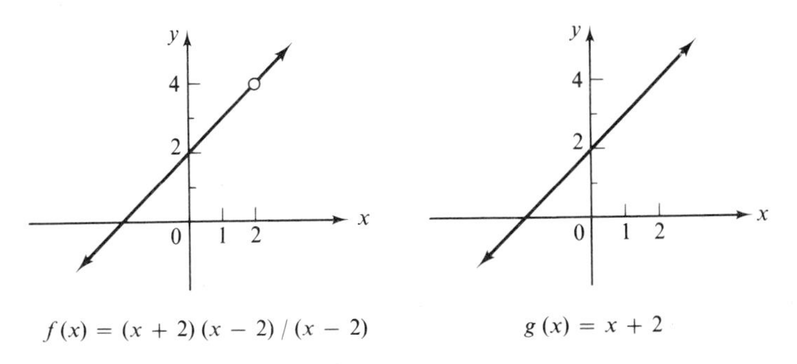

Figure 3-10

Example 3-37. Graph $y = x^2/[(x-1)(x+1)]$.

Note that the graph in Figure 3-11 does not touch the vertical lines $x = 1$ and $x = -1$. For values of x near 1 or -1, y is very large in absolute value. This is true because the variable y is a fraction whose denominator is getting closer and closer to zero while the numerator is getting closer and closer to 1 as x takes on values closer and closer to 1. These vertical lines $x = 1$ and $x = -1$ are called *vertical asymptotes* of the curve. Note that the curve cannot cross these vertical asymptotes since these values of x are not in the domain of the function.

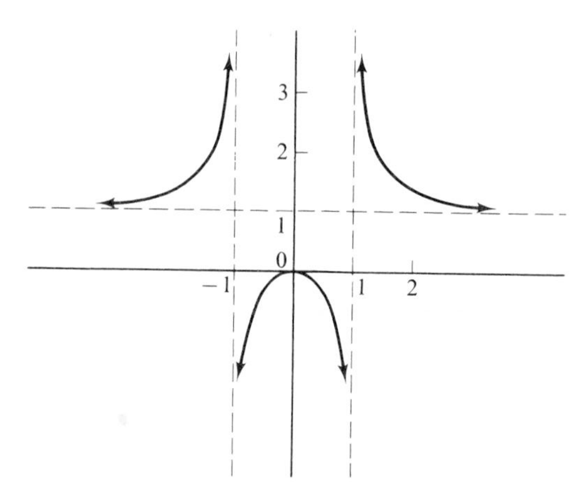

Figure 3-11

In Example 3-37 the line $y = 1$ is a *horizontal asymptote* for the graph. For large x, y is very close to 1. If the numerator and denominator polynomials have the same degree, and a is the leading coefficient for the numerator polynomial while b is the leading coefficient for the denominator polynomial, then $y = a/b$ is a horizontal asymptote.

We can see why this is true for Example 3-37 by writing

$$y = \frac{x^2}{(x^2-1)} = 1 + \frac{1}{(x^2-1)}$$

Clearly for large values of x the fraction $1/(x^2-1)$ will be very small; hence, y will be very close to 1. If the degree of the denominator polynomial is greater than the degree of the numerator polynomial, then $y = 0$ is a horizontal asymptote. If the degree of the numerator is greater than that of the denominator, then the graph has no horizontal asymptote.

In summary:

1. The graph of a rational function $R(x)$ will have a vertical asymptote $x=a$ if $x-a$ is a factor of the denominator polynomial, but $x-a$ is not a factor of the numerator polynomial.
2. Suppose

$$R(x)=\frac{a_nx^n+\ldots+a_1x+a_0}{b_mx^m+\ldots+b_1x+b_0}$$

   (i) If $m=n$ (i.e., the degree of the numerator and denominator polynomials is the same), then $y=a_n/b_m$ is a horizontal asymptote.
   (ii) If $m>n$, $y=0$ is a horizontal asymptote.
   (iii) If $n>m$, there are no horizontal asymptotes.

Example 3-38. The graph of $y=3x/(x-1)$ has vertical asymptote $x=1$ and horizontal asymptote $y=3$.

Example 3-39. The graph of $y=x/[(x-1)(x+3)]$ has two vertical asymptotes, $x=1$ and $x=-3$. It has the horizontal asymptote $y=0$.

Example 3-40. The graph of $y=x^3/x(x-2)$ has vertical asymptote $x=2$. Since x is a factor of the numerator, $x=0$ is not a vertical asymptote; however, $x=0$ is not in the domain of the function. This graph has no horizontal asymptote.

The *roots* of a rational function $R(x)=P(x)/Q(x)$ are those values of x for which $P(x)=0$ and $Q(x)\neq 0$.

Example 3-41. The only root of $R(x)=[(x+1)(x-2)]/(x+1)$ is $x=2$, since $R(2)=3\cdot 0/3=0/3=0$. When $x=-1$, $R(-1)=0\cdot(-3)/0=0/0$, which is undefined; therefore, $x=-1$ is not a root.

Example 3-42. The equation $y=3/(x-1)$ has no roots.

In graphing a rational function, the following information will be helpful:

1. Find the domain of the function. This tells us where the curve will be found as well as where it will not. In Example 3-37, we know that the curve between -1 and 1 cannot be the whole thing, since values of x greater than 1 and less than -1 are in the domain of the function.

2. Find the horizontal and vertical asymptotes.
3. Find the points where the curve cuts the x- and the y-axes. These are called the x- and y-intercepts.
4. Make use of any symmetries the graph may have. It will be symmetric with respect to the y-axis if $R(-x) = R(x)$, and it will be symmetric with respect to the origin if $R(-x) = -R(x)$. (See Section 2-7.)
5. Finally, plot some points on the graph. Points close to and on either side of the vertical asymptotes will usually be helpful.

Example 3-43. Graph $y = (2x + 1)/(x - 1)$.

Solution:
1. The domain of the function is the set of all real numbers except $x = 1$.
2. A vertical asymptote is $x = 1$ and a horizontal asymptote is $y = 2$.
3. When $x = 0$, $y = -1$, and when $y = 0$, $x = -1/2$.
4. The graph is not symmetric with respect to the y-axis or with respect to the origin. (See Figure 3-12.)
5.

x	1.5	2	3	4	5	.5	0	−.5	−1	−2	−3
y	8	5	3.5	3	2.75	−4	−1	0	.5	1	1.25

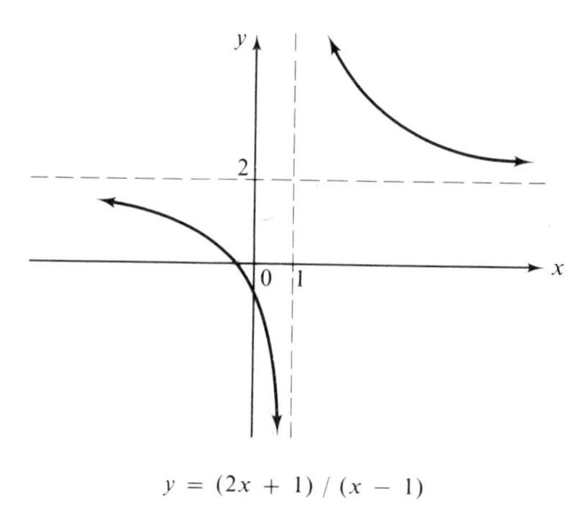

$y = (2x + 1)/(x - 1)$

Figure 3-12

Exercise 3-12

1. Give the domain of each of the following rational functions:
   (a) $y = (2x + 3)/[x(x + 1)]$
   (b) $y = (x - 1)/(x^2 + 2x - 3)$

(c) $y = (x^2 - 2x + 1)/(x^2 + 1)$
(d) $y = x^3/(x^2 + 3x)$
(e) $y = (3x + 1)/(2x - 1)$

2. Find the horizontal and vertical asymptotes for each of the functions in problem 1.

3. Are the functions $f(x) = x^2/x$ and $g(x) = x$ equal? Explain. What is the domain of each?

4. Are the functions $f(x) = x/x^2$ and $g(x) = 1/x$ equal? Explain. What is the domain of each?

5. Sketch the graph of each of the following rational functions. Give the domain of each and find all horizontal and vertical asymptotes. Determine any symmetries the graph may have.
(a) $y = x/(x^2 - 1)$
(b) $y = 2x/(x + 1)$
(c) $y = x^2/(x^2 - 1)$
(d) $y = 1/(x^2 + 1)$
(e) $y = 3x^2/(x^2 - 4)$

6. Find the roots of the following rational functions:
(a) $y = [(x + 1)(x + 3)]/[(2x - 1)(x + 3)]$
(b) $y = (x^3 - x)/(x + 1)$
(c) $y = 2/(x - 3)$
(d) $y = (4x^2 - 4x - 3)/(2x^4 - 3x^3)$
(e) $y = x/(x^2 + 1)$

7. A certain rational function has the following properties:
(a) The degree of both the numerator and denominator polynomials is one.
(b) The graph has the vertical asymptote $x = -1$ and the horizontal asymptote $y = 3$.
(c) The graph contains the point (0, 1).
Find a formula for this function.

8. A certain rational function has the following properties:
(a) The degree of the numerator polynomial is zero and the degree of the denominator polynomial is two.
(b) The graph has the vertical asymptotes $x = 1$ and $x = -1$, and $y = 0$ is a horizontal asymptote.
(c) The graph contains the point (0, −2).
Find a formula for this function.

Review Problems

1. Let $f(x) = ax + b$ and $g(x) = cx + d$ be two linear functions. Show that $g \circ f$ and $f + g$ are also linear functions.

2. Find the vertex of $y = -2x^2 + 3x - 1$ and graph this function. What is its range?

3. Choose values for the constant k so that the roots of $x^2 + 3x + k = 0$ are real and unequal; real and equal; complex.

4. State the Remainder Theorem, the Factor Theorem, the Fundamental Theorem of Algebra, and the Rational Root Theorem.

5. List all of the roots of the equation $x^2(2x + 1)(x - 3)^2(x^2 + 2x - 3) = 0$. What is the degree of this equation?

6. Find all of the roots of $x^3 - x^2 - x - 15 = 0$ if $-1 + 2i$ is a root of this equation.

7. Find an equation of lowest degree with rational coefficients having the roots -1, $1 + 2i$, $\sqrt{3}$.

8. If $P(x) = 3x^4 - 7x^2 + 3x - 9$, use synthetic division to find $P(-2)$. Use synthetic division to find the quotient and the remainder when $P(x)$ is divided by $x - 3$.

9. Given $2x^4 - 5x^3 + 7x^2 - 25x - 15 = 0$.
   (a) List all possible rational roots.
   (b) Find the smallest nonnegative integer and the largest nonpositive integer which our tests show to be an upper and a lower bound for the roots.
   (c) Find all of the roots of this equation.

*10. Use Descartes' Rule of Signs to classify the roots of the equation of problem 9.

*11. Given $2x^5 - x^4 + 4x^3 - x^2 - 16x + 12 = 0$. List all possible rational roots and test them two at a time by the method of Section 3-10 until all of the roots are found.

*12. Find the positive root of $x^3 + x^2 - 5x - 5 = 0$ correct to two significant digits.

13. Given $y = 2x/(x + 1)$.
   (a) Give the domain of this function.
   (b) Find its horizontal and vertical asymptotes.
   (c) Graph this function.

Bibliography

Bell, E. T., *Men of Mathematics*, Ch. 20. New York: Simon & Schuster, Inc., 1937.

Dickson, Leonard E., *New First Course in the Theory of Equations*. New York: John Wiley & Sons, Inc., 1939.

Eves, Howard, *An Introduction to the History of Mathematics*, Rev. Ed., Ch. 2, 3. New York: Holt, Rinehart & Winston, 1964.

Hardy, G. H., *A Course of Pure Mathematics*, Ch. 2. New York: The Macmillan Company, 1943.

Hooper, Alfred, *Makers of Mathematics*, Ch. 3. New York: Random House, 1948.

Kline, Morris, *Mathematics, A Cultural Approach*, Ch. 5. Reading, Massachusetts: Addison-Wesley, 1962.

———, *Mathematics in Western Culture*, Ch. 8. New York: Oxford University Press, 1953.

4

The Exponential and Logarithm Functions

4-1 Exponents

When we see the symbol 3^2, we know that this means $3 \cdot 3$. Similarly 3^3 is an abbreviation for $3 \cdot 3 \cdot 3$.

Definition 4-1. If a is some real number and n is a positive integer, a^n means

$$\underbrace{a \cdot a \cdot a \cdot \ldots \cdot a}_{n \text{ factors}}$$

or the product of n factors of a.

The number n is called the *exponent,* and the number a is called the *base.*

We have not always had this convenient notation for exponents. In the sixteenth century entirely different symbols were used for different powers of the variable. The following equation was written in 1559.

$$1 \diamond P\ 6\ \rho\ P\ 9\ [\ 1 \diamond P\ 3\ \rho\ P\ 24$$

The Greek letter ρ stands for the unknown, corresponding to our letter x. The diamond shaped symbol $\diamond$ stands for the second power of the unknown, or x^2. The letter P stands for plus and the bracket symbol, [, was used for equals. The equation inmodern notation would be written

$$1x^2 + 6x + 9 = 1x^2 + 3x + 24$$

As late as the eighteenth century, in American colleges x^2 was written $x \cdot x$ and x^3 as $x \cdot x \cdot x$.

From Definition 4-1 it is easy to see that $a^2 \cdot a^3 = (a \cdot a) \cdot (a \cdot a \cdot a) = a^5$, or in general, we have the following rule:

I. For two positive integers m and n, $a^n a^m = a^{n+m}$.

Example 4-1. $2^3 \cdot 2^5 = 2^8$

Since $a^3/a^2 = (a \cdot a \cdot a)/(a \cdot a) = a$, provided a is not zero, we have the following rule:

II. For two positive integers m and n, with $m > n$, and $a \neq 0$, $a^m/a^n = a^{m-n}$.

Example 4-2. $3^5/3^2 = 3^3$

These rules apply only if all of the bases are the same. Clearly we could not apply Rule I to $2^3 \cdot 3^2$ or Rule II to $2^5/5^3$ since the bases of the exponents are different.

Observing that $(a^2)^3 = a^2 \cdot a^2 \cdot a^2 = a \cdot a \cdot a \cdot a \cdot a \cdot a = a^6$, we state the third rule of exponents. For positive integers m and n, $(a^m)^n = a^{mn}$. Since multiplication is commutative, $mn = nm$, we have the following rule:

III. For positive integers m and n, $(a^n)^m = (a^m)^n = a^{mn}$.

Example 4-3. $(5^2)^3 = (5^3)^2 = 5^6$

In addition, we also state that for real numbers a and b and for any positive integer n, the following rules apply:

IV. $(a \cdot b)^n = a^n \cdot b^n$

V. $(a/b)^n = a^n/b^n, b \neq 0.$

Example 4-4. $(3 \cdot 7)^2 = 3^2 \cdot 7^2$
$(3/2)^5 = 3^5/2^5$

Rule IV says that the order of the operations of multiplication and of raising to powers can be interchanged. To evaluate $(3 \cdot 4)^2$ we can multiply first, then raise to the second power, $(3 \cdot 4)^2 = 12^2 = 144$, or we can square each factor first and then multiply, $(3 \cdot 4)^2 = 3^2 \cdot 4^2 = 9 \cdot 16 = 144$. Rule V tells us that the order of the operations of division and raising to powers can be interchanged.

It is perhaps worth noting here that there is *no* rule which states that $(a + b)^n = a^n + b^n$. This statement is not true, as an example will show: $(1 + 2)^2 = 3^2 = 9$; $1^2 + 2^2 = 1 + 4 = 5$.

The five laws of exponents given above follow directly from our definition of a^n, for n a positive integer. We would like to expand this definition so that an exponent might be zero, a negative integer, a fraction, or even an irrational number. In making these new definitions we will be concerned with preserving Rules I, II, and III.

For example, in defining $a^0 (a \neq 0)$ we will be motivated by the desire that Rule II will continue to hold when $m = n$. If so, we would have $a^m/a^m = a^{m-m} = a^0$. Since $a^m/a^m = 1$, we are forced to Definition 4-2.

Definition 4-2. $a^0 = 1, a \neq 0$

Example 4-5. $5^0 = 1$; $(-1)^0 = 1$; $100^0 = 1$

In the same way, if Rule II is to continue to hold when $m < n$, we would have, for example, $a^2/a^3 = a^{2-3} = a^{-1}$. Since $a^2/a^3 = (a \cdot a)/(a \cdot a \cdot a) = 1/a$, we must define $a^{-1} = 1/a$, and in general, for any integer n, we have the following definition:

Definition 4-3. $a^{-n} = 1/a^n, a \neq 0$

Example 4-6. $2^{-1} = 1/2$; $3^{-2} = 1/3^2$; $1/5^{-3} = 5^3$

In defining rational number exponents, we will be careful to do so in such a way that Rules I and III will still hold. For example, if Rule I is to apply for m and n rational numbers, then $2^{1/2} \cdot 2^{1/2} = 2^{1/2+1/2} = 2$, and $2^{1/2}$ must be a square root of 2. Similarly, $5^{1/3} \cdot 5^{1/3} \cdot 5^{1/3} = 5^{1/3+1/3+1/3} = 5$, and $5^{1/3}$ must be a cube root of 5.

Now any number has two square roots, three cube roots, and in general n nth roots, because the nth roots of a number a are the roots of the equation $x^n = a$. This equation has n different roots, except when a is zero. Not all of these roots will be real numbers, however. If a is positive and n is even, two of the roots will be real, one positive and one negative.

Example 4-7. The number 4 has two square roots, 2 and -2. The number 16 has four fourth roots. Two of them are real, 2 and -2; the other two are complex, $2i$ and $-2i$.

If a is either positive or negative and n is odd, then only one of the nth roots will be real.

Example 4-8. The number 8 has three cube roots, 2, $-1 + \sqrt{3}$, and $-1 - \sqrt{3}$. These are the roots of $x^3 = 8$. The number -1 has five fifth roots. Only one of these, -1, is real.

If a is negative and n is even, then a will have no real roots, they will all be complex.

Example 4-9. The number -4 has two square roots, $2i$ and $-2i$.

Since we want $a^{1/n}$ to be a unique real number, we will define this symbol as follows:

Definition 4-4. For a any real number and n a positive integer greater than 1,

$$a^{1/n} = \sqrt[n]{a}$$

where $\sqrt[n]{a}$ is chosen as follows:

(i) The number $a^{1/n}$ is a real number.
(ii) If a has two real nth roots, $a^{1/n}$ is chosen to be the positive one.

Example 4-10. $4^{1/2} = \sqrt{4} = 2$; $8^{1/3} = \sqrt[3]{8} = 2$; $(-1)^{1/3} = \sqrt[3]{-1} = -1$. The symbol $(-4)^{1/2} = \sqrt{-4}$ is undefined since there is no real number whose square is -4.

Since $p/q = (1/q)(p)$, if Rule III is to hold, then $a^{p/q} = a^{(1/q)(p)} = (a^{1/q})^p$. Moreover, $p/q = p(1/q)$, hence $a^{p/q} = a^{p(1/q)} = (a^p)^{1/q}$. Therefore, we have the following definition:

Definition 4-5. For a a real number and p and q positive integers,

$$a^{p/q} = (a^{1/q})^p = (\sqrt[q]{a})^p$$
$$= (a^p)^{1/q} = \sqrt[q]{a^p}$$

In other words, to find $a^{p/q}$ we can either find the qth root of a, then raise this number to the pth power, or equivalently we can raise a to the pth power, then find the qth root of this number.

Example 4-11 $4^{3/2} = (\sqrt{4})^3 = 2^3 = 8$. Equivalently, $4^{3/2} = \sqrt{4^3} = \sqrt{64} = 8$. $(-1)^{4/3} = (\sqrt[3]{-1})^4 = (-1)^4 = 1$, or $(-1)^{4/3} = \sqrt[3]{(-1)^4} = \sqrt[3]{1} = 1$.

Finally we can extend Definition 4-5 so that it holds for all rational number exponents, both positive and negative, by defining $a^{-p/q} = 1/a^{p/q}$.

We have now defined a^n for n a positive integer, for $n = 0$, for n a negative integer, and for n a rational number, provided the value of a is suitably restricted. The question now arises, can we define a^n for n *any* real number? Specifically, how could we define a^n for n an irrational number? What meaning can we assign to $2^{\sqrt{2}}$ for example? It seems reasonable that if the rational numbers 1, 1.4, 1.41, 1.414, 1.4142, etc., are successive approximations to $\sqrt{2}$, then 2^1, $2^{1.4}$, $2^{1.41}$, $2^{1.414}$, $2^{1.4142}$ will be successive approximations to $2^{\sqrt{2}}$. Since the exponents in this sequence are all rational, we can find any term in the sequence to any desired degree of accuracy. For example, $2^{1.4} = 2^{1+.4} = 2^1 \cdot 2^{4/10} = 2 \cdot \sqrt[10]{2^4} \simeq 2.639$.

Let us summarize. We now have six definitions:

Definition 4-1. For n a positive integer and a a real number,

$$a^n = \underbrace{a \cdot a \cdot a \cdot \ldots \cdot a}_{n \text{ factors}}$$

Definition 4-2. If $a \neq 0$, $a^0 = 1$.

Definition 4-3. For n a real number and $a \neq 0$,

$$a^{-n} = \frac{1}{a^n}$$

Definition 4-4. For n a positive integer greater than one, and for suitable values of a

$$a^{1/n} = \sqrt[n]{a}$$

Definition 4-5. For positive integers p and q, and for suitable values of a

$$a^{p/q} = \sqrt[q]{a^p} = (\sqrt[q]{a})^p$$

Definition 4-6. For r irrational, and for suitable values of a

$$a^r \simeq a^{p/q}$$

where $r \simeq p/q$.

We also have five Laws of Exponents:

For any real numbers m and n, and for any real numbers a and b for which the expression is defined,

I. $a^n \cdot a^m = a^{n+m}$.
II. $a^n/a^m = a^{n-m}$.
III. $(a^n)^m = (a^m)^n = a^{nm}$.
IV. $(a \cdot b)^n = a^n \cdot b^n$.
V. $(a/b)^n = a^n/b^n$.

Exercise 4-1

1. Use the laws of exponents to simplify the following expressions. Your answer should contain only positive exponents.
   (a) $5^2 \cdot 5^3$ (b) $3^{-2} \cdot 3^5 \cdot 3$
   (c) $7^2 \cdot 7^{-5}$ (d) $2^5 \cdot 2^{-2} \cdot 2^0$
   (e) $(3^2 \cdot 3^{-3} \cdot 3^0)/(3 \cdot 3^5 \cdot 3^{-2})$ (f) $(2^0 \cdot 2^3 \cdot 5^{-1})/(5^3 \cdot 2^{-4})$
2. Express as fractional exponents:
   (a) $\sqrt[3]{x^2}$ (b) $(\sqrt{x-y})^5$
   (c) $1/\sqrt[3]{a^7}$ (d) $\sqrt{10}/\sqrt[3]{10}$
   (e) $(\sqrt[5]{x^2+y^2})^3$ (f) $1/\sqrt[4]{(x-y)^5}$
3. Evaluate:
   (a) $4^{3/2}$ (b) $(-1)^{2/3}$

(c) $(-1)^{5/3}$ (d) $27^{-1/3}$
(e) $(-32)^{1/5}$ (f) $(1/9)^{-1/2}$
(g) $8^{-2/3}$ (h) $(4/25)^{-1/2}$
(i) $16^{3/4}$ (j) $(9/4)^{-3/2}$

4. Simplify each of the following expressions. Your answer should contain only positive exponents.
   (a) x^5/x^7 (b) a^{n+3}/a^{n-2}
   (c) x^{2m+1}/x^{m-4} (d) $z^{1/2}/z^{3/4}$
   (e) $(a+b)^{1/3}/(a+b)^{5/6}$ (f) $(x-y)^{-1/3}/(x-y)^{-1/2}$

5. Evaluate:
   (a) $2^3 + 2^0 + 2^{-1} + 2^{-2}$
   (b) $4^{3/2} + 4^{1/2} + 4^{-1/2} + 4^{-3/2}$
   (c) $10^3 + 10^2 + 10^1 + 10^0 + 10^{-1} + 10^{-2}$
   (d) $4 \cdot 10^2 + 1 \cdot 10^1 + 2 \cdot 10^0 + 8 \cdot 10^{-1} + 7 \cdot 10^{-2}$

6. Simplify (a) through (e). Your answer should contain only positive exponents.
   (a) $(x^{-1})^{-2}$
   (b) $(a^{2/3} \cdot b^{-1/6} \cdot c^{-3/2})/(a^{-1/3} \cdot b^{1/3} \cdot c^{3/2})$
   (c) $\sqrt{a^2/\sqrt{b}}$
   (d) $\sqrt[3]{x^{-2}}$
   (e) $(\sqrt{x} \cdot x^{-5/2})/(\sqrt[6]{x^5})$

7. Give an example to show that
   (a) $a^n + a^m \neq a^{n+m}$.
   (b) $(a-b)^n \neq a^n - b^n$.
   (c) $(a^n)^m \neq a^{(n^m)}$.

8. What is $1^{p/q}$ for any rational number p/q? What is $(-1)^{p/q}$ if p is even and q is odd? What is $(-1)^{p/q}$ if both p and q are odd?

4-2 The Exponential Function

An exponential function is a function of the form

$$y = a^x$$

where a is a positive real constant, $a \neq 1$.

Although superficially this may look very much like the polynomial function $y = x^n$, the two functions are quite different. In the polynomial the exponent is fixed and the base is the variable, e.g., $y = x^2$. For the exponential function the situation is just the reverse; the base is a constant and the exponent is the variable. Some examples of exponential functions are $y = 2^x$ and $y = 10^x$.

An interesting characteristic of the exponential function is that it increases very, very rapidly for values of the base greater than one, much more rapidly than any polynomial. Thus, although for small values of x, $y = a^x (a > 1)$ may be smaller than $y = x^n$, eventually there will be some value, for instance x_0, for which $a^x > x^n$ for all x greater than x_0, no matter how large n is. For example, for small values of x, $2^x < x^{10}$, but for $x = 100$, $x^{10} = 100^{10} = (10^2)^{10} = 10^{20}$, while $2^x = 2^{100} = (2^4)^{25} = 16^{25}$. Thus, for $x \geq 100$, $2^x > x^{10}$.

Since a^x, $a > 0$, was defined for all real values of x in Section 4-1, the domain of an exponential function is the set of all real numbers. The range of any exponential function is the set of positive real numbers.

From Figure 4-1 we can see that the graphs of $y = a^x$ for all $a > 0$, $a \neq 1$, have similar features. Since $a^0 = 1$, all of the exponential functions pass through the point (0, 1). Moreover, a^x is never zero and is never negative. This is reflected in the fact that the graph of $y = a^x$ always lies above the x-axis.

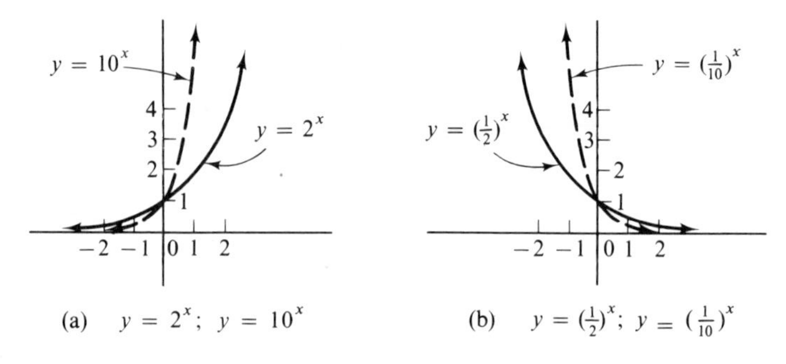

Figure 4-1

The exponential function

If a is greater than one, then the graph of $y = a^x$ rises rapidly as x increases. [See Figure 4-1 (a).] For negative values of x which are large in absolute value, a^x is very close to zero. For example, $10^{-2} = .01$; $10^{-5} = .00001$.

For values of a between zero and one, the graph of the function $y = a^x$ falls as x increases. [See Figure 4-1 (b).] In this case, for large positive values of x, a^x is very close to zero.

$$\left(\frac{1}{2}\right)^{10} = \frac{1}{1024} \simeq .001$$

The exponential functions in Figure 4-1 (a) are examples of a class of functions which are called strictly increasing functions. The functions in Figure 4-1 (b) are called strictly decreasing functions.

Definition 4-7. A function f is said to be *strictly increasing* if for any x_1, x_2 in the domain of f, whenever $x_1 < x_2$, it follows that $f(x_1) < f(x_2)$.

Definition 4-8. A function f is said to be *strictly decreasing* if for any x_1, x_2 in the domain of f, whenever $x_1 < x_2$, it follows that $f(x_1) > f(x_2)$.

It is not hard to see that any function which is strictly increasing or strictly decreasing must be one-to-one and therefore will have an inverse.

Exercise 4-2

1. If $y = (1/3)^{-x}$, complete the following table:

x	2	1	0	-1	-2
y					

   Graph the function; give its domain; give its range.

2. If $y = 2^{-x}$, complete the following table:

x	2	1	0	-1	-2
y					

   Graph the function; give its domain; give its range.

3. Graph each of the following functions:
   (a) $y = 3^x$ (b) $y = (1/3)^x$
   (c) $y = 2^{2x}$ (d) $y = (2/3)^x$
   (e) $y = 1^x$ (f) $y = 2^{x+1}$

4. Graph the inverse of the functions in problem 3 (a) and (b).

5. What kind of difficulties might you encounter in defining $y = a^x$ for a a negative number and in defining $y = a^x$ for $a = 0$?

6. Find x if:
   (a) $2^x = 1$ (b) $2^{x+1} = 8$
   (c) $2^{2x-1} = 1/4$ (d) $9^{x-1} = 27$
   (e) $10^{x-1} = .001$ (f) $17^x = 1$

7. Locate x between consecutive integers if:
   (a) $10^x = 3$ (b) $2^x = 17$
   (c) $10^x = .09$ (d) $2^x = .1$
   (e) $10^x = 127$ (f) $3^x = .2$

8. If $f(x) = a^x$, $g(x) = (1/a)^x$, $a > 0$, find:
   (a) $f \cdot g$ (b) $g \circ f$
   (c) $f(-x)$ (d) f/g

4-3 The Number *e* and the Function $y = e^x$

The base of an exponential function can be any positive number except one. A very commonly used base is the number designated by the symbol e.* The number e, like π, is irrational and is approximately equal to 2.718281828459045. Like π, the number e arises in many applications.

Banking. It is common to see banks advertise that they compound interest *continuously*. Suppose that you were fortunate enough to find a very generous bank that paid 100% interest on savings. If you invested \$1.00 and the bank compounded interest annually, at the end of the year your savings would have increased to \$2.00. If the bank compounded twice a year, at the end of six months you would have \$1.50. Interest on this for six months more would be 75¢, so at the end of a year your savings would be \$2.25, or $(1 + 1/2)^2$. If the bank compounded quarterly, at the end of the year you would have $(1 + 1/4)^4$, or \$2.44. In general, if the bank compounded interest n times a year, your dollar would accumulate to $(1 + 1/n)^n$ at the end of one year at interest of 100%.

Now suppose the bank compounded interest *continuously*. You might think that at the end of a year you would be a millionaire (or even an *infinitaire*) and the bank would be out of business. Not at all. At the end of one year compounded continuously at 100%, your investment would have increased to e dollars, or approximately \$2.72. To put it another way, the sequence of numbers

$$(1 + 1)^1, (1 + 1/2)^2, (1 + 1/3)^3, \ldots, (1 + 1/n)^n, \ldots$$

does not increase without bound, but gets closer and closer to the number e as n gets larger and larger. We say that the *limit* of this sequence is e.

If the bank is not so generous and pays interest at 5%, for instance, compounded continuously, then at the end of one year an investment of \$100 would increase to $\$100 \cdot e^{.05}$, or approximately \$105.13. In general, if the interest is designated by the letter r, then a principle of P dollars compounded continuously would amount to $P \cdot e^r$ at the end of one year. This is of course an example of an exponential function whose base is e.

Biology. Suppose that a colony of bacteria has ample food and no poisons are present to inhibit their growth. Under such circumstances their growth will be extremely rapid, in fact so rapid that in order to find a formula giving the number of bacteria x present at any time t, we

* The letter e comes from Euler, the Swiss mathematician who devised this symbol.

must use the exponential function

$$x = x_0 e^{kt}$$

Here x_0 is the number of bacteria present at the beginning of the experiment (when $t = 0$), and k is a positive constant which depends on the type of bacteria.

Chemistry. Suppose that some substance A is being converted by a chemical reaction into another substance B. At the beginning of the reaction ($t = 0$), no B is present and the amount of A present is x_0. The function which gives x, the amount of A present at any time t, is

$$x = x_0 e^{-kt}$$

where again k is a positive constant which depends on the chemicals involved. The minus sign occurs because this is a decreasing function. As t increases, x, the amount of A present, decreases.

Archeology. The archeologist can calculate the age of certain ancient objects by a method called carbon 14 dating. This method depends on the rate of decay of carbon 14, a radioactive isotope of ordinary carbon 12. Both forms of carbon are found in all living things and their proportion remains constant during the life of the organism. When the organism dies, the ratio of carbon 14 to carbon 12 begins to change. The amount of carbon 12 stays the same, but the unstable carbon 14 begins to break down. The radioactive decay of carbon 14 occurs at a predictable rate and can be measured. Thus, the rate of decay of carbon 14 in an ancient bone or piece of wood can be measured and used to calculate its age. The formula giving the amount x of carbon 14 present at any time t is an exponential function,

$$x = x_0 e^{-kt}$$

where x_0 stands for the amount of carbon 14 present at the time the organism died, and $k \simeq .0001203$. Using this method the date of the Dead Sea Scrolls has been set at about 20 B.C.

In general, whenever we find a substance acting in such a way that its rate of change is proportional to the amount of the substance present, then the equation giving the amount of the substance at any time will involve an exponential function.

4-4 The Logarithm Function

Since every exponential function $y = a^x$, $a > 0$, $a \neq 1$, is one-to-one, each of these functions has an inverse. The inverse of an exponential function is called a *logarithm function*. Specifically, if $y = a^x$, then $x =$

$\log_a y$. (Read "x is the logarithm of y to the base a.") To put it another way, if $f(x) = a^x$, then $f^{-1}(x) = \log_a x$. The constant a is called the *base* of the logarithm, just as it was called the base of the exponential.

We say that

$$y = a^x \quad \text{if and only if } x = \log_a y \tag{4-1}$$

This relationship is easy to remember if we note that (1) the base of the exponential is the base of the logarithm, and (2) the logarithm is the exponent. The logarithm of y to the base a is the exponent to which a must be raised to give the number y.

Example 4-12. Since $2^3 = 8$, then $\log_2 8 = 3$. Since $10^{-2} = .01$, then $\log_{10} .01 = -2$. If $\log_{10} x = 2$, then $10^2 = x$ and $x = 100$. If $\log_3 9 = x$, then $3^x = 9$ and $x = 2$.

Since the domain of these exponential functions is the set of all real numbers and the range is the set of positive real numbers, then it follows that the domain of every logarithm function is the set of all positive real numbers while its range is the set of all real numbers. We can draw the graph of a logarithm function by reflecting the graph of the corresponding exponential function across the line $y = x$. In Figure 4-2 we have sketched the graphs of $y = 2^x$ and its inverse $y = \log_2 x$, and of $y = 10^x$ and its inverse $y = \log_{10} x$.

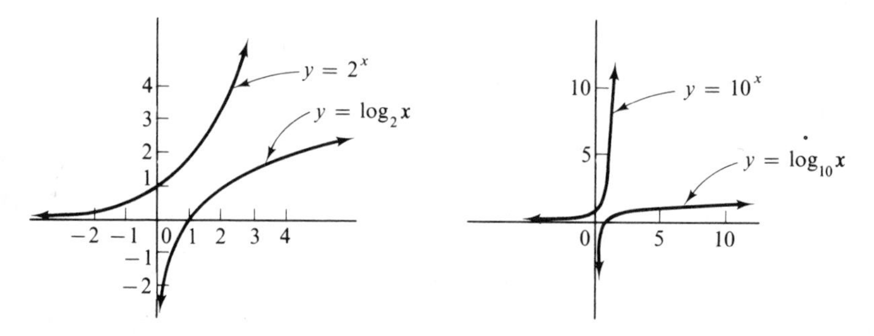

Figure 4-2

Just as every exponential function contains the point (0, 1), so every logarithm function contains the point (1, 0). That is, $\log_a 1 = 0$ for any base a, because $a^0 = 1$. For $a > 1$, the logarithm function is strictly increasing, and if $x > 1$, $\log_a x$ is positive. If $0 < x < 1$, then $\log_a x$ is negative. Since the domain of the logarithm function is the set of positive real numbers, $\log_a 0$ is undefined, as is the logarithm of a negative number.

Example 4-13. $\log_{10}100 = 2; \log_2(1/4) = -2$

The base e is so important that sometimes $y = e^x$ is referred to as *the* exponential function, and is sometimes written $y = \exp x$. Its inverse, $y = \log_e x$, is called the *natural logarithm* function and is usually designated by the symbol $y = \ln x$. (The letters ln stand for natural logarithm.)

Recall that the composite of a function and its inverse is the identity function (Section 2-9). Thus, $(f \circ f^{-1})(x) = f(f^{-1}(x)) = x$, for all x in the domain of f^{-1} and $(f^{-1} \circ f)(x) = f^{-1}(f(x)) = x$, for all x in the domain of f. Since the exponential and logarithm functions are inverses of each other, it follows that

$$\log_a(a^x) = x, \text{ for all real numbers } x$$

and

$$a^{\log_a x} = x \qquad x > 0$$

If $a = e$, then we have $\ln(e^x) = x$ for all x, and $e^{\ln x} = x; x > 0$.

We have seen that a logarithm is an exponent, and the behavior of the logarithm function reflects this. By using the Laws of Exponents we can prove the following three theorems which are called the Laws of Logarithms.

For $a > 0, a \neq 1, M, N > 0$, we have the following:

Theorem 4-1. $\log_a(M \cdot N) = \log_a M + \log_a N$

Theorem 4-2. $\log_a(M/N) = \log_a M - \log_a N$

Theorem 4-3. $\log_a M^p = p \log_a M$

These laws can be proved by using the relation (4-1) between the logarithm and the exponential functions.

Proof of Theorem 4-1: Let $\log_a M = x$ and $\log_a N = y$. Then $a^x = M$ and $a^y = N$. Now $M \cdot N = a^x \cdot a^y = a^{x+y}$ by the first Law of Exponents. Since $M \cdot N = a^{x+y}$, by relation (4-1) $\log_a(M \cdot N) = x + y$, and thus, $\log_a(M \cdot N) = \log_a M + \log_a N$. ■

The proofs of the other two Laws of Logarithms are left for the student (Exercise 4-4, problems 7 and 8).

Example 4-14.

$$\log_a(xyz) = \log_a x + \log_a y + \log_a z$$
$$\log_a x^2 = 2 \log_a x$$
$$\log_a(xy/z) = \log_a x + \log_a y - \log_a z$$
$$\log_a(\sqrt{x} \cdot y^3)/z^2 = (1/2)\log_a x + 3 \log_a y - 2 \log_a z$$

From Figure 4-2, we note that the logarithm function is one-to-one, and this leads us to two useful, if obvious, conclusions:

(1) If $M = N$, $(M, N > 0)$, then $\log_a M = \log_a N$.
(2) If $\log_a M = \log_a N$, then $M = N$.

Observation (1) allows us to perform such operations as taking the logarithm of both sides of an equation (as long as both sides are positive), and observation (2) allows us to reverse this process.

Example 4-15. Solve for x: $5^x = 3$.

Solution: Taking the logarithm of both sides to the base a, we have $\log_a 5^x = \log_a 3$. By the third Law of Logarithms this can be written $x \log_a 5 = \log_a 3$, hence $x = \log_a 3/\log_a 5$.

Example 4-16. Solve for x: $\log_a(x + 1) = \log_a 5$.

Solution: By observation (2), $x + 1 = 5$, and $x = 4$.

Exercise 4-4

1. Graph $y = e^x$ ($e \simeq 2.71$) and its inverse $y = \ln x$ on the same set of axes.

2. Rewrite each of the following in exponential form:
   (a) $\log_2 4 = 2$ (b) $\log_3(1/9) = -2$
   (c) $\log_{10} 1000 = 3$ (d) $\log_8 2 = 1/3$
   (e) $\log_4(1/2) = -1/2$ (f) $\log_{10} .001 = -3$

3. Rewrite each of the following in logarithmic form:
   (a) $4^2 = 16$ (b) $2^{-1} = 1/2$
   (c) $5^3 = 125$ (d) $2^5 = 32$
   (e) $27^{-1/3} = 1/3$ (f) $64^{1/6} = 2$
   (g) $10^{-1} = .1$ (h) $16^{3/2} = 64$

4. Find x in each of the following equations:
   (a) $\log_3 9 = x$ (b) $\log_2 x = 3$
   (c) $\log_4 2 = x$ (d) $\log_8(1/2) = x$
   (e) $\log_{10} x = 3$ (f) $\log_2 x = -2$
   (g) $\log_x 25 = 2$ (h) $\log_x(1/27) = -3$
   (i) $\log_{16} 8 = x$ (j) $\log_x 4 = 1/2$

5. Evaluate:
   (a) $\log_2(2^3)$ (b) $\log_{10}(10^{-1})$
   (c) $10^{\log_{10}5}$ (d) $2^{\log_2 10}$
   (e) $\ln(e^3)$ (f) $e^{\ln 4.2}$
   (g) $\log_a a$ (h) $\log_a 1$
   (i) $\log_a(1/a)$ (j) $\log_a(a^x)$

6. Compute $\log_{10}.1$; $\log_{10}.01$; $\log_{10}.001$; $\log_{10}.0001$. What happens to $\log_{10}x$ as x gets closer and closer to zero? Do you think that the other logarithm functions behave in this way?

7. Prove Theorem 4-2.

8. Prove Theorem 4-3.

9. Use the Laws of Logarithms to rewrite the following expressions as a sum or difference of logarithms:
   (a) $\log_a(x^2y^2)$ (b) $\log_a(xy/uv)$
   (c) $\log_a\sqrt{xyz}$ (d) $\log_a\sqrt[3]{x+y}$
   (e) $\log_a\sqrt{x^2y^{-1/2}z^{2/3}}$ (f) $\log_a\sqrt{x^3/\sqrt[3]{y^2}}$
   (g) $\log_{10}\sqrt{s(s-a)(s-b)(s-c)}$
   (h) $\ln(2\pi\sqrt{l/g})$

10. Rewrite each of the following as a single logarithm with coefficient one:
    (a) $\log_a p + \log_a q - \log_a r$ (b) $2\ln x - 3\ln y$
    (c) $(1/2)(\log_a 3 - 2\log_a 5)$ (d) $\ln 7 + 2\ln 10$
    (e) $(1/2)(\log_a 8 - 1/2\log_a 21 + \log_a 17)$
    (f) $\log_{10}(x+1) + \log_{10}x - \log_{10}2$

11. If $\ln 2 = .6931$ and $\ln 3 = 1.0986$, use the Laws of Logarithms to find:
    (a) $\ln 4$ (b) $\ln 2/3$
    (c) $\ln 6$ (d) $\ln 1/3$
    (e) $\ln 9/2$ (f) $\ln 8/9$

12. Solve for x:
    (a) $\log_a 3 + \log_a x = \log_a 21$ (b) $\log_a 2 + \log_a x = \log_a 2$
    (c) $\log_a(x^2) - \log_a x = \log_a 2$ (d) $2\log_a x = \log_a 4$
    (e) $2\log_a x - 3\log_a x = \log_a(1/3)$

13. Solve for x:
    (a) $5^x = 10$ (b) $3^{2x} = 2$
    (c) $2^{x+1} = 3$ (d) $3^x = 5^{x+1}$

14. Show that:
    (a) $10^{(2\log_{10}(x+y))} = x^2 + 2xy + y^2$
    (b) $a^{(3\log_a 2)} - b^{(1/2\log_b 4)} = 6$
    (c) $10^{(\log_{10}5 + \log_{10}2)} = 10$
    (d) $\log_{10}(\log_2(\log_3 9)) = 0$
    (e) $\log_{10}(10^{x+1}) - \log_{10}(10^x) = 1$

15. A fourth Law of Logarithms states that

$$\log_b a = \log_c a / \log_c b$$

for any bases a, b, and c greater than one. Prove this law.

16. Prove that

$$\log_N b = 1/\log_b N \quad \text{for } N, b > 1$$

17. Prove that

$$\log_N b = \log_{1/N}(1/b) \quad \text{for } N, b > 1$$

18. The function $y = e^{-x^2}$ is important in statistics, where it is used in describing the standard normal curve. Graph this function. What is its range?

19. A *catenary* is the curve assumed by a perfectly flexible, nonstretchable cable of uniform density hanging freely from two supports. An equation for the catenary is

$$y = \frac{e^x + e^{-x}}{2}$$

Graph this function.

4-5 Common Logarithms

The logarithms whose base is ten are called common or Briggsian logarithms. This base is particularly useful since our number system also has ten for its base. It is so widely used that when the base is omitted, it is understood that the base is ten. Thus, log 15 usually means $\log_{10} 15$.

Every number can be written as the product of a number between 1 and 10 and a power of ten. This is called *scientific notation*. Thus,

$$4638 = 4.638 \times 10^3$$
$$463.8 = 4.638 \times 10^2$$
$$46.38 = 4.638 \times 10^1$$
$$4.638 = 4.638 \times 10^0$$
$$.4638 = 4.638 \times 10^{-1}$$
$$.04638 = 4.638 \times 10^{-2} \text{ etc.}$$

The exponent on the 10 indicates how many places the decimal is to be moved to the right or to the left.

Recall that $\log_{10} 10^n = n$, thus, $\log_{10} 10^2 = 2$; $\log_{10} 10^{-1} = -1$, and so on. Now $\log 4638 = \log(4.638 \times 10^3)$, and by the first Law of Logarithms this is $\log 4.638 + \log 10^3 = \log 4.638 + 3$. Moreover, $\log(.04638) = \log(4.638 \times 10^{-2}) = \log 4.638 + \log 10^{-2} = \log 4.638 - 2$.

Thus, the common logarithm of any number can be written as the logarithm of a number between 1 and 10 plus an integer. The integer is called the *characteristic* of the logarithm and its value is determined by the position of the decimal in the original number.

For example, 5231 has characteristic 3, since $5231 = 5.231 \times 10^3$. The number .1011 has characteristic -1, since $.1011 = 1.011 \times 10^{-1}$. The number 3.2 has characteristic zero, since this number is already between 1 and 10.

With a little practice the characteristic can be found by inspection without rewriting the number in scientific notation. Simply locate the position immediately to the right of the first nonzero digit in the number, then count the number of places over to the decimal. If you move to the right, the characteristic is positive; if to the left, it is negative.

Example 4-17. The characteristic of 30,241 is 4.

$$3_{\wedge}0\,2\,4\,1.$$

1 2 3 4

The characteristic of .00124 is -3.

$$.0\,0\,1_{\wedge}2\,4$$

3 2 1

The logarithm of a number between 1 and 10 is always some decimal between 0 and 1, since $\log 1 = 0$ and $\log 10 = 1$, and this number is called the *mantissa* of the logarithm. Mantissas are found in the table of logarithms (Table 1) in the back of the book.

Since any number we will be looking up in the table of logarithms will be between 1 and 10, and every mantissa will be a decimal between 0 and 1, the decimals are often left off in the table. Thus, in looking for the mantissa of log 7.28, we look for 728, and if the mantissa is given as 8621, then we must supply the decimal, making it .8621.

The values we find in Table 1 are four-place *approximations* to the logarithm, since, in general, a logarithm will be an irrational number, and as such will have an infinite nonrepeating decimal. If greater precision is desired, tables to five or more places are available.

Example 4-18. Find (a) log 728 and (b) log .00132.

Solution: (a) The characteristic is 2, and the mantissa from Table 1 is .8621. Thus, $\log 728 = .8621 + 2 = 2.8621$. (b) This time the characteristic is -3 and the mantissa is .1206. Thus, $\log .00132 = .1206 - 3$, a negative number. The usual practice in this case is to write the logarithm as $7.1206 - 10$ instead of carrying out the subtraction.

We could just as easily write the logarithm as $17.1206 - 20$ or $27.1206 - 30$, and we will see later that sometimes we may want to use one of these forms for computation purposes. If we write negative characteristics in this way, then the logarithm of every number consisting of the digits 132 (no matter where the decimal place is) will have the same decimal part or mantissa. Only the characteristic will be different.

Example 4-19.

$$\begin{aligned} \log 132 &= 2.1206 \\ \log 13.2 &= 1.1206 \\ \log 1.32 &= 0.1206 \\ \log .132 &= 9.1206 - 10 \\ \log .0132 &= 8.1206 - 10 \end{aligned}$$

We will frequently want to reverse this procedure, that is, given the common logarithm of a number, to find the number.

Example 4-20. Suppose we know that $\log x = 3.8376$. Now 3 is the characteristic and tells us where the decimal place will be. The mantissa is .8376, and if we look this up in the body of Table 1, we find that it corresponds to the number 6.88. Thus, $x = 6.88 \times 10^3 = 6{,}880$.

Example 4-21. Find x if $\log x = 9.7412 - 10$.

Solution: The mantissa here is .7412 and looking in Table 1 we find that this corresponds to the number 5.51. The characteristic is $9 - 10 = -1$, hence $x = 5.51 \times 10^{-1} = .551$.

Looking at Table 1, you will notice that not every mantissa is listed. For example, if $\log x = 0.4208$, you will not find 4208 in the table. You will find the mantissas 4200 and 4216, corresponding to the numbers 2.63 and 2.64, respectively. Thus, x must be some number between 2.63 and 2.64. To find this number we use a method called *linear interpolation*. In this method, we assume that for values of x very close together, the graph of $y = \log x$ is a straight line between these two values. This is, of course, not true, but if the values of x are very close together, the error in making this assumption will be quite small. If we make this assumption, then the x-value corresponding to $y = .4208$ will be halfway between 2.63 and 2.64, and we can write $x = 2.635$ (approximately). (See Figure 4-3.) In linear interpolation we assume that the change in y will be proportional to the change in x.

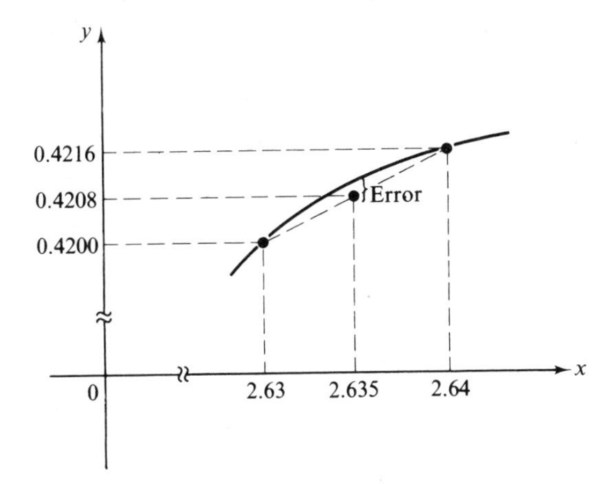

Figure 4-3

Linear interpolation

Example 4-22. Find x if $\log x = 1.3204$.

Solution: Looking in Table 1, we find 3204 is between 3201 and 3222 and these mantissas correspond to 209 and 210 respectively. Arranging this information in a table, we compute the differences between the entries.

$$1\left\{\begin{matrix} d\left\{\begin{matrix} 209 & 3201 \\ x & 3204 \end{matrix}\right\}3 \\ \begin{matrix} 210 & 3222 \end{matrix} \end{matrix}\right\}21$$

Letting d stand for the difference between 209 and the number we are looking for, we equate the ratios of the differences:

$$d/1 = 3/21 \quad \text{or} \quad d \simeq .14$$

Rounding off to one digit, we have $d \simeq .1$ and $x = 209 + .1 = 209.1$. The position of the decimal here of course has no significance in our final answer. Its only function was to tell us how to add the difference d to 209. The position of the decimal in the final answer is determined by the characteristic, which is 1. This tells us that the decimal is one place to the right of the first nonzero digit, thus, $x = 20.91$.

We can also use linear interpolation to find the mantissa of a four digit number.

Example 4-23. Find log 3204.

Solution: Since Table 1 gives mantissas only for 320 and 321, we must interpolate to find the mantissa of 3204.

$$10\left\{\begin{matrix} 4\left\{\begin{matrix} 3200 & 5051 \\ 3204 & y \end{matrix}\right\} d \\ \quad 3210 \qquad 5065 \end{matrix}\right\} 14$$

Equating the ratio of the differences, we have

$$d/14 = 4/10 \quad \text{or} \quad d \simeq 5.6$$

Thus, $y = 5051 + 5.6 = 5056.6 \simeq 5057$. Since the characteristic of 3204 is 3, $\log 3204 = 3.5057$.

Exercise 4-5

1. Write in scientific notation:
   (a) 2371 (b) .0124
   (c) .00001 (d) 10.7
   (e) 2.43 (f) 541,000,000

2. Write as a decimal:
   (a) 1.71×10^3 (b) 2.04×10^{-2}
   (c) 5.41×10^{-1} (d) 1.0×10^5
   (e) 8.02×10^{-3} (f) 7.92×10^{-8}

3. Find:
   (a) log 1.54 (b) log 153
   (c) log .00831 (d) log 11,000
   (e) log 25.8 (f) log .0124

4. Find x if:
   (a) $\log x = 0.9722$ (b) $\log x = 3.8014$
   (c) $\log x = 8.6599 - 10$ (d) $\log x = 1.8609$
   (e) $\log x = 7.4425 - 10$ (f) $\log x = 9.1367 - 10$

5. Use linear interpolation to find x if:
   (a) $\log x = 2.7200$ (b) $\log x = 9.5219 - 10$
   (c) $\log x = 0.8827$ (d) $\log x = 1.7999$
   (e) $\log x = 3.2420$ (f) $\log x = 7.7122 - 10$

6. Use linear interpolation to find:
   (a) log .03214 (b) log 1924
   (c) log 4.023 (d) log 1.632×10^{-5}
   (e) log 77.24 (f) log 1.003×10^{14}

7. Find x if:
   (a) $10^x = 432$ (b) $10^{\log 3} = x$
   (c) $x = 10^{.00699}$ (d) $\log 10^{1.723} = x$

8. From Table 1, log 2 = 0.3010 and log $2^2 = 2 \log 2 = 0.6020$. However, the table gives log 4 = 0.6021. Explain this discrepancy.

9. By the fourth Law of Logarithms (Exercise 4-4, problem 15),

$$\log_e a = \log_{10} a / \log_{10} e$$

   Use this conversion formula to find:
   (a) ln 10 (b) ln 1.02
   (c) ln 2 (d) ln .00722
   ($e \simeq 2.718$)

10. Find x if:
   (a) $\ln x = 3.1$ (b) $\ln x = 10$
   (c) $\ln x = 0.12$ (d) $\ln x = -1.4$

11. Consider the function $y = \log(\log x)$. What is the domain of this composite function? Is this an increasing function? Graph the function.

4-6 Applications of Logarithms

Many computations can be greatly simplified by the use of logarithms. Using the Laws of Logarithms, problems of multiplication and division become transformed into problems involving addition and subtraction, while problems of raising to powers or extracting roots are changed into problems of multiplication or division. All that is needed is a table of mantissas.

Example 4-24. Use logarithms to compute (4.32)(.00271)/520.

Solution: If $N = (4.32)(.00271)/520$, then $\log N = \log 4.32 + \log .00271 - \log 520$.

$$\begin{aligned} \log 4.32 &= 0.6355 \\ + \log .00271 &= \underline{7.4330 - 10} \\ &\ 8.0685 - 10 \\ - \log 520 &= \underline{2.7160} \\ \log N &= 5.3525 - 10 \end{aligned}$$

$$N = .0000225$$

Example 4-25. Find $\sqrt[3]{427}$.

Solution: If $N = \sqrt[3]{427}$, then $\log N = (1/3)\log 427$.

$$\log 427 = 2.6304$$
$$(1/3)\log 427 = 0.8768$$
$$N = 7.53$$

Example 4-26. Find $\sqrt[3]{.575}$.

Solution: If $N = \sqrt[3]{.575}$, then $\log N = (1/3)\log .575$.

$$\log .575 = 9.7597 - 10$$
$$(1/3)\log .575 = ?$$

In this case we run into a problem, because when we try to divide 10 by 3, we get a decimal, but the characteristic must be an integer. To avoid this, we write the characteristic, which is -1, as $29 - 30$, instead of $9 - 10$. The division by 3 then proceeds without difficulty.

$$\log .575 = 29.7597 - 30$$
$$(1/3)\log .575 = 9.9199 - 10$$
$$N = .831$$

The modern slide rule is based on the use of logarithms. A slide rule is marked off in such a way that the lengths of segments are proportional to the logarithms of various numbers. Thus, in Figure 4-4, if

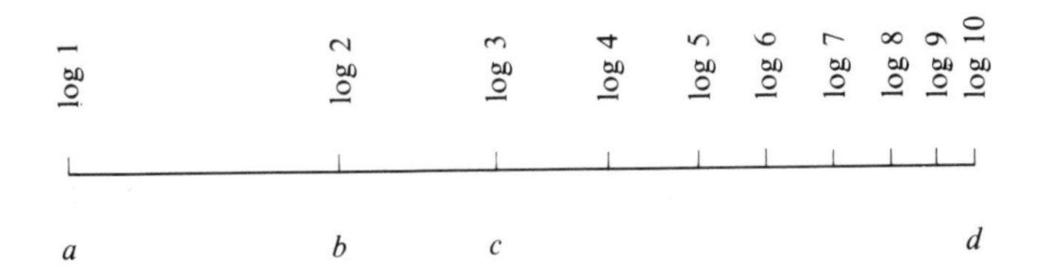

Figure 4-4

segment *ad* is taken to be one unit, then segment *ab* has length $\log 2 = .301$; segment *ac* has length $\log 3 = .477$, and so on. A slide rule has two of these rulers, arranged in such a way that one slides along the other. To find the product of 2 and 3, for example, we align the rulers so that segments *ab* and *ac* lie end to end. Their combined length will be $\log 2 + \log 3 = \log 6$.

Exercise 4-6

Use the Laws of Logarithms and the table of mantissas to compute the following, giving the answer to three digits:

1. $(23.8)(.0187)$
2. $4.21/723$
3. $\sqrt{(17.7)(92.1)}$
4. $(.00171)^2/(24.8)^3$
5. $\sqrt[3]{191}$
6. $\sqrt[4]{.0478} \cdot \sqrt[5]{121}$
7. $(2.09)^{10}/134$
8. $\sqrt{402}$
9. $(79.2)(.184)/9.92$
10. $\sqrt{(.821)/(105)}$
11. $(14.1)^2(.008)/(692)$
12. $\sqrt{(8.09)^3}$
13. $(4.32)^{1.02}$
14. $(\sqrt{2})^{\sqrt{2}}$
15. 2^{π}
16. $\sqrt[3]{(17.1)^2(180)/(32{,}000)}$
17. If $2^x = 313$, find x to three digits.
18. If $x^{1.02} = 17.1$, find x to three digits.
19. Heron's formula gives the area of a triangle in terms of the lengths of its sides.

$$A = \sqrt{s(s-a)(s-b)(s-c)}$$

where a, b, and c are the lengths of the sides of the triangle, and $s = (1/2)(a + b + c)$. Use this formula and logarithms to find the area of a triangle whose sides measure 4.31 inches. 7.18 inches, and 8.01 inches.

20. The period T of a simple pendulum is given by the formula

$$T = 2\pi\sqrt{l/g}$$

where T is measured in seconds, l is the length of the pendulum in feet, and $g \simeq 32$ ft/sec². Use this formula and logarithms to find the period of a pendulum 1/2 foot long. (Take $\pi \simeq 3.1416$.)

21. The volcano that formed Crater Lake in Oregon killed a tree whose charcoal gave a count of 44.5% of the amount of carbon 14 found in a living tree. In using the method of carbon 14 dating, we arrive at the equation

$$.445 = e^{-.0001203\,t}$$

where t is the age of the tree in years. Use logarithms to solve for t.

22. It has been hypothesized that the population growth of the United States can be predicted by the following formula*

$$N = 210/(1 + 51.5\, e^{-0.03\,t})$$

where N is measured in millions and t is measured in years (A.D.) Find N for $t = 1980$.

23. According to Newton's law of cooling, if an object whose temperature is T_1 is surrounded by air at temperature T_0, then its temperature T at anytime t will be given by the formula

$$T = T_0 + (T_1 - T_0)e^{-kt}$$

where k is a constant which depends on the object involved. If the temperature of your cup of coffee is 180°, how long will it take for it to get cool enough to drink (130°) if the surrounding air is at 70°? (Take $k = .1216$, then t will be given in minutes.)

4-7 Historical Note

Most notable mathematical achievements are the work of many men. Usually one man has the germ of an idea and successive generations of mathematicians advance the theory step by step until it reaches full maturity years or even centuries later. An exception to this is the invention of logarithms. Logarithms were conceived and developed by one man, John Napier, of Edinburgh, Scotland.

Napier's little book of 147 pages, published in 1614, was the result of 20 years of intensive labor. His achievement is all the more remarkable when we consider that he did not have the advantage of using our modern algebraic symbols. The exponential notation we now use was unknown to Napier.

After Napier's book was published, he was visited by Henry Briggs, a well-known professor of geometry at Gresham College in London. Together they agreed to alter Napier's tables so that the logarithm of one would be zero and the logarithm of ten would be some power of ten. This led to the tables of logarithms to the base ten that we use

* R. Pearl and L. J. Reed, *Proc. Nat. Acad. Sci.*, 6: 275 (1920).

today. They are sometimes called Briggsian logarithms in honor of Henry Briggs.

The word *logarithm* was invented by Napier from two Greek words meaning *ratio number*. Briggs suggested the word *mantissa,* a Latin term originally meaning an addition. Briggs also introduced the term characteristic.

Napier's wonderful invention was enthusiastically adopted throughout Europe. It saved scientists, astronomers, navigators, and engineers countless hours of drudgery. Laplace stated that this invention "by shortening the labors doubled the life of the astronomer."

Review Exercise

1. Simplify. Your answer should contain only positive exponents.
   (a) $(x^{-1} \cdot y^{2/3})/(x^2 \cdot y^{-1/3})$ (b) $(\sqrt{a^3} \cdot b^2)/(a^{-2} \cdot \sqrt[3]{b})$

2. Evaluate:
   (a) $16^{-3/4}$ (b) $(-8)^{2/3}$ (c) $27^{1/3} - 3^0$

3. Graph $y = 2^x$. What is the domain of this function? What is its range? What is the inverse of this function? Graph the inverse.

4. Rewrite in logarithmic form:
   (a) $3^{-4} = 1/81$ (b) $\sqrt{25} = 5$

5. Rewrite in exponential form:
   (a) $\log_3 9 = 2$ (b) $\log_8(1/2) = -1/3$

6. Find x in each of the following:
   (a) $\log_3(1/3) = x$ (b) $\log_x 100 = 2$
   (c) $\ln e^{-5} = x$ (d) $e^{2 \ln 7} = x$

7. State the three Laws of Logarithms.

8. Use the Laws of Logarithms to rewrite the following as a sum or difference of logarithms:
   (a) $\log \sqrt[3]{x^2 y/z^4}$ (b) $\log(\sqrt{a} \cdot \sqrt[3]{b^2}/\sqrt{z^3})$

9. Solve for x.
   (a) $5^x = 7^{2x+1}$ (b) $\log(x + 1) - \log 2 = \log 5$

10. Rewrite as a single logarithm with coefficient one:
    (a) $2 \log x - (1/2)\log 3x$ (b) $(1/2)(3 \ln x + 5 \ln y - \ln z)$

11. Interpolate to find x if:
    (a) $\log .003257 = x$ (b) $\log x = 1.8725$

12. Use the Laws of Logarithms and the table of mantissas to compute the following correct to three digits:

$$\sqrt[5]{1.02/.0751}$$

13. If $\ln 2 = .6931$ and $\ln 3 = 1.0986$, find $\ln 6/27$.

14. If $x = x_0 e^{-kt}$, find t if $x = 500$, $x_0 = 100$, and $k = .5$.

Bibliography

Brand, Louis, *Differential and Difference Equations,* pp. 33–39. New York: John Wiley & Sons, Inc., 1966.

Eves, Howard, *An Introduction to the History of Mathematics,* Rev. Ed., Ch. 9. New York: Holt, Rinehart & Winston, 1964.

Hooper, Alfred, *Makers of Mathematics,* Ch. 5. New York: Random House, 1948.

5

The Trigonometric Functions

5-1 Angles and Their Measure

In geometry we define an angle to be the set of points on two rays which have a common end point, and we measure angles with a protractor. The protractor is a semicircle which is divided into 180 equal parts. Using this we can assign to every angle a unique number between zero and 180, inclusive, called its *degree* measure.

In trigonometry we will extend this notion of measurement of angles so that *any real number whatever* may be the measure of some angle. We do this by defining a *directed* angle. If we choose one of the rays of an angle and call it the initial side and call the other ray the terminal side, then we can assign a measure to an angle by thinking of rotating the initial side until it coincides with the terminal side. If the rotation is counterclockwise, the measure will be positive, if clockwise, the measure will be negative. The measure of a rotation through a complete circle will be taken to be 360°.

It is not hard to see that an angle can have more than one measure. In Figure 5-1, the angle can be assigned the measures 30°, −330°, or $360° + 30° = 390°$. In fact every angle has infinitely many measures, for if t is some measure of an angle, then $t + 360°$, $t - 360°$, $t + 2 \cdot 360°$, etc., will also be measures for this angle. In general, if t is any measure for the angle, then $t + 360° \cdot k$ will also be a measure for any integer k ($k = 0, \pm 1, \pm 2, \ldots$).

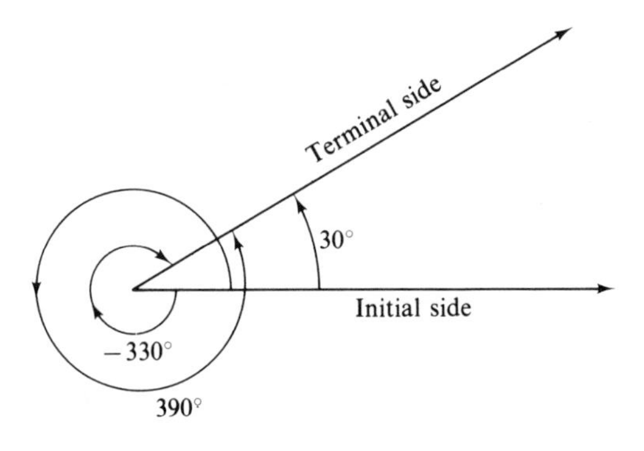

Figure 5-1

In order to decide which side of the angle will be the initial side, we will place our angle on a coordinate system with the vertex of the angle at the origin and the initial side lying along the positive x-axis. (See Figure 5-2.) This is called *standard position*.

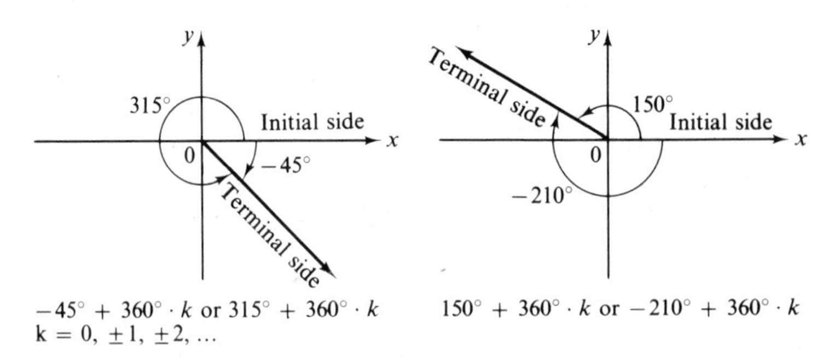

Figure 5-2

When an angle is placed in this position its terminal side may fall in any one of the four quadrants, or it may fall on the x- or y-axis. An angle whose terminal side falls on one of the axes is called a *quadrantal angle*.

Example 5-1. In what quadrant will the terminal side of an angle having measure 486° fall?

Solution: If we place the initial side of the angle on the positive x-axis, then we can think of it revolving in a positive direction (counterclockwise) through $486° = 360° + 126°$. This is one complete revolution, plus 126° more. The terminal side will thus lie in the second quadrant. (See Figure 5-3.)

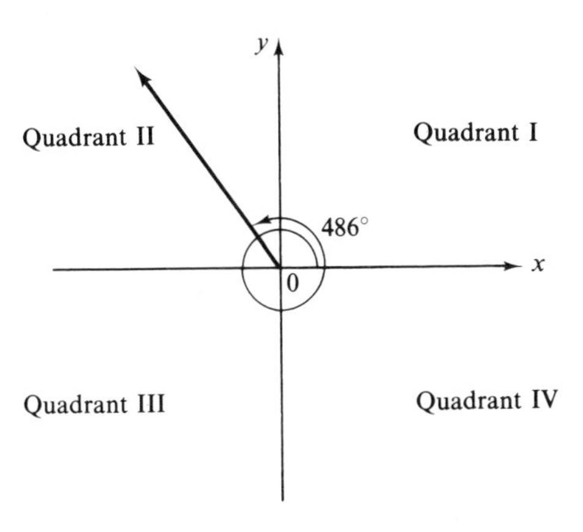

Figure 5-3

Example 5-2. List all of the measures of the angle whose terminal side lies on the positive y-axis.

Solution: The positive measures would be $90°$, $90° + 360° = 450°$, $90° + 2 \cdot 360° = 810°$, etc. The negative measures would be $-270°$, $-270° - 360° = -630°$, etc. A formula that will give all of these measures would be $90° + 360° \cdot k$, $k = 0, \pm 1, \pm 2, \ldots$.

The degree is not the only unit of measure for angles. Just as lengths may be measured in inches or centimeters, and temperature may be measured in degrees Fahrenheit or Celsius, so angles may be measured in degrees or in *radians*. The radian is a unit of measure which is used extensively in calculus, so it is important that the student become familiar with this concept.

To describe radian measure, let us start with a number line and with a circle whose radius is one unit and whose center is at the origin of a coordinate system. (This is called a *unit circle*.) (See Figure 5-4.) Now think of the number line as being flexible, like a piece of string, and let us wrap it around the unit circle without stretching so that zero on the number line lies on the point (1, 0); the positive part of the number line is wrapped counterclockwise and the negative part of the number line is wrapped clockwise. This wrapping will assign to every number on the number line a unique point on the unit circle. Of course every point on the unit circle will have many (infinitely many) real numbers assigned to it, since the real line is infinite in extent.

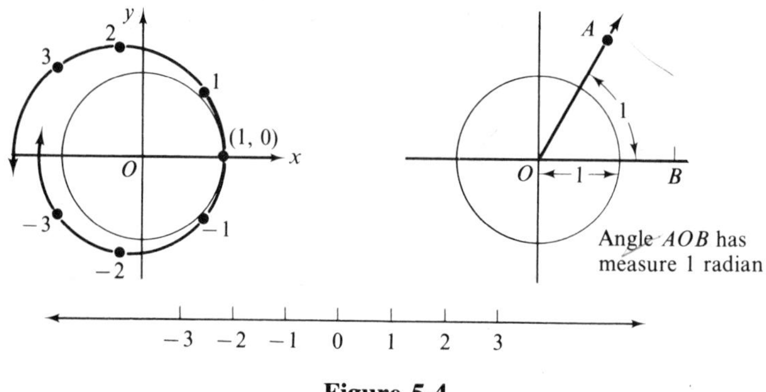

Figure 5-4

Now the terminal side of an angle in standard position on the coordinate axes will cut the unit circle in a point. Any real number assigned to this point by the wrapping will be called a *radian measure* of the angle. Consider for example the angle whose terminal side cuts the circle at the point corresponding to one on the number line (Figure 5-4). We say its measure is one radian. Thus, a radian measure of an angle is found by measuring the length of arc on the unit circle determined by the sides of the angle. Since the circumference of a circle with radius one unit is 2π units, $360°$ is equivalent to 2π radians and $180°$ is equivalent to π radians. To convert from one measure to the other, we note that

$1°$ is equivalent to $\pi/180$ radians.

1 radian is equivalent to $180/\pi$ degrees.

Since $\pi \simeq 3.1416$, 1 radian is approximately $57°$.

Every angle will have infinitely many radian measures, just as it has infinitely many measures in degrees. If an angle has measure r radians, then $r + 2\pi k$, $k = 0, \pm 1, \pm 2, \ldots$ will also be a radian measure of that angle.

Example 5-3. To change $\pi/2$ radians to degrees, we multiply by $180/\pi$ degrees/rad.

$$(\pi/2)(180/\pi) = 90°$$

To change 270° to radians, we multiply by $\pi/180$ rad/degree.

$$(270°)(\pi/180) = 3\pi/2 \text{ radians}$$

Radian measure is so important that it will be used extensively in the sections that follow. If degree measure is used it will be distinguished by the degree symbol. If no symbol is used, the student may assume that the measure is given in radians. Thus, if we refer to the measure of an angle as 3, it will be understood that this means 3 radians. If degree measure were meant, we would write 3°.

Exercise 5-1

1. In what quadrant will the terminal side of an angle having each of the following measures fall? Sketch each.
   (a) 372° (b) −192°
   (c) 1022° (d) − 96°
   (e) 271° (f) −524°
   (g) 200° (h) −200°

2. If the terminal side of an angle whose measure is $k°$ falls in the first quadrant, in what quadrant will the terminal side of the angle having measure $-k°$ fall? Answer this question if the terminal side of the angle whose measure is $k°$ falls in the second quadrant, in the third quadrant, in the fourth quadrant. Give examples to illustrate each case.

3. Give all of the degree measures of the angle whose terminal side lies on the negative x-axis, on the negative y-axis, on the positive x-axis.

4. Find the smallest positive degree measure of an angle if another measure of that angle is:
   (a) 1075° (b) −124°
   (c) 483° (d) −324°
   (e) −180° (f) −842°

5. Change to radian measure:
   (a) 30° (b) −180°
   (c) 225° (d) 390°
   (e) −450° (f) 60°
   (g) −45° (h) 18°

6. Change to degree measure:
   (a) $\pi/6$ (b) $-4\pi/3$

(c) $3\pi/4$ (d) $\pi/9$
(e) -6π (f) 3
(g) $14\pi/5$ (h) $-2\ 1/2$

7. Find the smallest positive radian measure of an angle if another measure of that angle is:
(a) $7\pi/2$ (b) $-3\pi/8$
(c) $9\pi/4$ (d) $-4\pi/3$
(e) -20π (f) $37\pi/2$

8. In what quadrant will the terminal side of an angle having each of the following measures fall? Sketch each.
(a) $2\pi/3$ (b) $5\pi/4$
(c) $17\pi/8$ (d) $7\pi/4$
(e) $13\pi/3$ (f) 15

9. A wheel is turning at the rate of 48 revolutions per minute (rpm). Express this in degrees per second, in radians per second.

10. If the real line is wrapped around the unit circle as described in this section, what is the smallest positive number that will be assigned to the point (0, 1)? to the point (−1, 0)? to the point (0, −1)? to the point $(1/\sqrt{2}, 1/\sqrt{2})$? What is the largest negative number that will be assigned to each of these points?

11. Construct a radian protractor and mark on it in both degrees and radians the points corresponding to 0°, 30°, 45°, 60°, 90°, 120°, 135°, 150°, 180°, 1 radian, 2 radians.

5-2 The Trigonometric Functions

Let AOB be an angle in standard position, and let (x, y) be any point [except (0, 0)] on its terminal side. If we drop a perpendicular from (x, y) to the x-axis (Figure 5-5), we will have a right triangle whose legs have length $|x|$ and $|y|$ respectively. By the Pythagorean Theorem the length of the hypotenuse is $\sqrt{x^2 + y^2}$. We will call this number r. Clearly $r > 0$.

The ratios of any two of these numbers, x, y, and r, will be the same no matter what point we pick on the terminal side. For suppose (x', y') is some other point on the terminal side. If we drop a perpendicular to the x-axis as before, we have a second right triangle which is similar to the first, since they are both right triangles and share the angle AOB. From geometry we know that if two triangles are similar, then ratios of corresponding sides are equal. Thus, $x/y = x'/y'$; $y/r = y'/r'$; and so on.

Since these ratios depend only on the angle, we can define a function

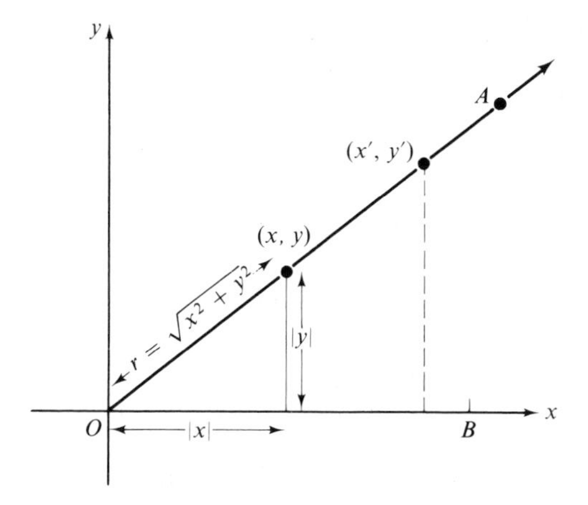

Figure 5-5

which associates with each angle one of these ratios, for example y/x. It will be much more convenient, however, if we define our function so that the domain is a set of real numbers, rather than a set of angles, for then we can graph the function. Therefore, we will take the domain of the function to be the set of real numbers which are the measures of the angles in radians. The function which associates the ratio y/x with each such measure will be called the *tangent function,* and it is abbreviated *tan.* Thus, we will write $\tan \theta = y/x$, where θ is the angle measure in radians. Suppose, for example $\theta = \pi/4$ (45°). (See Figure 5-6.) The terminal side of this angle is the line $y = x$. If we choose any point on this line, for instance (1, 1), we find that

$$\tan \frac{\pi}{4} = \frac{1}{1} = 1$$

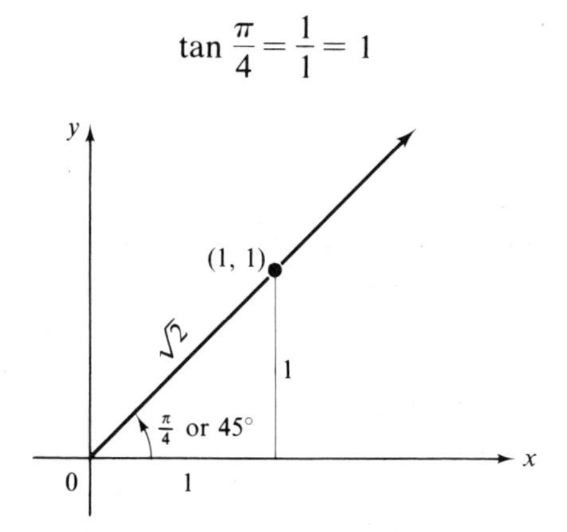

Figure 5-6

The *sine function* (abbreviated *sin*), and the *cosine function* (abbreviated *cos*) are defined similarly.

$$\sin \theta = \frac{y}{r}$$

$$\cos \theta = \frac{x}{r}$$

Since for $\theta = \pi/4$, $x = y$, if we take $x = 1$, $y = 1$, then $r = \sqrt{1^2 + 1^2} = \sqrt{2}$, and

$$\sin \frac{\pi}{4} = \frac{1}{\sqrt{2}} \simeq .707$$

$$\cos \frac{\pi}{4} = \frac{1}{\sqrt{2}} \simeq .707$$

We can get three more ratios by taking the reciprocals of the ratios y/r, x/r, and y/x, and these define three more trigonometric functions called the *cosecant* (abbreviated *csc*), the *secant (sec)*, and the *cotangent (cot)* respectively.

Definition 5-1. Let θ be any real number. Let (x, y) be any point except $(0, 0)$ on the terminal side of an angle in standard position whose measure in radians is θ, and let $r = \sqrt{x^2 + y^2}$. Then the six trigonometric functions are defined as follows:

$$\sin \theta = \frac{y}{r} \qquad\qquad \csc \theta = \frac{r}{y} \quad y \neq 0$$

$$\cos \theta = \frac{x}{r} \qquad\qquad \sec \theta = \frac{r}{x} \quad x \neq 0$$

$$\tan \theta = \frac{y}{x} \quad x \neq 0 \qquad\qquad \cot \theta = \frac{x}{y} \quad y \neq 0$$

These definitions are basic and should be memorized.

From these definitions it follows that since r is never zero, the sine and cosine functions are defined for all real numbers θ. On the other hand, the tangent and secant functions are undefined when $x = 0$, that is, when the terminal side of the angle falls on the y-axis. This occurs when θ is $\pi/2$, $3\pi/2$, or any odd integral multiple of $\pi/2$. Similarly, the cosecant and cotangent functions are not defined when $y = 0$. This happens when $\theta = 0$, π, or any integral multiple of π.

The sine and cosecant functions are called *reciprocal functions* since $\sin \theta = 1/\csc \theta$. Similarly, the cosine and secant functions and the tangent and cotangent functions are also called reciprocal functions since $\cos\theta = 1/\sec\theta$ and $\tan\theta = 1/\cot\theta$.

Although we have defined θ to be the measure of an angle in *radians* it will sometimes be useful to let it stand for the measure in degrees as well. Thus, we may write sin 45°. We will take this to be equivalent to $\sin \pi/4$, and, therefore, $\sin 45° = \sin \pi/4 = 1/\sqrt{2}$.

From our definition, r is always positive, however, x and y may be either positive or negative, depending on the quadrant in which the terminal side of the angle falls. Thus, in the first quadrant (Figure 5-7) $x > 0$ and $y > 0$, and all of the trigonometric functions are positive for angles whose terminal side lies in this quadrant.

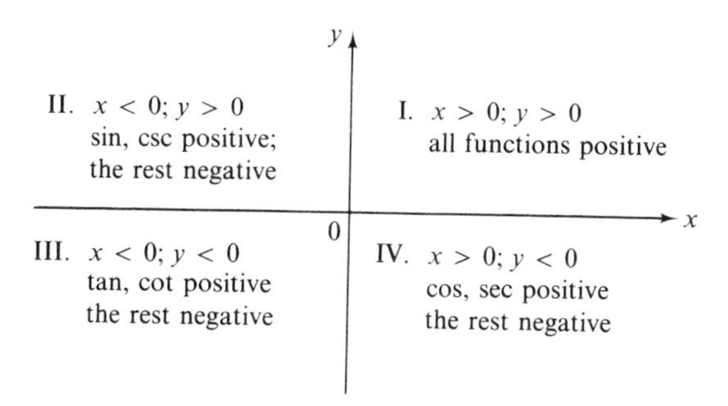

Figure 5-7

If the terminal side of the angle falls on the x- or y-axis (a quadrantal angle), then the values of the trigonometric functions will be 0, 1, -1, or the function may be undefined. For example, suppose $\theta = \pi/2$ (90°). The terminal side then is the positive y-axis, and if we choose the point (0, 1) on this side, then $x = 0$, $y = 1$, and $r = 1$. Thus, for $\theta = \pi/2$,

$$\sin\frac{\pi}{2} = \frac{1}{1} = 1 \qquad \csc\frac{\pi}{2} = \frac{1}{1} = 1$$

$$\cos\frac{\pi}{2} = \frac{0}{1} = 0 \qquad \sec\frac{\pi}{2} = \frac{1}{0} \text{ (which is undefined)}$$

$$\tan\frac{\pi}{2} = \frac{1}{0} \text{ (which is undefined)} \qquad \cot\frac{\pi}{2} = \frac{0}{1} = 0$$

(Recall that $a/b = k$ means that $b \cdot k = a$. Thus, $0/1 = k$ means k is some number such that $k \cdot 1 = 0$. Clearly, k must be zero for this equation to hold. On the other hand, if $1/0 = k$, then $k \cdot 0 = 1$. But any number multiplied by zero will always give zero, never 1; hence, there is no number k satisfying this equation, and we say that $1/0$ is undefined.)

Example 5-4. If $\sin\theta = 2/3$, find the values of the other trigonometric functions.

Solution: Since $\sin\theta = y/r$, we can take $y = 2$ and $r = 3$. To find x, we use the relation $x^2 + y^2 = r^2$, or $x^2 = 3^2 - 2^2 = 5$. Since x can be either positive or negative, we find that $x = \pm\sqrt{5}$. We thus have two cases to consider. If $y = 2$ and $x = \sqrt{5}$, then the angle terminates in quadrant I (see Figure 5-8), and

$$\sin\theta = \frac{2}{3} \qquad \csc\theta = \frac{3}{2}$$

$$\cos\theta = \frac{\sqrt{5}}{3} \qquad \sec\theta = \frac{3}{\sqrt{5}}$$

$$\tan\theta = \frac{2}{\sqrt{5}} \qquad \cot\theta = \frac{\sqrt{5}}{2}$$

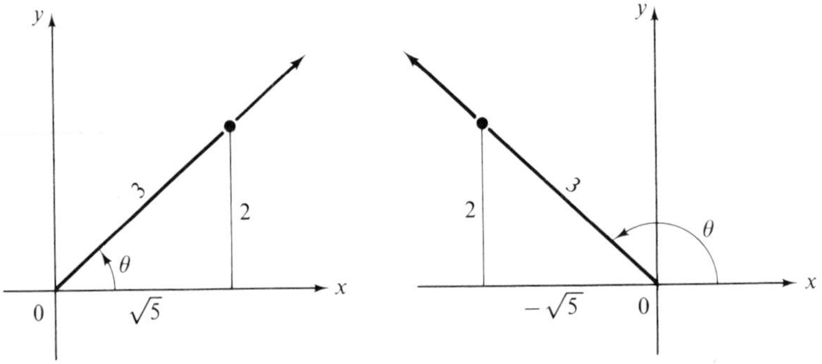

Figure 5-8

On the other hand, if $x = -\sqrt{5}$, then the angle terminates in the second quadrant (see Figure 5-8), and

$$\sin\theta = \frac{2}{3} \qquad \csc\theta = \frac{3}{2}$$

$$\cos\theta = \frac{-\sqrt{5}}{3} \qquad \sec\theta = \frac{-3}{\sqrt{5}}$$

$$\tan\theta = \frac{-2}{\sqrt{5}} \qquad \cot\theta = \frac{-\sqrt{5}}{2}$$

Example 5-5. If $\tan\theta = -1/2$, find $\cos\theta$ and $\sin\theta$ if $\sin\theta$ is positive.

Solution: Since $\tan\theta = y/x = -1/2$, then either $y = -1$ and $x = 2$, or

$y = 1$ and $x = -2$. However, we are told that $\sin \theta$ is positive. Since $\sin \theta = y/r$, and r is always positive, then y must be positive also, and we must have $y = 1, x = -2$. Since $r = \sqrt{x^2 + y^2} = \sqrt{(-2)^2 + 1^2} = \sqrt{5}$, then $\sin \theta = 1/\sqrt{5}$ and $\cos \theta = -2/\sqrt{5}$. (See Figure 5-9.)

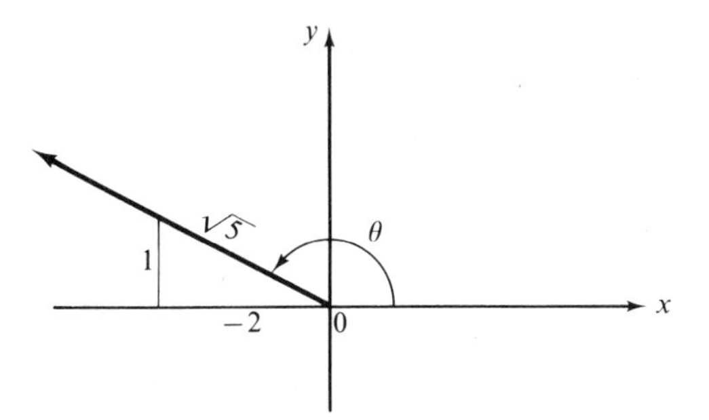

Figure 5-9

We can get an interesting picture of the trigonometric functions as lengths of line segments if we choose an appropriate point on the terminal side of the angle. For example, suppose we choose a point on the unit circle. Its distance from the origin will be 1 unit. Since $r = 1$, then $\sin \theta = y/r = y$, and $\cos \theta = x/r = x$. Thus, the two legs of the right triangle have lengths $\cos \theta$ and $\sin \theta$ respectively. (See Figure 5-10.)

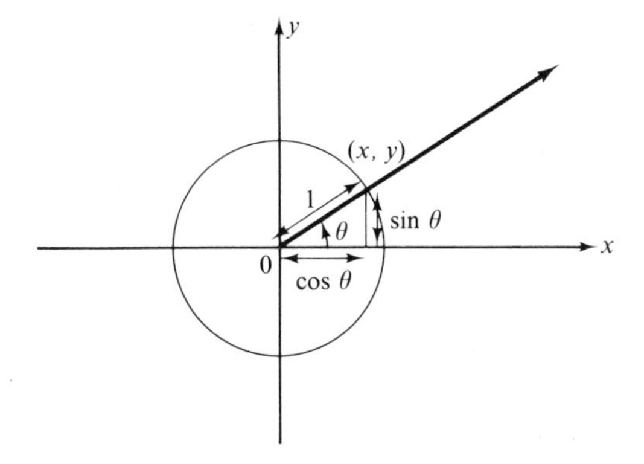

Figure 5-10

Next, suppose we pick a point on the terminal side for which $x = 1$. Then $\tan \theta = y/x = y$, and $\sec \theta = r/x = r$. Thus, the legs of the right triangle have lengths 1 and $\tan \theta$, and the hypotenuse has length $\sec \theta$. (See Figure 5-11.)

Finally, let us choose a point on the terminal side for which $y = 1$.

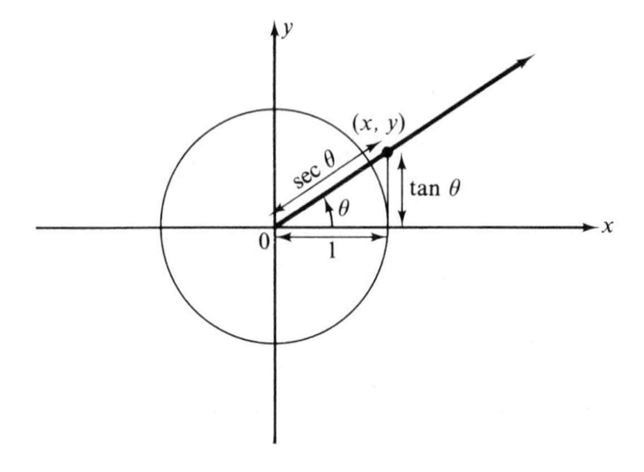

Figure 5-11

Then $\cot \theta = x/y = x$ and $\csc \theta = r/y = r$, and the legs of the right triangle have lengths $\cot \theta$ and 1, and the hypotenuse has length $\csc \theta$. (See Figure 5-12.)

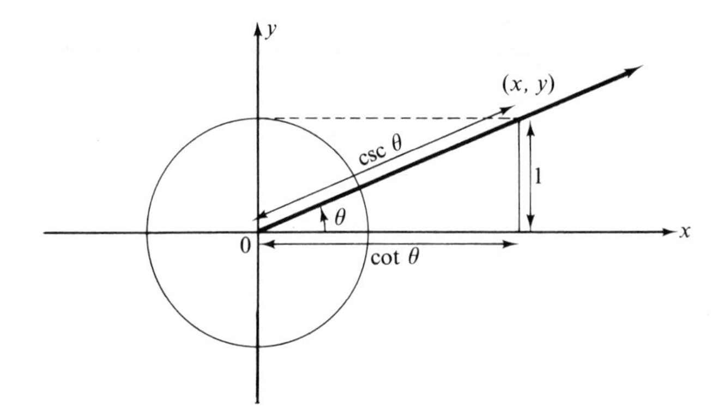

Figure 5-12

Using the Pythagorean Theorem these three right triangles give us the following interesting relationships:

$$\sin^2\theta + \cos^2\theta = 1$$
$$1 + \tan^2\theta = \sec^2\theta$$
$$\cot^2\theta + 1 = \csc^2\theta$$

We will discuss these in more detail in Section 5-9.

Exercise 5-2

1. Give the domain of each of the trigonometric functions.

2. If P is a point on the terminal side of an angle whose measure is θ, find the values of the trigonometric functions of θ for the following:
   (a) $P(2,1)$ (b) $P(-3,4)$
   (c) $P(-1,-1)$ (d) $P(1,-3)$

3. In what quadrant will the terminal side of an angle whose measure is θ fall if
   (a) $\sin\theta$ and $\cos\theta$ are both positive?
   (b) $\sin\theta$ and $\cos\theta$ are both negative?
   (c) $\sin\theta$ and $\tan\theta$ are both negative?
   (d) $\csc\theta$ is positive and $\sec\theta$ is negative?
   (e) $\tan\theta$ is positive and $\cos\theta$ is negative?
   (f) $\tan\theta$ is negative and $\csc\theta$ is positive?

4. In what quadrants may the terminal side of an angle whose measure is θ fall if
   (a) $\sin\theta$ is negative? (b) $\tan\theta$ is positive?
   (c) $\cos\theta$ is positive? (d) $\sec\theta$ is negative?
   (e) $\cot\theta$ is negative? (f) $\csc\theta$ is positive?

5. Which of the trigonometric functions will have the same sign and what will that sign be in
   (a) Quadrants I and II? (b) Quadrants I and III?
   (c) Quadrants I and IV? (d) Quadrants II and III?
   (e) Quadrants II and IV? (f) Quadrants III and IV?

6. If $\sin\theta = 1/2$, find the values of $\cos\theta$ and $\tan\theta$. (There are two cases to consider.) Sketch.

7. If $\cos\theta = -5/6$, find $\sin\theta$ and $\tan\theta$. Sketch.

8. If $\tan\theta = -3/2$, find $\sin\theta$ and $\cos\theta$. Sketch.

9. If $\tan\theta = 3$ and $\cos\theta < 0$, find the values of the remaining trigonometric functions.

10. If $\sec\theta = 3$ and $\tan\theta < 0$, find the values of the remaining trigonometric functions.

11. Complete the following table:

θ	$\sin\theta$	$\cos\theta$	$\tan\theta$	$\csc\theta$	$\sec\theta$	$\cot\theta$
0 $(0°)$						
$\pi/2$ $(90°)$	1	0	und.	1	und.	0
π $(180°)$						
$3\pi/2$ $(270°)$						

12. Why is $\sin \theta$ always less than or equal to 1?

13. Why is $\sec \theta$ always greater than or equal to 1?

14. If $0 < \theta < \pi/2$ (i.e., the terminal side of the angle lies in Quadrant I) for what value of θ will $\tan \theta$ be equal to 1? For what values of θ will $\tan \theta$ be less than 1? Greater than 1?

15. Why is $\sin \theta$ always less than or equal to $\csc \theta$?

16. If $0 \leq \theta \leq \pi/2$, for what value of θ is:
    (a) $\sin \theta = \cos \theta$ (b) $\sin \theta = \csc \theta$
    (c) $\tan \theta = \cot \theta$ (d) $\sin \theta = \tan \theta$
    (e) $\cos \theta = \sec \theta$ (f) $\cos \theta = \cot \theta$

5-3 Special Angles

If we wanted to find $\sin 1° = \sin .01745$ (radians), we could construct an angle whose measure was 1°, drop a perpendicular to the x-axis from the terminal side, measure the sides of the right triangle very carefully, and compute the appropriate ratio. Actually, these ratios are computed using more sophisticated mathematical methods and are available to us in tables. (See Table 2.) The values in the table are for the most part approximations, since in general the values of the trigonometric functions will be irrational numbers. (Recall that $\sin \pi/4 = 1/\sqrt{2}$.) However, for some special angles these ratios can be found exactly.

Before we discuss this, let us note that if an angle is *acute* (i.e., its measure is between 0 and $\pi/2$), then we can construct a right triangle one of whose angles is congruent to this acute angle and define the trigonometric functions in terms of the lengths of the sides of the triangle. Historically, this definition preceeds Definition 5-1 and it has the advantage of being independent of a coordinate system.

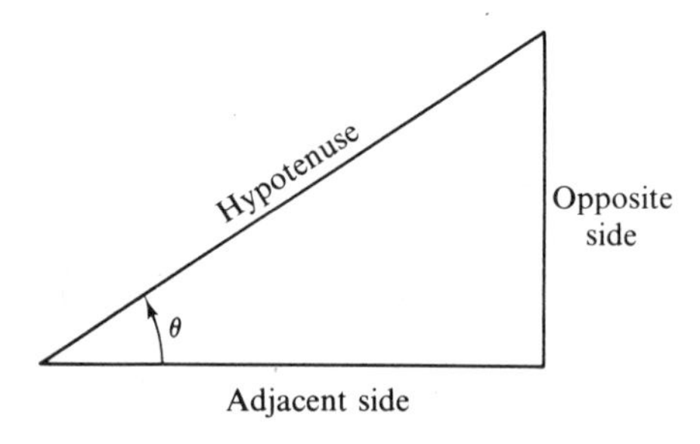

Figure 5-13

The three sides of a right triangle can be described rel angle as the hypotenuse, the side opposite the angle, and jacent to the angle.

Definition 5-2. If θ is a real number, $0 < \theta < \pi/2$ ($0° < \theta < 90°$), then θ is the measure of an acute angle of a right triangle and

$$\sin\theta = \frac{\text{length of the opposite side}}{\text{length of the hypotenuse}}$$

$$\cos\theta = \frac{\text{length of the adjacent side}}{\text{length of the hypotenuse}}$$

$$\tan\theta = \frac{\text{length of the opposite side}}{\text{length of the adjacent side}}$$

The reciprocal functions are defined similarly.

If $\theta = \pi/4$, then the other acute angle of the triangle also has measure $\pi/4$, since the acute angles of a right triangle are complementary. Thus, the right triangle will be isosceles and we can take the lengths of its congruent sides to be one unit. The length of the hypotenuse will then be $\sqrt{1^2 + 1^2} = \sqrt{2}$. From a sketch of this right triangle we can read off the values of all of the trigonometric functions for $\theta = \pi/4$. (See Figure 5-14.)

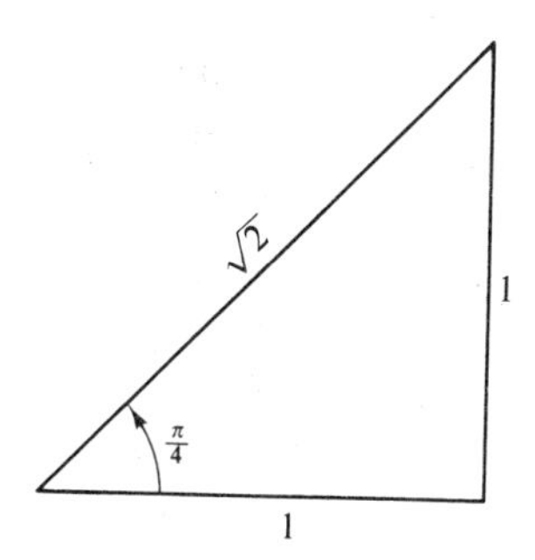

Figure 5-14

$$\sin\frac{\pi}{4} = \frac{\text{opposite}}{\text{hypotenuse}} = \frac{1}{\sqrt{2}} \qquad \csc\frac{\pi}{4} = \frac{\text{hypotenuse}}{\text{opposite}} = \frac{\sqrt{2}}{1} = \sqrt{2}$$

$$\cos\frac{\pi}{4} = \frac{\text{adjacent}}{\text{hypotenuse}} = \frac{1}{\sqrt{2}} \qquad \sec\frac{\pi}{4} = \frac{\text{hypotenuse}}{\text{adjacent}} = \frac{\sqrt{2}}{1} = \sqrt{2}$$

$$\tan\frac{\pi}{4} = \frac{\text{opposite}}{\text{adjacent}} = \frac{1}{1} = 1 \qquad \cot\frac{\pi}{4} = \frac{\text{adjacent}}{\text{opposite}} = \frac{1}{1} = 1$$

Next let us consider an equilateral triangle. Each of its angles has measure $\pi/3$ (60°). Constructing a perpendicular from a vertex to the opposite side gives us two congruent right triangles. If we take the length of one of the sides of the equilateral triangle to be two units, then one side of the right triangle will have length one unit and the other will have length $\sqrt{3}$, since $1^2 + (\sqrt{3})^2 = 2^2$. The other acute angle will have measure $\pi/6$ (30°).

From the $\pi/3 - \pi/6$ right triangle (Figure 5-15) we can find the values of the trigonometric functions for both $\pi/3$ and $\pi/6$.

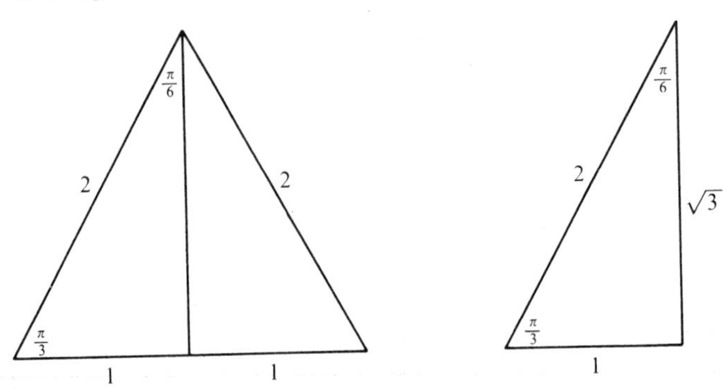

Figure 5-15

$$\sin\frac{\pi}{3} = \frac{\text{opposite}}{\text{hypotenuse}} = \frac{\sqrt{3}}{2}$$

$$\cos\frac{\pi}{3} = \frac{\text{adjacent}}{\text{hypotenuse}} = \frac{1}{2}$$

$$\tan\frac{\pi}{3} = \frac{\text{opposite}}{\text{adjacent}} = \frac{\sqrt{3}}{1}$$

$$\sin\frac{\pi}{6} = \frac{\text{opposite}}{\text{hypotenuse}} = \frac{1}{2}$$

$$\cos\frac{\pi}{6} = \frac{\text{adjacent}}{\text{hypotenuse}} = \frac{\sqrt{3}}{2}$$

$$\tan\frac{\pi}{6} = \frac{\text{opposite}}{\text{adjacent}} = \frac{1}{\sqrt{3}}$$

Note that $\sin \pi/3 = \cos \pi/6$ and $\cos \pi/3 = \sin \pi/6$. This is not a coincidence. Since the side opposite to one acute angle is adjacent to the other acute angle, and the two acute angles are complementary, we have the following rule:

Any trigonometric function of a number θ, $0 < \theta < \pi/2$, is equal to the cofunction of $\pi/2 - \theta$.

The sine and cosine are cofunctions as are the tangent and cotangent, and the secant and cosecant.

Example 5-6. $\cos \frac{\pi}{5} = \sin \frac{3\pi}{10}$

$$\tan \frac{\pi}{4} = \cot \frac{\pi}{4}$$

$$\csc 1 = \sec \left(\frac{\pi}{2} - 1\right) \simeq \sec .5708$$

$$\sin 50° = \cos 40°$$

$$\cot 10° = \tan 80°$$

$$\sec 15° = \csc 75°$$

Do not confuse reciprocal functions and cofunctions. The sine and cosecant functions are reciprocal functions since $\sin \theta = 1/\csc \theta$. The sine and cosine are cofunctions as their names would indicate. Only the tangent and cotangent functions are both reciprocal functions and cofunctions.

We can find the values of the trigonometric functions exactly for some other angles closely related to those whose measures are $\pi/4$, $\pi/3$, and $\pi/6$. Consider for example $\theta = 3\pi/4$. The terminal side of this angle lies in quadrant II, and if we drop a perpendicular to the x-axis we get an isosceles triangle whose sides can be taken to have the lengths 1, 1, $\sqrt{2}$.

If we label the sides of this triangle with the *directed distances* 1, -1, and $\sqrt{2}$, then we have $y = 1$, $x = -1$, $r = \sqrt{2}$, and we can find the functions of $3\pi/4$ using Definition 5-1.

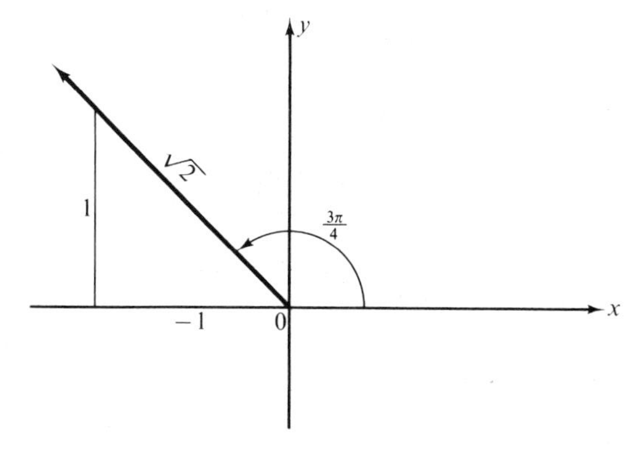

Figure 5-16

$$\sin\frac{3\pi}{4} = \frac{1}{\sqrt{2}} \qquad \csc\frac{3\pi}{4} = \sqrt{2}$$

$$\cos\frac{3\pi}{4} = \frac{-1}{\sqrt{2}} \qquad \sec\frac{3\pi}{4} = -\sqrt{2}$$

$$\tan\frac{3\pi}{4} = -1 \qquad \cot\frac{3\pi}{4} = -1$$

Example 5-7. Find the sine, cosine, and tangent of $7\pi/6$.

Solution: Since $7\pi/6 = \pi + \pi/6$, if we sketch this angle and drop a perpendicular to the x-axis, we will find that we have a $\pi/3$-$\pi/6$ right triangle with $y = -1$, $x = -\sqrt{3}$, and $r = 2$. (See Figure 5-17.) Consequently,

$$\sin\frac{7\pi}{6} = -\frac{1}{2}$$

$$\cos\frac{7\pi}{6} = -\frac{\sqrt{3}}{2}$$

$$\tan\frac{7\pi}{6} = \frac{1}{\sqrt{3}}$$

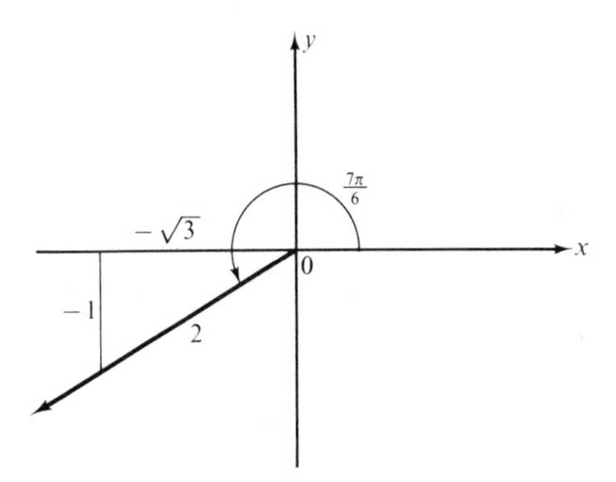

Figure 5-17

A third, more modern, way of defining the trigonometric functions uses the unit circle. We have already observed that for any point on the unit circle $r = 1$, hence $\sin\theta = x$ and $\cos\theta = y$.

Definition 5-3. Let θ be any real number and let (x, y) be the point of intersection of the terminal side of an angle in standard position whose measure is θ with the unit circle. Then

$$\sin\theta = y \qquad \csc\theta = \frac{1}{y} \quad y \neq 0$$

$$\cos\theta = x \qquad \sec\theta = \frac{1}{x} \quad x \neq 0$$

$$\tan\theta = \frac{y}{x} \quad x \neq 0 \qquad \cot\theta = \frac{x}{y} \quad y \neq 0$$

Because of this definition, the trigonometric functions are frequently called the *circular functions.*

This definition suggests a device for remembering the values of the functions of the quadrantal angles. If we draw a unit circle (Figure 5-18) and label the points where the circle cuts the axes, then the coordinates of these points will give us the values we seek.

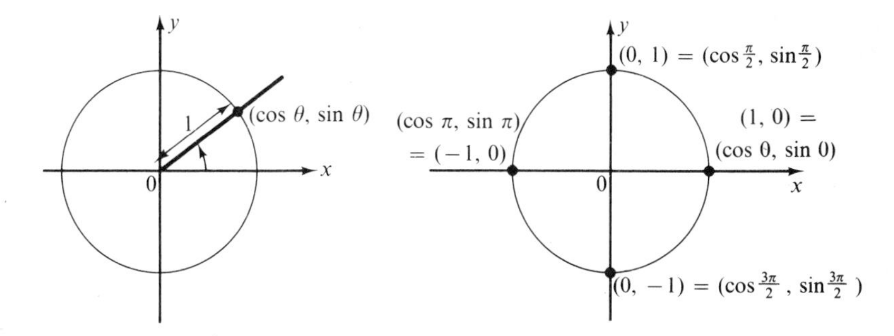

Figure 5-18

The x-coordinate of each point is the value of the cosine; thus, $\cos 0 = 1$, $\cos \pi/2 = 0$, $\cos \pi = -1$, and $\cos 3\pi/2 = 0$. The y-coordinate of each point is the value of the sine function; hence, $\sin 0 = 0$, $\sin \pi/2 = 1$, $\sin \pi = 0$, and $\sin 3\pi/2 = -1$. The value of the tangent function will be the ratio of the y-coordinate to the x-coordinate, and we have $\tan 0 = 0$, $\tan \pi/2$ undefined, $\tan \pi = 0$, and $\tan 3\pi/2$ undefined. The values of the reciprocal functions are of course the reciprocals of these values.

Exercise 5-3

1. Find the values of the trigonometric functions of A and B if the right triangle has the dimensions indicated in Figure 5-19.
2. Find the values of the trigonometric functions of A and B if the right triangle has the dimensions indicated in Figure 5-20.

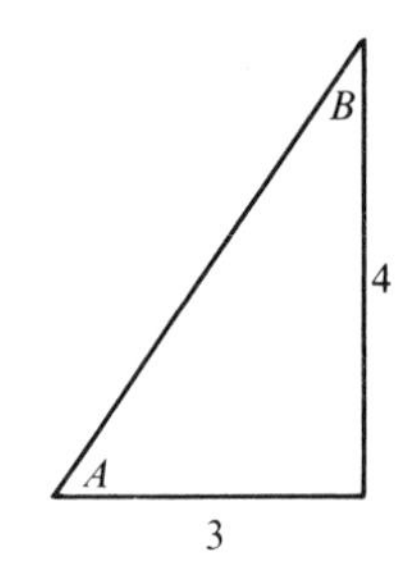

Figure 5-19

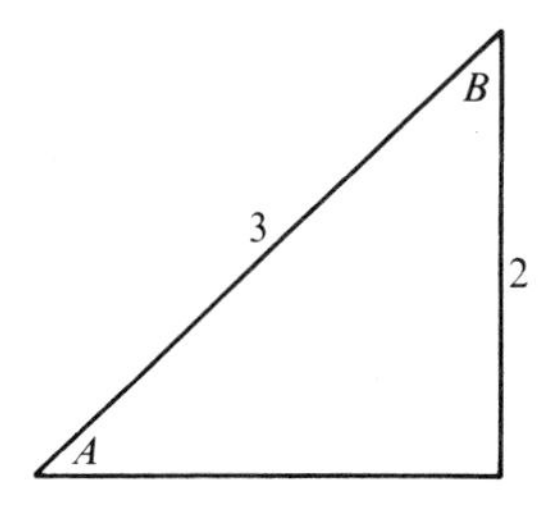

Figure 5-20

3. If $0 < \theta < \pi/2$, and $\sin\theta = 2/3$, find the value of the reciprocal function, of the cofunction.

4. If $0 < \theta < \pi/2$, and $\sec\theta = 3$, find the value of the reciprocal function, of the cofunction.

5. Evaluate:
   (a) $\cos \pi/2 - \sin \pi/6$
   (b) $\tan \pi/4 + \cos \pi$
   (c) $\sin 90° + \cos 0° - \tan 45°$
   (d) $\sin 90° \cdot \cos 30° + \cos 90° \cdot \sin 30°$
   (e) $\cot \pi/6 + \tan \pi/3$

6. Evaluate:
   (a) $\cos \pi \cdot \sin \pi/2 - \sin \pi \cdot \cos \pi/2$
   (b) $\dfrac{\tan 30° - \tan 60°}{1 + \tan 30° \tan 60°}$
   (c) $\sin 3\pi/2 \cdot \cos \pi/6 + \cos 3\pi/2 \cdot \sin \pi/6$
   (d) $\tan \pi/6 - \cot \pi/3$
   (e) $\csc 45° + \cot 30° - \sec 60°$

7. Find the values of the sine, cosine, and tangent functions for each of the values listed. Sketch.
   (a) $2\pi/3$ (b) $4\pi/3$
   (c) $7\pi/4$ (d) $11\pi/6$

(e) $5\pi/6$ (f) $5\pi/4$
(g) $5\pi/3$

8. Express as a function of an angle whose measure is between 0 and $\pi/4$ ($0°$ and $45°$):
   (a) $\csc 3\pi/8$ (b) $\sec 3\pi/7$
   (c) $\cot 4\pi/9$ (d) $\sin 86°$
   (e) $\cos 72°$ (f) $\tan 46°$

9. Find the value of θ if $0 \leq \theta \leq \pi/2$ and
   (a) $\sin \theta = 1/\sqrt{2}$ (b) $\sin \theta = \sqrt{3}/2$
   (c) $\tan \theta = 1$ (d) $\cos \theta = 1/2$
   (e) $\tan \theta = 1/\sqrt{3}$ (f) $\cos \theta = 0$

10. Find the values of θ if $0 \leq \theta < 2\pi$ and
    (a) $\cos \theta = 1$ (b) $\sin \theta = -1$
    (c) $\tan \theta$ is undefined (d) $\cos \theta = -1$
    (e) $\csc \theta = 1$ (f) $\sec \theta$ is undefined

11. Find the values of θ, $0 \leq \theta < 2\pi$ for each of the following. There will be two answers. Sketch each.
    (a) $\tan \theta = -1$ (b) $\cos \theta = 1/2$
    (c) $\sin \theta = -1/\sqrt{2}$ (d) $\tan \theta = -1/\sqrt{3}$

12. Use Definition 5-3 of the sine and cosine functions to tell whether these functions are increasing or decreasing on the intervals given below:

	sine	cosine
$0 < \theta < \pi/2$		
$\pi/2 < \theta < \pi$		
$\pi < \theta < 3\pi/2$		
$3\pi/2 < \theta < 2\pi$		

5-4 Periodic Functions and Functions of a Negative Number

From the definitions of the trigonometric functions, their domain is the set of all real numbers, with some exclusions in the case of the tangent, cosecant, secant, and cotangent functions. However, the trigonometric functions are not one-to-one, as are the exponential and logarithm functions. Instead, the trigonometric functions are *periodic,* that is, a trigonometric function takes on a certain set of values, then these values are repeated over and over again.

Definition 5-4. A function f is said to be *periodic* if for some number $p > 0, f(x+p) = f(x)$ for all x in the domain of f. The smallest positive number p for which this is true is called the *period* of the function.

We have already seen that the angle whose measure is $\theta + 2\pi$ has the same terminal side as the angle whose measure is θ. It follows that the values of the trigonometric functions are the same for these two numbers.

$$\sin(\theta + 2\pi) = \sin \theta \qquad \csc(\theta + 2\pi) = \csc \theta$$
$$\cos(\theta + 2\pi) = \cos \theta \qquad \sec(\theta + 2\pi) = \sec \theta$$
$$\tan(\theta + 2\pi) = \tan \theta \qquad \cot(\theta + 2\pi) = \cot \theta$$

The period of the sine and cosine functions and their reciprocal functions is 2π, since this is the smallest positive number for which these equations hold. However, it is also true that

$$\tan(\theta + \pi) = \tan \theta \quad \text{and} \quad \cot(\theta + \pi) = \cot \theta$$

The period of the tangent and cotangent functions is π.

This property of the trigonometric functions has useful results in the construction and use of tables. It is unnecessary, for instance, to tabulate values for $\sin 5\pi$ or $\cos 480°$, since these will be the same as $\sin \pi$ and $\cos 120°$ respectively.

Functions of a negative number can be changed to functions of a positive number by the following formulas. For all real numbers θ,

$$\sin(-\theta) = -\sin \theta \qquad \csc(-\theta) = -\csc \theta$$
$$\cos(-\theta) = \cos \theta \qquad \sec(-\theta) = \sec \theta$$
$$\tan(-\theta) = -\tan \theta \qquad \cot(-\theta) = -\cot \theta$$

These formulas tell us that the cosine and secant are *even* functions and the other trigonometric functions are *odd.* (See Section 2-7.)

To see why these formulas hold, first let us consider the case in which $0 < \theta < \pi/2$ (Figure 5-21).

If we construct two congruent triangles, one in the first and one in the fourth quadrant, then corresponding to a point (x, y) on the terminal side of the angle whose measure is θ, there will be a point $(x, -y)$ on the terminal side of the angle whose measure is $-\theta$. Thus,

$$\sin(-\theta) = \frac{-y}{r} = -\sin\theta \qquad \csc(-\theta) = \frac{r}{-y} = -\csc\theta$$

$$\cos(-\theta) = \frac{x}{r} = \cos\theta \qquad \sec(-\theta) = \frac{r}{x} = \sec\theta$$

$$\tan(-\theta) = \frac{-y}{x} = -\tan\theta \qquad \cot(-\theta) = \frac{x}{-y} = -\cot\theta$$

For the case in which $\pi/2 < \theta < \pi$, we construct two congruent triangles, one in the second quadrant and one in the third (Figure 5-22).

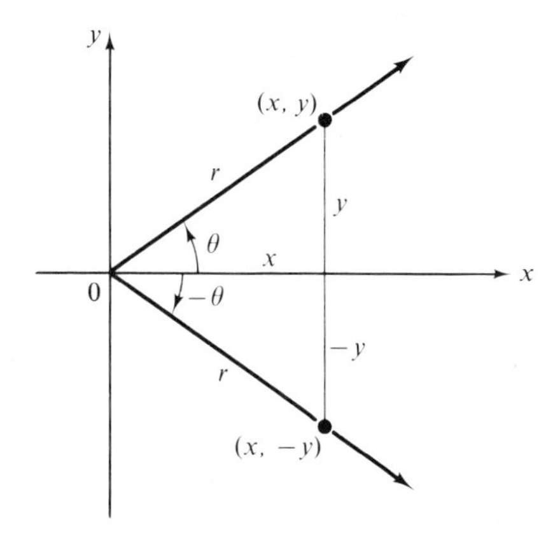

Figure 5-21

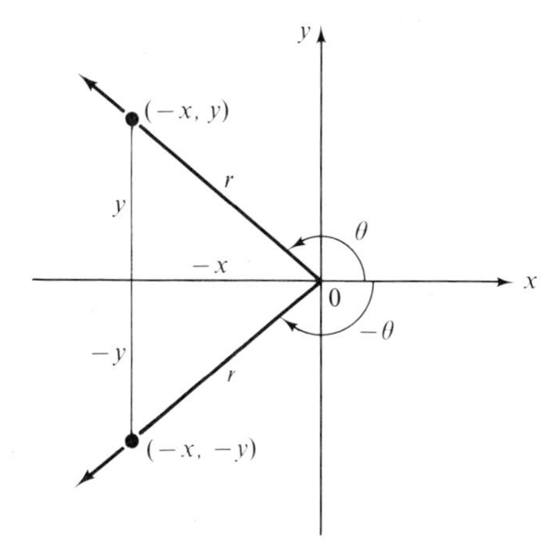

Figure 5-22

Corresponding to a point $(-x, y)$ on the terminal side of the angle whose measure is θ, there will be a point $(-x, -y)$ on the terminal side of the angle whose measure is $-\theta$. Then

$$\sin(-\theta) = \frac{-y}{r} = -\sin\theta$$

$$\cos(-\theta) = \frac{-x}{r} = \cos\theta$$

$$\tan(-\theta) = \frac{-y}{-x} = \frac{y}{x} = -\tan\theta$$

$$\cot(-\theta) = \frac{-x}{-y} = \frac{x}{y} = -\cot\theta$$

$$\csc(-\theta) = \frac{r}{-y} = -\csc\theta$$

$$\sec(-\theta) = \frac{r}{-x} = \sec\theta$$

The argument is similar for $\pi < \theta < 3\pi/2$, and for $3\pi/2 < \theta < 2\pi$ and will be left for the student (Exercise 5-4, problem 5). Except for those cases in which the functions are undefined the formulas also hold for $\theta = 0$, $\pi/2$, π, and $3\pi/2$ (Exercise 5-4, problem 6).

Since the formulas hold for all θ, $0 \leq \theta \leq 2\pi$, the periodic nature of these functions tells us that they hold for *all* real numbers θ.

Exercise 5-4

1. Express as a function of a number between 0 and 2π (or 0° and 360°).
   (a) $\sin 27\pi$ (b) $\sin 500°$
   (c) $\cos 8\pi/3$ (d) $\cos(-921°)$
   (e) $\tan 16\pi/7$ (f) $\tan(-28\pi/3)$
   (g) $\csc(-480°)$ (h) $\sec(-1421°)$
   (i) $\sin 16$ (j) $\cos(-21)$

2. Find the exact value of
   (a) $\sin 11\pi/4$ (b) $\cos 9\pi/4$
   (c) $\cos(-17\pi)$ (d) $\sin(-11\pi/2)$
   (e) $\tan(-7\pi)$ (f) $\sin 15\pi/6$

3. Find the exact value of
   (a) $\sin(-13\pi/2)$ (b) $\cos 15\pi/2$
   (c) $\tan(-20\pi/3)$ (d) $\cos 21\pi/4$
   (e) $\tan 14\pi/6$ (f) $\sec 181\pi$

4. In Figure 5-23 the arrows indicate that the function continues forever in the manner indicated. Which are graphs of periodic functions? Give the period of those which are periodic functions.

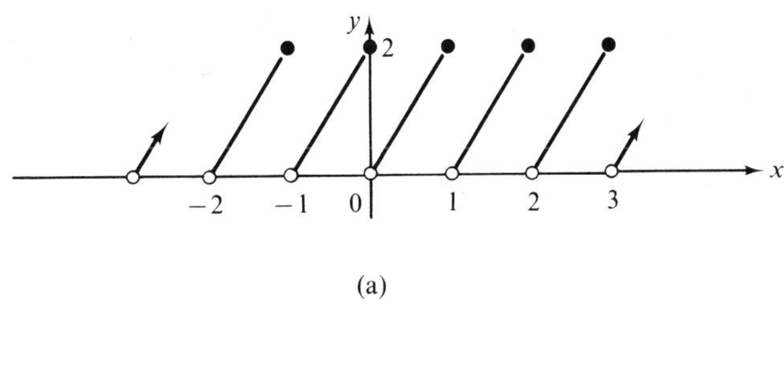

(a)

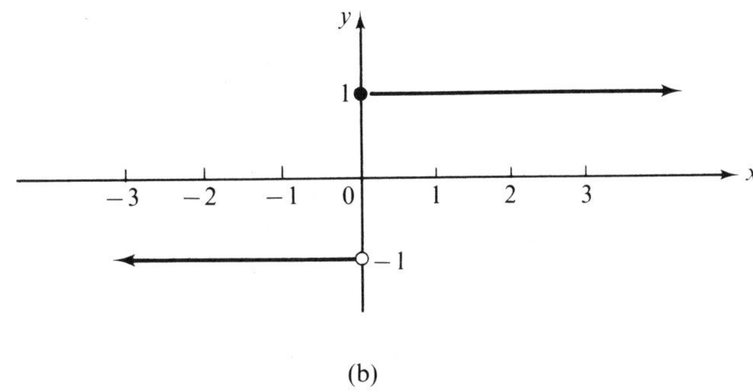

(b)

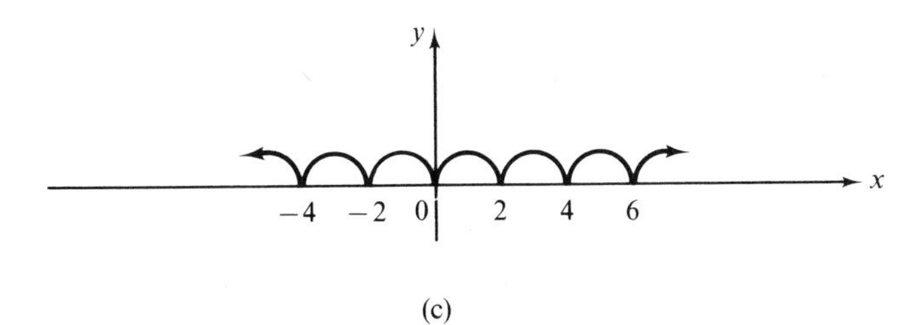

(c)

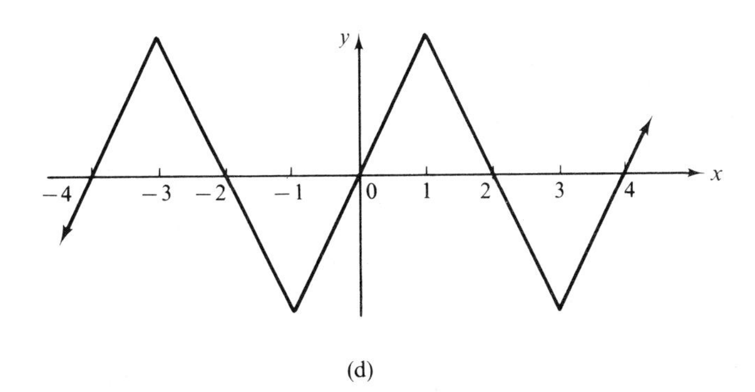

(d)

Figure 5-23

5. Verify the formulas for the functions of a negative number for $\pi < \theta < 3\pi/2$; for $3\pi/2 < \theta < 2\pi$. Sketch.

6. Verify the formulas for a function of a negative number for $\theta = 0$; $\pi/2$; π; and $3\pi/2$.

7. Show that a constant function is periodic. Does it have a period?

5-5 Graphs of the Trigonometric Functions

From Definition 5-3, Section 5-3, the values of the sine and cosine functions are given by the y- and x-coordinates respectively of points on the unit circle. Thus, for $\theta = 0$, $\sin\theta = 0$, and as θ increases to $\pi/2$, the value of the sine function increases to one. (See Figure 5-18.) As θ increases from $\pi/2$ to π, the value of $\sin\theta$ decreases from one to zero. It continues to decrease as θ goes from π to $3\pi/2$, where it takes on its smallest value, -1. From $3\pi/2$ to 2π $\sin\theta$ increases again until $\sin 2\pi = 0$. This behavior is pictured in the graph of the sine function (Figure 5-24).

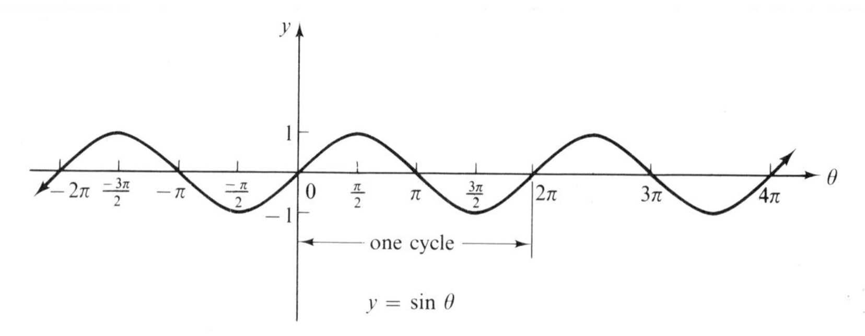

Figure 5-24

The graph of the cosine function is very similar. From $\theta = 0$ to $\theta = \pi$, $\cos\theta$ decreases from 1 to -1, and as θ goes from π to 2π, $\cos\theta$ increases from -1 to 1. (See Figure 5-25.)

The graphs clearly show the periodic nature of these functions. The portion of the curve between $\theta = 0$ and $\theta = 2\pi$ is repeated over and over. This portion of the graph is called a *cycle*. The domain of the sine and cosine functions is the set of all real numbers; the range is the set of real numbers between -1 and 1, inclusive. The cosine function is an *even* function, since $\cos(-\theta) = \cos\theta$; and this means that the graph is symmetric with respect to the vertical axis. Similarly, the fact that the sine function is an *odd* function ($\sin(-\theta) = -\sin\theta$) means its graph is symmetric with respect to the origin. (See Section 2-7.)

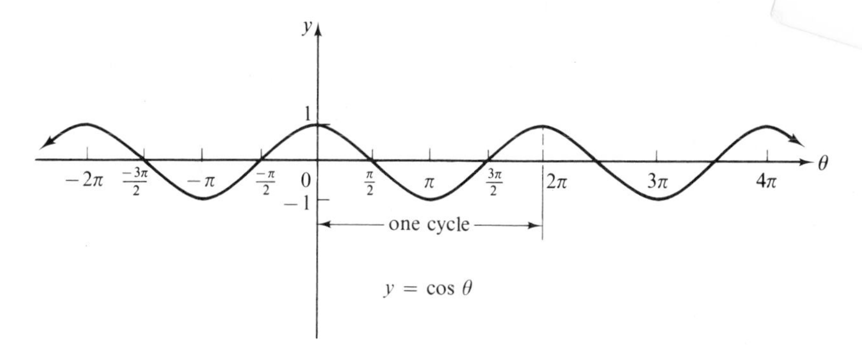

Figure 5-25

The tangent function is zero for $\theta = 0$ and undefined when $\theta = \pi/2$. What does this function do between these two values? If we think of $\tan\theta$ as y/x, where (x, y) is a point on the unit circle, we can see that as θ increases from 0 to $\pi/2$, x is getting smaller and smaller, while y is getting closer and closer to one. If the denominator of a fraction gets smaller while the numerator does not change, the fraction itself will become larger. Thus, if $x = .1$, $y = \sqrt{99/100} \simeq 1$, and $\tan\theta \simeq 10$. If $x = .01$, $y = \sqrt{9999/10000} \simeq 1$, and $\tan\theta \simeq 100$. The tangent function then gets larger and larger as θ gets closer and closer to $\pi/2$. Since the tangent function is an odd function ($\tan(-\theta) = -\tan\theta$), the graph is symmetric with respect to the origin, and for $-\pi/2 < \theta < 0$, $\tan\theta$ is a negative number whose absolute value becomes larger and larger as θ gets closer and closer to $-\pi/2$. (See Figure 5-26.)

From the graph we can see that the domain of the tangent function is all of the real numbers except odd integral multiples of $\pi/2$ ($\theta \neq (2k + 1)(\pi/2)$, $k = 0, \pm 1, \pm 2, \ldots$), while the range is the set of all real num-

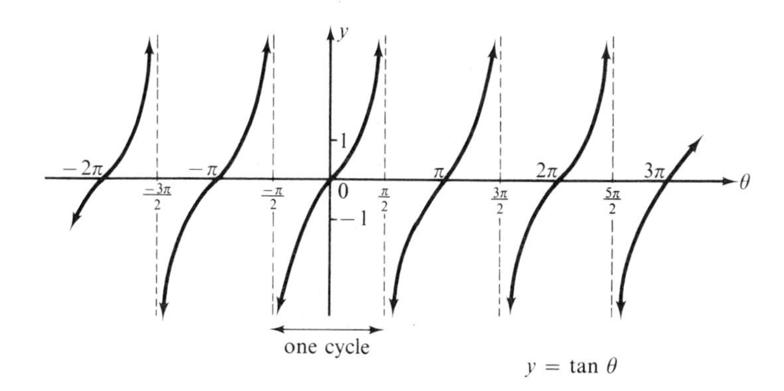

Figure 5-26

bers. The period is π, and we see that the portion of the graph from $-\pi/2$ to $\pi/2$ is repeated again and again. This is a cycle of this function. The dotted lines at $\theta = \pm\pi/2$; $\theta = \pm 3\pi/2$, etc., are called asymptotes of the graph. The curve approaches closer and closer to these lines but never crosses them.

The graph of the cosecant function can best be drawn by referring to the graph of its reciprocal function, the sine function. When $\sin\theta = \pm 1$, $\csc\theta = \pm 1$, and when $\sin\theta = 0$, $\csc\theta$ is undefined. The two functions are shown on the same set of axes in Figure 5-27. From this graph we see that the domain of the cosecant function is the set of all real numbers except $k\pi$, $k = 0, \pm 1, \pm 2, \ldots$, and the range is the set of all real numbers which are greater than or equal to 1 or less than or equal to -1. The period is 2π.

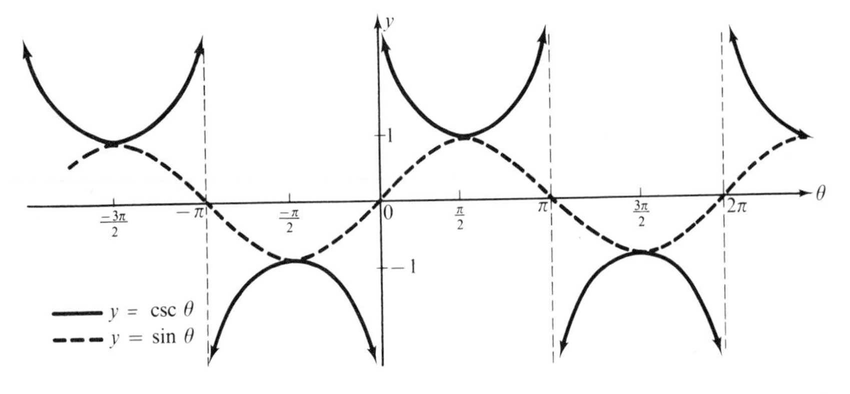

Figure 5-27

The graphs of the secant and cotangent functions are left for the student (Exercise 5-5, problems 2 and 3).

If we modify the sine or cosine function by multiplying it by a constant, the effect on the graph is to change the height of the "wave." Thus, if $f(\theta) = 2\sin\theta$, then when $\theta = \pi/2, f(\theta) = 2$, and the change in the graph is pictured in Figure 5-28.

This height is called the *amplitude* of the sine (or cosine) function. More formally, the amplitude of a periodic function is one-half of the difference between the maximum and the minimum value of the function. In the case of the function $y = A\sin x$, or $y = A\cos x$, the amplitude is $|A|$.

If we change the trigonometric functions by multiplying the measure of the angle by a constant, then the period of the function will be changed. In the function $f(\theta) = \sin 2\theta$, for example, as θ goes from

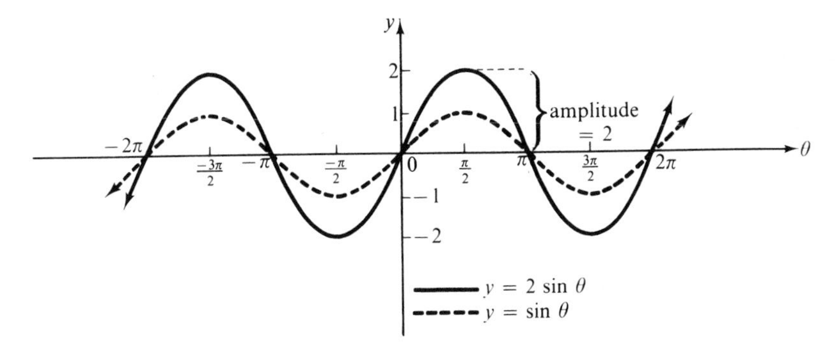

Figure 5-28

zero to π, 2θ goes from zero to 2π. The function thus completes one cycle in the interval from zero to π and the period is cut in half. (See Figure 5-29.)

In general, if the measure of the angle is multiplied by a nonzero constant B, then the normal period is changed by the factor $1/|B|$. For example, $y = \cos 3x$ has period $2\pi \cdot (1/3) = 2\pi/3$; $y = \sin(1/2)x$ has period $2\pi \cdot 2 = 4\pi$; and $y = \tan 2x$ has period $\pi \cdot (1/2) = \pi/2$, since the period of $y = \tan x$ is π.

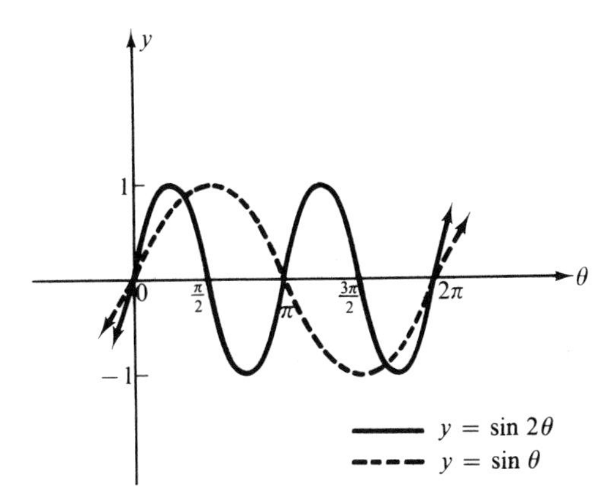

Figure 5-29

Adding a constant to the measure of the angle causes the graph to be shifted to the right or left on the horizontal axis. The amount of shift is called the *phase shift* and is found by setting the entire angle measure equal to zero and solving for the independent variable. For example, if $y = \sin(2x + \pi)$, we set $2x + \pi$ equal to zero and find $x =$

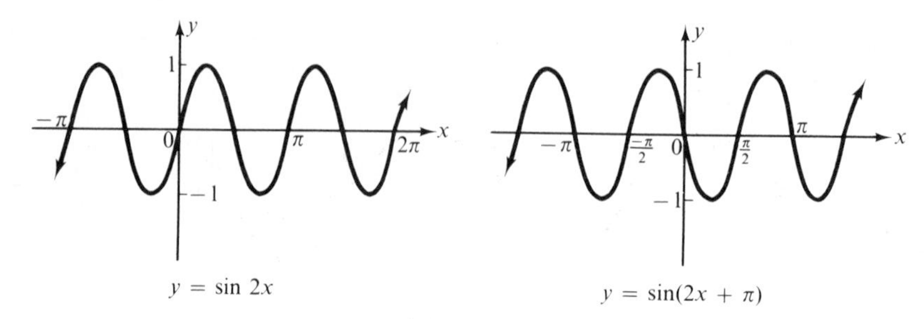

$y = \sin 2x$ $y = \sin(2x + \pi)$

Figure 5-30

$-\pi/2$. This means that the graph of this function will be the graph of $y = \sin 2x$ shifted to the left by $\pi/2$ units. (See Figure 5-30.)

Let us summarize these observations. If $y = A \text{ fcn}(Bx + C); A \neq 0;$ $B \neq 0$; where "fcn" is one of the trigonometric functions, then the following are true:

1. The amplitude is $|A|$ if the function is the sine or the cosine. The amplitude is not defined for the other functions since they do not have maximum or minimum values.
2. The period will be the product of $1/|B|$ and the normal period of of the function. The normal period of the sine and cosine functions and their reciprocal functions is 2π; the normal period of the tangent and cotangent functions is π.
3. The phase shift is $-C/B$.

Example 5-8. Find the amplitude, period, and phase shift of $y = 2 \cos(3x - \pi)$ and graph this function.

Solution: The amplitude is 2 and the period is $2\pi \cdot (1/3) = 2\pi/3$. To find the phase shift, we set $3x - \pi = 0$ and find $x = \pi/3$. To graph this function, we first sketch $y = 2 \cos 3x$, then shift the graph to the right $\pi/3$ units. (See Figure 5-31.)

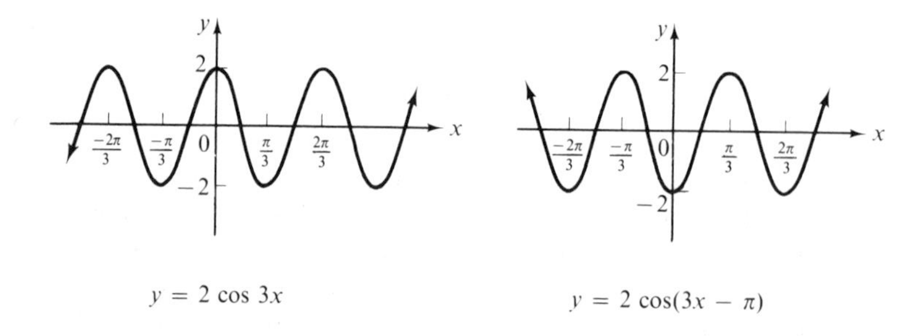

$y = 2 \cos 3x$ $y = 2 \cos(3x - \pi)$

Figure 5-31

Exercise 5-5

1. Complete the following table for each of the functions listed:

	Domain	Range	Period
sine			
cosine			
tangent			
cosecant			
secant			
cotangent			

2. Graph $y = \sec x$.

3. Graph $y = \cot x$.

4. Graph $y = \cos x$ and $y = 2 \cos x$ on the same set of axes.

5. Graph $y = \cos x$ and $y = \cos 2x$ on the same set of axes.

6. Graph $y = \sin x + 2$. In general what effect does adding a constant to a function have on the graph?

7. Graph $y = -\sin x$. In general what effect does multiplying a function by -1 have on the graph?

8. Find the amplitude (where it applies), the period, and the phase shift of each of the following:
   (a) $y = \cos(3x + \pi/2)$ (b) $y = 3 \tan(2x + \pi/4)$
   (c) $y = 3 \cos[(1/3)x - \pi/2]$ (d) $y = 2 \sin[(2/3)x + \pi/6]$
   (e) $y = \sqrt{2} \cos(2x - 1)$ (f) $y = 4 \sin(2x + \pi/4)$
   (g) $y = \tan \pi x/2$ (h) $y = \tan(3x + \pi)$

9. Find the amplitude (where it applies), the period, and the phase shift of each of the following and graph:
   (a) $y = 3 \sin 2x$ (b) $y = 2 \cos 3x$
   (c) $y = \tan 2x$ (d) $y = 2 \sin[(1/2)x - \pi]$
   (e) $y = 4 \sin(x/2)$ (f) $y = (1/2)\cos(x - 1)$
   (g) $y = \sin(x + \pi/4)$ (h) $y = 2 \cos(x - \pi/3)$

10. Consider the function $y = A \sin (Bx + C)$; $A \neq 0$; $B \neq 0$. What is the effect on the graph if
    (a) $|A|$ is increased? (b) $|B|$ is increased?
    (c) $C > 0$; C is increased and B is unchanged?
    (d) $|A|$ is decreased? (e) $|B|$ is decreased?
    (f) $C < 0$; C is decreased and B is unchanged?

11. Graph $y = \sin x$ and $y = \cos(x - \pi/2)$. What do you observe about these two functions?

12. Make a table of values and use this to graph the function $y = \tan x \cdot \cos x$; $0 \leq x \leq 2\pi$.

13. Make a table of values and use this to graph the function $y = \sin^2 x + \cos^2 x$; $0 \leq x \leq 2\pi$.

14. Let us define functions which we shall call the *square* functions as follows (Figure 5-32):

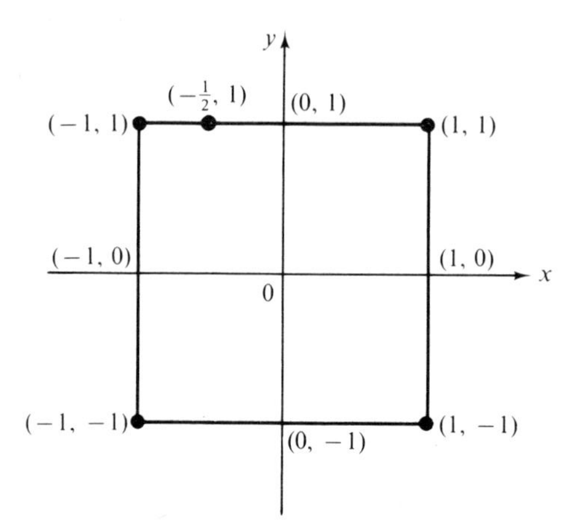

Figure 5-32

On a square whose side is 2 units long, let θ be the distance measured along the square starting at the point (1, 0). If θ is positive, the distance is measured counterclockwise; if θ is negative, it is measured clockwise. Thus $\theta = 1$ corresponds to the point (1, 1) while $\theta = 2\frac{1}{2}$ corresponds to the point $(-1/2, 1)$. We will define two functions of θ, cosq and sqin to be the x- and y-coordinates of these points respectively. Thus, for example, cosq $0 = 1$; sqin $0 = 0$; cosq $2\frac{1}{2} = -1/2$; sqin $2\frac{1}{2} = 1$.

(a) Complete the following table:

	0	1	2	3	4	5	6	7	8	−1	−2	−3
cosq												
sqin												

(b) Are these functions periodic? If so, what is their period? What is the domain and range of each?
(c) Graph each of these functions.
(d) Are these functions even, odd, or neither? Explain.

5-6* Graphing the Sum of Two Functions

It is frequently useful to graph a function such as

$$y = \sin x + \sin 2x$$

This function is the sum of two simpler functions

$$y_1 = \sin x \quad \text{and} \quad y_2 = \sin 2x$$

each of which is easy to graph. We can graph the sum by first graphing the simpler functions which are its terms and then use these to sketch the sum by a method called *addition of ordinates.*

The *ordinate* of a point (x, y) is its y-coordinate and can be thought of as the perpendicular distance from that point to the x-axis. To graph $y = \sin x + \sin 2x$, we first graph $y_1 = \sin x$ and $y_2 = \sin 2x$ on the same set of axes. (See Figure 5-33.)

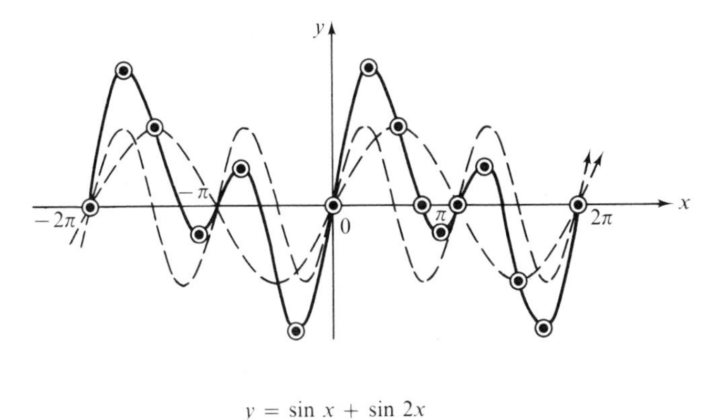

$y = \sin x + \sin 2x$

Figure 5-33

For any value of x, the ordinate of a point on the graph of $y = \sin x + \sin 2x$ can be found by adding the ordinates of the corresponding points on $y_1 = \sin x$ and $y_2 = \sin 2x$. Of course these ordinates must be added algebraically. If they have the same sign, they are added, if they have opposite signs, they are subtracted. This addition (or sub-

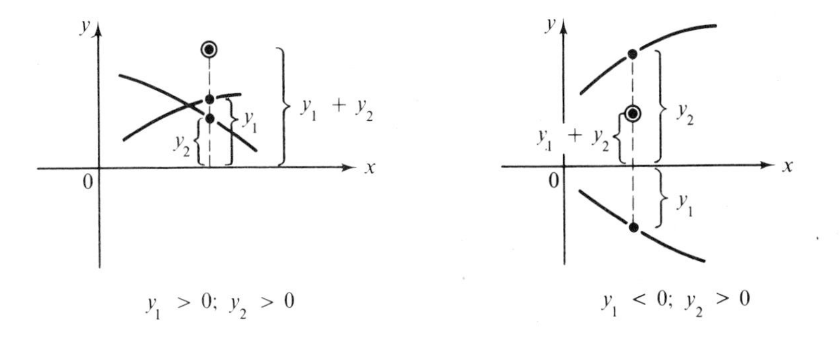

Figure 5-34

traction) can be done geometrically with the aid of a straight-edge (Figure 5-34). We simply add the distances from the x-axis, where this distance is taken to be positive if the curve is above the x-axis and negative if it is below the x-axis.

The sum of two periodic functions is not necessarily periodic. It will be periodic if there exists some number which is a positive integral multiple of each of the two periods. The smallest such integral multiple will be the period of the sum function.

Example 5-9. The function $y = \sin 2x + \cos 3x$ is periodic. The period of $\sin 2x$ is π, and that of $\cos 3x$ is $2\pi/3$. The number 2π is the smallest positive integral multiple of both π and $2\pi/3$; thus, the period of $y = \sin 2x + \cos 3x$ is 2π. It is not hard to see why this is the case. At $x = 2\pi$, the function $y = \sin 2x$ will have completed two cycles, while $y = \cos 3x$ has completed three cycles, thus one cycle of the sum will be completed.

Example 5-10. The function $y = \sin x + \sin \pi x$ is not periodic, because $y = \sin x$ has period 2π, while $y = \sin \pi x$ has period 2. There is no number which is a positive integral multiple of both 2π and 2, since one is irrational and the other is rational.

Exercise 5-6

Graph each of the following by the method of addition of ordinates. Which are periodic? Find the periods of those which are periodic.

1. $y = \sin x + 2$
2. $y = \sin x + x$
3. $y = \sin x + \cos x$
4. $y = \cos x + \cos 2x$
5. $y = \sin x + |x|$
6. $y = |x| + x$

Find the period of each of the following functions:

7. $y = \sin(1/2)x + \sin 2x$
8. $y = \sin(1/2)x + \cos(1/3)x$
9. $y = \cos 3x + \cos(1/2)x$
10. $y = \sin(2/3)x + \cos(1/3)x$

11. Suppose that f and g are periodic functions with the same period p. Prove that $f + g$ is periodic with period p. Prove that $f \cdot g$ is periodic with period p.

12. Let f and g be functions, and suppose g is periodic but f is not. Prove that $f \circ g$ is periodic.

5-7 Mathematics and Music

The great mathematician Fourier (1768–1830) showed that all musical sounds, vocal or instrumental, can be described in mathematical terms. The functions needed to describe these sounds are the trigonometric functions we have been studying.

First let us consider the sound given off when a tuning fork is struck. If a pen is attached to one prong of the tuning fork and this is allowed to trace a mark on a roll of paper moving at a constant rate, the result will be the curve shown in Figure 5-35. This looks remarkably like a sine or cosine wave. These vibrations of the tuning fork are transferred to the molecules of the air and this causes the sound we hear.

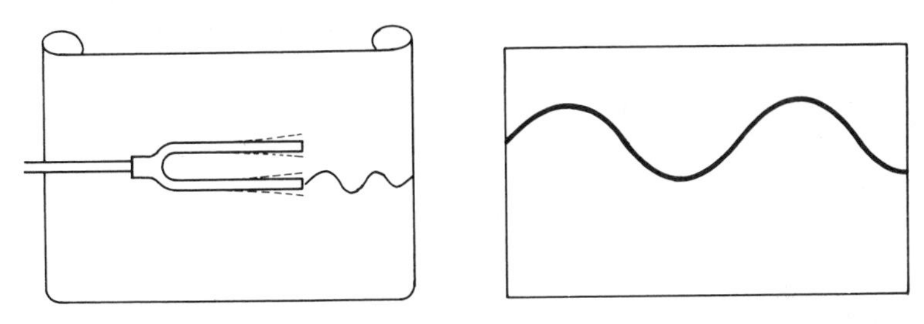

Figure 5-35

Every musical instrument produces sounds in this way, by inducing vibrations in the molecules of air. This motion may be initiated by vibrating strings, as in the violin or piano, by vibrating columns of air, in the clarinet or the organ, or by the vibration of other materials in the case of the cymbal or the drum.

If the tuning fork is struck harder, so as to give off a louder sound, the amplitude of the curve will be increased. The tuning fork will still give off the same note, however, and the period of the graph will remain the same.

In discussing the graphs of sounds, it is useful to speak of the *frequency* rather than the period. The frequency of a graph is the reciprocal of its period when the period is measured in seconds, and frequency is measured in cycles per second. The frequency of 261 cycles per second is associated with the note of middle C, not only for a tuning fork, but for any musical sound, no matter how it is produced. For instance, if the wings of a mosquito vibrate at the rate of 261 beats per second, the whine we hear will be middle C. If a chain saw runs at such a rate that 261 teeth cut into the wood in one second, the note we hear will again be middle C. The pitch of the sound we hear is related to the frequency (and hence the period) of its graph. If the frequency is increased, the note we hear will be higher in pitch. Doubling the frequency increases the pitch by one octave.

The graph of the note of middle C given off by a tuning fork can be described by the function

$$y = k \sin 522\pi t$$

where the independent variable t is measured in seconds and k, the amplitude, will vary with the loudness of the sound.

Not all musical sounds are as simple as those given off by tuning forks. The note of middle C played on a violin and that same note sung by a human voice, for instance, sound quite different, even to the untrained ear. How does the mathematician explain these differences? Fourier proved that the function which represents any periodic sound is a *sum* of simpler functions of the form $a \sin bx$. Moreover, the frequencies of these sine terms are all integral multiples of the smallest frequency. The term with the smallest frequency is called the fundamental tone.

For example, a note played by a violin may be represented by the function

$$y = .06 \sin 1{,}000\pi t + .02 \sin 2{,}000\pi t + .01 \sin 3{,}000\pi t$$

Note that the period of the first term is $2\pi/1{,}000\pi = 1/500$, thus its frequency is 500 cycles per second. The frequency of the second term is twice this, or 1,000, and the frequency of the third is 1,500 cycles per second. The graph of each of these terms together with the graph of their sum is shown in Figure 5-36.

A musical sound has three characteristics, loudness, pitch, and quality, and each has a mathematical counterpart. The louder of two sounds has a graph with a greater amplitude. Sounds with the same pitch produce graphs having the same frequency, while the graphs of high pitched sounds have a greater frequency than those of low pitched sounds. The frequency of the graph of a complex sound will be that of the fundamental tone. The frequency of the graph of the violin note in Figure 5-36 is 500.

The quality of a musical sound is related to the shape of its graph,

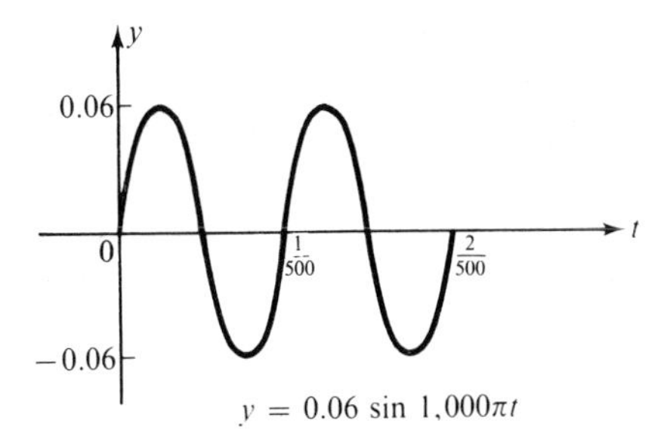

$y = 0.06 \sin 1{,}000\pi t$

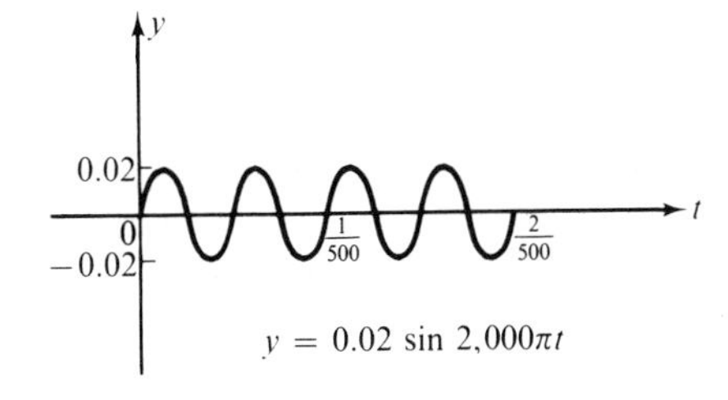

$y = 0.02 \sin 2{,}000\pi t$

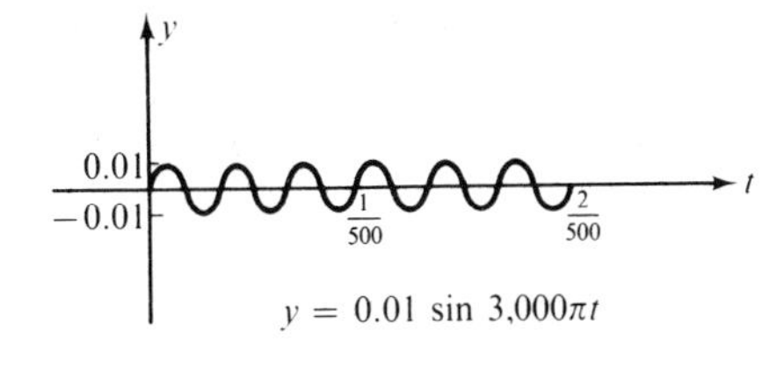

$y = 0.01 \sin 3{,}000\pi t$

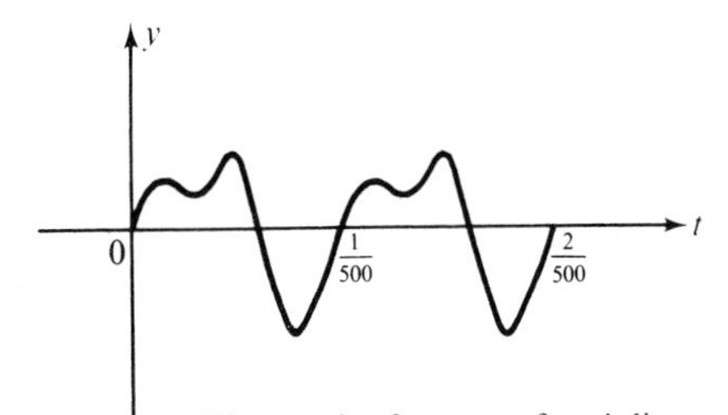

The graph of a note of a violin
$y = 0.06 \sin 1{,}000\pi t + 0.02 \sin 2{,}000\pi t + 0.01 \sin 3{,}000\pi t$

Figure 5-36

and this is determined by the different terms in the function which represents it. The graphs for different notes on the same instrument will have the same general shape, though the amplitude and frequency will vary.

A complex tone, such as that of the violin, is made up of simpler tones such as those produced by the tuning fork. A tone practically indistinguishable from the note of the violin graphed in Figure 5-36 could be produced by striking simultaneously, with suitable relative loudness, three tuning forks whose frequencies are 500, 1,000, and 1,500 vibrations per second. It is theoretically possible to play Tchaikovsky's First Piano Concerto entirely with tuning forks!

5-8 The Inverse Trigonometric Functions

The trigonometric functions are not one-to-one and therefore cannot have inverses. However, it is possible to choose a small portion of the domain of a trigonometric function in such a way that the function defined on this small portion will be one-to-one and will therefore have an inverse. Let us see how this can be done with the cosine function.

If we restrict the domain so that $0 \leq x \leq \pi$, then the cosine function will be one-to-one on this new domain and will have an inverse (Figure 5-37). We will call this inverse function the arccosine function, abbreviated arccos, and if $y = \cos x$, and $0 \leq x \leq \pi$, then $x = \arccos y$. We can think of this as saying "x is the number whose cosine is y," or "x is the length of arc whose cosine is y." The length of the arc along the unit circle is, of course, the measure of the angle in radians.

Since the domain of $y = \cos x$, $0 \leq x \leq \pi$, is $\{x \mid 0 \leq x \leq \pi\}$ and its range is $\{y \mid -1 \leq y \leq 1\}$, it follows that the domain of $y = \arccos x$ is $\{x \mid -1 \leq x \leq 1\}$ and its range is $\{y \mid 0 \leq y \leq \pi\}$.

Example 5-11. Find arccos (-1).

Solution: We want to find the number whose cosine is -1. Now there are infinitely many numbers that satisfy this condition, $\pi, -\pi, 3\pi$, etc.; however, we must choose the one that is in the range of the arccosine function, i.e., a number between 0 and π, inclusive. Hence, $\arccos(-1) = \pi$.

If we try to define the inverse of the sine function in the same way, we run into difficulties, because the sine function is not one-to-one on the interval from 0 to π. It is, however, one-to-one between $-\pi/2$ and $\pi/2$. (It is also one-to-one between $\pi/2$ and $3\pi/2$, however it is customary to choose the portion between $-\pi/2$ and $\pi/2$.) Accordingly, we define the arcsine function (abbreviated arcsin) to be the inverse of $y = \sin x; -\pi/2 \leq x \leq \pi/2$. (See Figure 5-38.) The domain of $y = \arcsin x$ is $\{x \mid -1 \leq x \leq 1\}$; its range is $\{y \mid (-\pi/2) \leq y \leq (\pi/2)\}$.

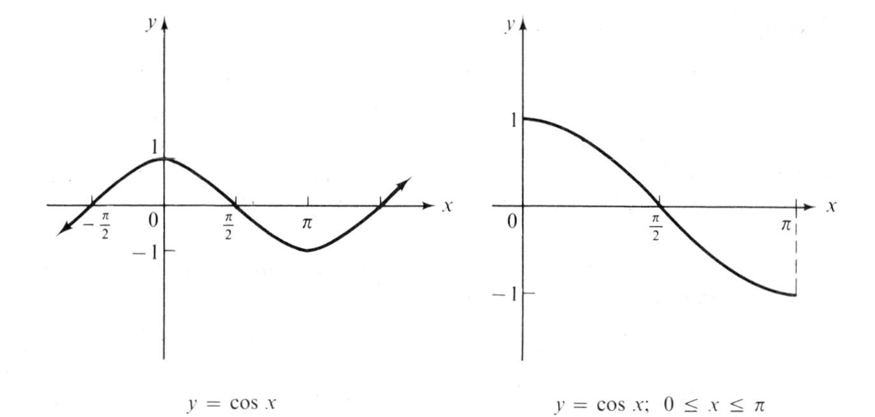

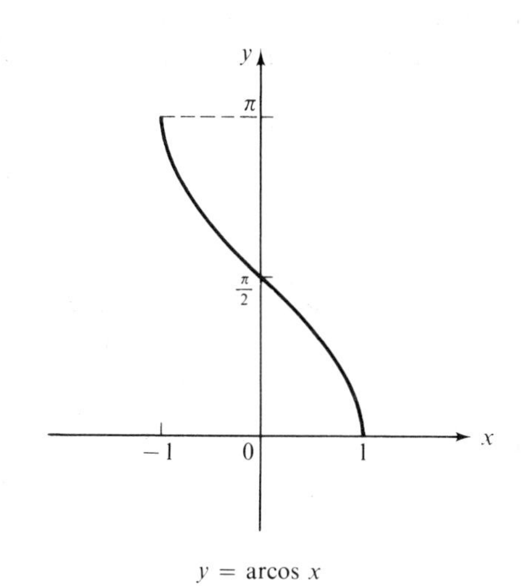

Figure 5-37

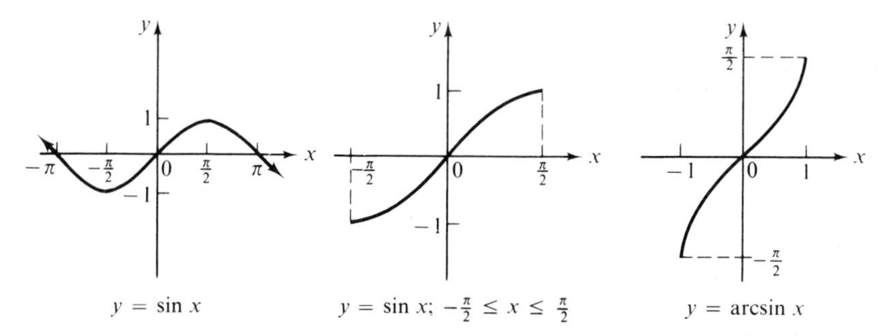

Figure 5-38

Example 5-12. Find arcsin $(-1/\sqrt{2})$.

Solution: We are looking for the number whose sine is $-1/\sqrt{2}$. The sine function is negative in Quadrants III and IV, however the range of the arcsine function is $\{y \mid (-\pi/2) \leq y \leq (\pi/2)\}$, consequently we must choose the angle whose terminal side lies in the fourth quadrant and its measure must be negative. Thus, arcsin $(-1/\sqrt{2}) = -\pi/4$.

The inverse tangent function may be defined similarly. The restricted tangent function $y = \tan x$; $-\pi/2 < x < \pi/2$, is one-to-one, so we can define the arctangent function (arctan) to be the inverse of this function. The domain of $y = \arctan x$ is the set of all real numbers, while its range is $\{y \mid (-\pi/2) < y < (\pi/2)\}$. Recall that $y = \tan x$ is undefined for $x = \pi/2$ and $x = -\pi/2$; hence, these numbers are not in the domain of the tangent function. If they are not in the domain of the tangent function, then they cannot be in the range of its inverse. (See Figure 5-39.)

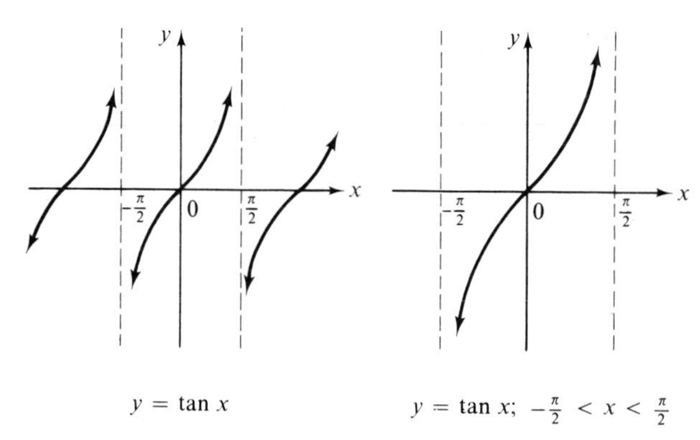

$y = \tan x$ $\qquad$ $y = \tan x;\ -\frac{\pi}{2} < x < \frac{\pi}{2}$

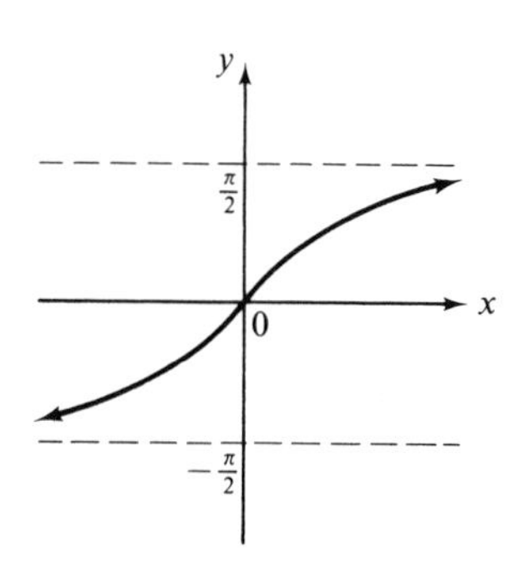

$y = \arctan x$

Figure 5-39

Example 5-13. Find arctan $\sqrt{3}$.

Solution: We need to find the number whose tangent is $\sqrt{3}$. This number suggests the $\pi/3-\pi/6$ right triangle, and a quick sketch tells us that tan $\pi/3 = \sqrt{3}$. Since $\pi/3$ is in the range of the arctangent function, arctan $\sqrt{3} = \pi/3$.

Definition 5-5. $y = \arcsin x$ if and only if $x = \sin y$; $-\pi/2 \leq y \leq \pi/2$

Definition 5-6. $y = \arccos x$ if and only if $x = \cos y$; $0 \leq y \leq \pi$

Definition 5-7. $y = \arctan x$ if and only if $x = \tan y$; $-\pi/2 < y < \pi/2$

The symbols $\sin^{-1}$, $\cos^{-1}$, and $\tan^{-1}$ are also used for the inverses of the sine, cosine, and tangent functions. Thus, you may see $y = \sin^{-1}x$ instead of $y = \arcsin x$.

Note that as a result of these definitions the arcsine, arccosine, and arctangent of a *positive* number will always be a number between 0 and $\pi/2$. The arcsine and arctangent of a *negative* number will be a number between $-\pi/2$ and 0, while the arccosine of a negative number will be a number between $\pi/2$ and π.

We could define the arccosecant, arcsecant, and arccotangent functions in much the same way as we have defined the arcsine, arccosine, and arctangent functions; however, it is simpler to define the former in terms of the latter.

If $y = \csc x$, then $\sin x = 1/y$, and $x = \text{arccsc } y$ and $x = \arcsin 1/y$. Thus arccsc $y = \arcsin 1/y$. The other two functions can be defined similarly.

Definition 5-8.
$$y = \text{arccsc } x = \arcsin 1/x$$
$$y = \text{arcsec } x = \arccos 1/x$$
$$y = \text{arccot } x = \arctan 1/x$$

Since the domain of the arcsine function is the set of numbers between -1 and 1, inclusive, the domain of the arccosecant function must be those values of x for which $-1 \leq 1/x \leq 1$. If $x > 0$ and $1/x \leq 1$, then $x \geq 1$. If $x < 0$ and $-1 \leq 1/x$, then $x \leq -1$. Thus, the domain of the arccosecant function will be those real numbers x such that $x \leq -1$ or $x \geq 1$. The range will be the same as the range of the arcsine function, $\{y \mid (-\pi/2) \leq y \leq (\pi/2)\}$.

It is worth emphasizing that the *inverse* of a trigonometric function and the *reciprocal* of a trigonometric function are not the same at all. The function which is the inverse of $y = \sin x$ is $y = \arcsin x$, and the function which is its reciprocal is $y = \csc x$. A glance at the graphs of these two functions (Figures 5-27 and 5-38) should convince the student that they are totally different.

We have seen that the composite of a function and its inverse is the identity function $y = x$. It follows then, for example, that $\sin(\arcsin x) = x$, for $-1 \leq x \leq 1$, and $\arcsin(\sin x) = x$, for $-\pi/2 \leq x \leq \pi/2$.

Example 5-14. Find $\tan(\arccos 4/5)$.

Solution: We are asked here to find the tangent of the number whose cosine is 4/5. We can do this without ever finding this number. If $x = \arccos 4/5$, then $\cos x = 4/5$, and x is the measure of an acute angle. Drawing a right triangle and labeling one of its angles x, we see that the lengths of its sides must be 3, 4, and 5, and $\tan x = \tan(\arccos 4/5) = 3/4$. (Compare this problem with problems 6–10 in Exercise 5-2.)

Example 5-15. If $y = 3 \arcsin(x + 1)$, solve for x.

Solution: Since $y/3 = \arcsin(x + 1)$, then $\sin y/3 = x + 1$ and $x = \sin y/3 - 1$.

Exercise 5-8

1. Find:
   (a) $\arccos(-1/\sqrt{2})$ (b) $\arcsin(1/\sqrt{2})$
   (c) $\arctan 1/\sqrt{3}$ (d) $\arcsin(-\sqrt{3}/2)$
   (e) $\arctan(-1/\sqrt{3})$ (f) $\arccos(-1/2)$
   (g) $\arcsin(-1/\sqrt{2})$ (h) $\arccos(1/2)$
   (i) $\arccos(-\sqrt{3}/2)$ (j) $\arcsin \sqrt{3}/2$
2. Find the arcsine, arccosine, and arctangent of 0, 1, and -1.
3. Find:
   (a) $\sin(\arccos 1/4)$ (b) $\cos(\arctan 3)$
   (c) $\tan(\arccos 2/3)$ (d) $\sin(\arccos(-1/4))$
   (e) $\cos(\arctan(-3))$ (f) $\tan(\arccos(-2/3))$
4. Find:
   (a) $\sin(\arctan(-5))$ (b) $\cos(\arctan(-5))$
   (c) $\tan(\arctan(-5))$ (d) $\tan(\arccos 1)$
   (e) $\cos(\arccos 1/2)$ (f) $\sin(\arccos .1)$
5. Show the following:
   (a) $-\arccos 0 + \arccos(-1) = \arcsin 1$
   (b) $\arccos 0 + \arctan(-1) = \arctan 1$
   (c) $\arccos(-1/2) + \arcsin(-\sqrt{3}/2) = \arctan \sqrt{3}$
   (d) $-\arctan 1 + \arccos(-1) = \arccos(-1/\sqrt{2})$

6. Explain why the following are true:
   (a) $\arcsin(-x) = -\arcsin x; -1 \le x \le 1$
   (b) $\arctan(-x) = -\arctan x$
   (c) $\arccos(-x) = \pi - \arccos x; -1 \le x \le 1$

7. Graph $y = \text{arccsc } x$.

8. What is the domain of the arcsecant function? What is its range? Graph this function.

9. What is the domain of the arccotangent function? What is its range? Graph this function.

10. Solve for x:
    (a) $y = 2 \arccos(2x)$
    (b) $y = 2 \arctan(x + \pi)$
    (c) $y = 3 \arcsin(x/2)$
    (d) $y = \pi + \arccos(2x + 1)$
    (e) $y = 2/3 \arctan(4x - \pi)$
    (f) $y = 2 + 3 \arcsin(3x - 1)$

11. Find values for x that satisfy the following equations:
    (a) $\arcsin(2x + 1) = \pi/2$
    (b) $\arcsin(x - 2) = \pi/6$
    (c) $\arcsin(x + 2) = \arccos 2/3$

12. In the equation $c^2 = a^2 + b^2 - 2ab \cos C$, solve for C.

13. (a) Show that $\sin(\arccos x) = \sqrt{1 - x^2}$ for $-1 \le x \le 1$.
    (b) Show that $\tan(\arccos x) = \sqrt{1 - x^2}/x$ for $-1 \le x \le 1; x \ne 0$.

14. (a) Show that $\sin(\arctan x) = x/\sqrt{x^2 + 1}$ for all x.
    (b) Show that $\cos(\arctan x) = 1/\sqrt{x^2 + 1}$ for all x.

15. (a) Show that $\cos(\arcsin x) = \sqrt{1 - x^2}$ for $-1 \le x \le 1$.
    (b) Show that $\tan(\arcsin x) = x/\sqrt{1 - x^2}$ for $-1 < x < 1$.

16. (a) Show that $\sec(\arccos x) = 1/x; x \ne 0; -1 \le x \le 1$.
    (b) Show that $\cot(\arctan x) = 1/x; x \ne 0$.
    (c) Show that $\csc(\arcsin x) = 1/x; x \ne 0; -1 \le x \le 1$.

17. Show that if $0 < \theta < \pi/2$, then the following are true:
    (a) $\arccos(\sin \theta) = \pi/2 - \theta$
    (b) $\arctan(\cot \theta) = \pi/2 - \theta$
    (c) $\text{arcsec}(\csc \theta) = \pi/2 - \theta$

18. Evaluate each of the following:
    (a) $\arcsin(\cos \pi/8)$
    (b) $\arctan(\cot 3\pi/7)$
    (c) $\arccos(\sin 1)$
    (d) $\arcsin(\cos \pi/5)$

19. Prove that $\arcsin x + \arccos x = \pi/2$ for $0 \le x \le 1$.

20. For what values of x do the following equations hold?
    (a) $\tan(\arctan x) = x$
    (b) $\arctan(\tan x) = x$
    (c) $\cos(\arccos x) = x$
    (d) $\arccos(\cos x) = x$

21. Find:
    (a) $\tan(\arctan 2)$
    (b) $\arccos(\cos 2\pi/5)$
    (c) $\sin(\arcsin 1/4)$
    (d) $\arctan(\tan(-\pi/4))$
    (e) $\arcsin(\sin \pi)$
    (f) $\arccos(\cos(-\pi/4))$
    (g) $\arctan(\tan 2\pi)$
    (h) $\arcsin(\sin 3\pi/2)$

5-9 Basic Identities

A trigonometric identity is an equation involving trigonometric functions that is true for all values of the variable for which the expressions are defined. Thus,

$$\csc \theta = \frac{1}{\sin \theta}$$

is an identity, since it holds for all values of θ in the domain of the cosecant function. On the other hand

$$\sin \theta = \cos \theta$$

is not an identity since this equation holds only for certain values of θ, ($\theta = \pi/4 + k\pi$, $k = 0, \pm 1, \pm 2, \ldots$) and is not true for others ($\theta = 0$, for instance).

A problem that arises frequently in calculus involves changing an expression involving trigonometric functions to another, usually simpler, form. To do this the student must be familiar with the basic trigonometric identities.

The student is already familiar with the first set of identities we will list, the *reciprocal identities:*

$$\csc \theta = \frac{1}{\sin \theta} \qquad \theta \neq k\pi \qquad k = 0, \pm 1, \pm 2, \ldots$$

$$\sec \theta = \frac{1}{\cos \theta} \qquad \theta \neq \pi/2 + k\pi \qquad k = 0, \pm 1, \pm 2, \ldots$$

$$\cot \theta = \frac{1}{\tan \theta} \qquad \theta \neq k\pi/2 \qquad k = 0, \pm 1, \pm 2, \ldots$$

The last identity, for example, holds for all values of θ for which $\tan \theta$ is defined and $\tan \theta \neq 0$. Since $\tan \theta$ is not defined for $\theta = \pm\pi/2$, $\pm 3\pi/2, \ldots$, and $\tan \theta = 0$ when $\theta = 0, \pm\pi, \pm 2\pi, \ldots$, this identity will be valid for all $\theta \neq k\pi/2$; $k = 0, \pm 1, \pm 2, \ldots$.

Two more useful identities are called the *quotient identities:*

$$\tan \theta = \frac{\sin \theta}{\cos \theta}$$

$$\cot \theta = \frac{\cos \theta}{\sin \theta}$$

The first of these can be verified by using Definition 5-1. Since $\sin \theta = y/r$ and $\cos \theta = x/r$, then $\sin \theta/\cos \theta = y/r \div x/r = y/x = \tan \theta$. The second identity follows from the reciprocal identity for tangent and cotangent.

The third set of basic identities is called the *Pythagorean identities* since they are derived from the Pythagorean relation $y^2 + x^2 = r^2$. If we divide both sides of this equation by r^2, we get $(y/r)^2 + (x/r)^2 = 1$, or

$$\sin^2\theta + \cos^2\theta = 1$$

Note that $(\sin \theta)^2$ is written $\sin^2\theta$.

Using the same Pythagorean relation and dividing through by x^2, we get

$$\tan^2\theta + 1 = \sec^2\theta$$

Dividing by y^2 gives us

$$1 + \cot^2\theta = \csc^2\theta$$

These identities were verified geometrically in Section 5-2.

To summarize, the basic identities are the following:

The reciprocal identities

$$\csc \theta = \frac{1}{\sin \theta}$$

$$\sec \theta = \frac{1}{\cos \theta}$$

$$\tan \theta = \frac{1}{\cot \theta}$$

The quotient identities

$$\tan \theta = \frac{\sin \theta}{\cos \theta}$$

$$\cot \theta = \frac{\cos \theta}{\sin \theta}$$

The Pythagorean identities

$$\sin^2\theta + \cos^2\theta = 1$$
$$\tan^2\theta + 1 = \sec^2\theta$$
$$1 + \cot^2\theta = \csc^2\theta$$

Complete familiarity with these basic identities is essential. The student should memorize them.

Example 5-16. Express each of the trigonometric functions in terms of the sine function.

Solution: From the first Pythagorean identity, we get $\cos\theta = \pm\sqrt{1 - \sin^2\theta}$, and from the first quotient identity, we have $\tan\theta = \sin\theta/\pm\sqrt{1-\sin^2\theta}$. The reciprocal identities then give us $\csc\theta = 1/\sin\theta$, $\sec\theta = 1/\pm\sqrt{1-\sin^2\theta}$, and $\cot\theta = \pm\sqrt{1-\sin^2\theta}/\sin\theta$. The plus or minus sign in each case will be chosen according to the quadrant in which the terminal side of the angle falls. In the second quadrant, for example, the sine function is positive and the tangent, cotangent, and secant functions are negative, so the negative sign will be correct.

Verifying an identity consists in transforming one side of an expression by means of the basic identities until it looks exactly like the other side. Thus, we have the rule—work on one side of the identity only. This will usually, but not always, be the more complicated side.

In addition to having the basic identities at his fingertips, there are several other techniques that may be helpful to the student in tackling identities.

(1) *Perform the indicated operations.*

If the more complicated side is a sum of fractions, find the common denominator and add them. If it is the product of binomials or trinomials, multiply these factors out. Sometimes factoring will be helpful.

Example 5-17. Verify

$$\frac{\sin x}{1 + \cos x} + \frac{1 + \cos x}{\sin x} = 2 \csc x$$

Solution: Working with the left-hand side, we get a common denominator and add.

$$\text{L.H.S.} = \frac{\sin^2 x + (1 + \cos x)^2}{\sin x(1 + \cos x)}$$

$$= \frac{\sin^2 x + 1 + 2 \cos x + \cos^2 x}{\sin x (1 + \cos x)}$$

$$= \frac{(\sin^2 x + \cos^2 x) + 1 + 2 \cos x}{\sin x (1 + \cos x)}$$

By the first Pythagorean identity, this is

$$= \frac{2 + 2 \cos x}{\sin x (1 + \cos x)}$$

$$= \frac{2(1 + \cos x)}{\sin x (1 + \cos x)}$$

$$= \frac{2}{\sin x}$$

$$= 2 \csc x \quad \text{(which is the right-hand side)}$$

(2) *Since the student will find that he is more familiar with the basic identities involving the sine and cosine functions, it is often helpful to rewrite one side of the identity so that it is in terms of sines and cosines alone.*

Example 5-18. Verify

$$\sec^2\theta + \csc^2\theta = \sec^2\theta \cdot \csc^2\theta$$

Solution: Rewriting the left-hand side in terms of sines and cosines, we have

$$\text{L.H.S.} = \frac{1}{\cos^2\theta} + \frac{1}{\sin^2\theta}$$

adding gives us

$$= \frac{\sin^2\theta + \cos^2\theta}{\cos^2\theta \cdot \sin^2\theta}$$

$$= \frac{1}{\cos^2\theta \cdot \sin^2\theta}$$

$$= \sec^2\theta \cdot \csc^2\theta$$

(3) *Practice.*

A problem which arises in calculus involves transforming a single term such as $\sin^3 x$ or $\tan^5 x$ to a seemingly more complicated form which is better suited to the purpose of the problem.

Example 5-19. Transform the left-hand side of the identity

$$\cos^3 x = \cos x - \sin^2 x \cos x$$

into the right-hand side.

Solution: Rewriting the left-hand side as $\cos^2 x \cdot \cos x$, we replace $\cos^2 x$ by $1 - \sin^2 x$ and have

$$\begin{aligned} \text{L.H.S.} &= \cos^2 x \cdot \cos x \\ &= (1 - \sin^2 x)\cos x \\ &= \cos x - \sin^2 x \cdot \cos x \end{aligned}$$

Another problem the student will encounter in calculus involves transforming an algebraic expression into a simple trigonometric expression by means of a trigonometric substitution.

Example 5-20. Simplify $1/\sqrt{1 - x^2}$ by means of a trigonometric substitution.

Solution: If we set $x = \sin\theta$, then $1 - x^2 = 1 - \sin^2\theta = \cos^2\theta$, and $1/\sqrt{1-x^2} = 1/\sqrt{\cos^2\theta} = 1/\cos\theta = \sec\theta$.

Algebraic expressions involving $a^2 + x^2$, $a^2 - x^2$, or $x^2 - a^2$ may be simplified by making the substitutions $x = a\tan\theta$, $x = a\sin\theta$, or $x = a\sec\theta$, respectively, and using the first two Pythagorean identities.

Exercise 5-9

1. Give the set of real numbers θ for which the quotient identities are defined.

2. Give the set of real numbers θ for which the Pythagorean identities are defined.

3. Express each of the trigonometric functions in terms of:
   (a) the cosine function
   (b) the tangent function
   (c) the cosecant function
   (d) the secant function
   (e) the cotangent function

4. Express each of the following in terms of the sine and cosine functions alone. Simplify.
   (a) $\sec x + \tan x$
   (b) $\sec\theta\tan\theta + \csc\theta\cot\theta$
   (c) $\dfrac{\tan\alpha - \cot\alpha}{\sin\alpha}$
   (d) $\dfrac{\sec A - \csc A}{\sec A + \csc A}$

(e) $\dfrac{\tan y \sec y}{1 - \sec y \csc y}$

(f) $\tan \phi \cot \phi - \sec \phi \csc \phi$

5. Verify each of the following identities:

(a) $\dfrac{1 - \cos^2\theta}{\sin \theta} = \sin \theta$

(b) $\csc A \sin A = 1$

(c) $\csc B \tan B = \sec B$

(d) $\cot x \tan x = 1$

(e) $\cot y \sin y = \cos y$

(f) $(\tan^2 s + 1) \cos^2 s = 1$

(g) $(1 - \sin^2 x)(1 + \tan^2 x) = 1$

(h) $\sin^2 y \cot^2 y = \cos^2 y$

(i) $\dfrac{\sin \alpha \sec \alpha}{\tan \alpha} = 1$

(j) $(1 - \cos^2 x)(1 + \cot^2 x) = 1$

6. Verify each of the following identities:

(a) $\cot x \cos x = \dfrac{1 - \sin^2 x}{\sin x}$

(b) $\sec \theta \cos^2\theta = \cos \theta$

(c) $\sin y(\csc y - \cot y) = 1 - \cos y$

(d) $\sin \alpha \sec \alpha \cot \alpha = 1$

(e) $\cos \beta \csc \beta \tan \beta = 1$

(f) $\sin^2 x(1 + \cot^2 x) = 1$

(g) $\sec x - \sec x \sin^2 x = \cos x$

(h) $\dfrac{1 + \tan^2 y}{\tan^2 y} = \csc^2 y$

(i) $\dfrac{\cos s - \sin s}{\cos s} = 1 - \tan s$

(j) $\dfrac{1 + \tan^2 A}{\csc^2 A} = \tan^2 A$

7. Verify each of the following identities by performing the indicated operation on the left-hand side:

(a) $(1 - \cos A)(1 + \sec A)\cos A = \sin^2 A$

(b) $\dfrac{\sin x}{\cos x} + \dfrac{\cos x}{\sin x} = \sec x \csc x$

(c) $(\tan x + 1)^2 - 2 \tan x = \sec^2 x$

(d) $\dfrac{1}{1 - \cos \theta} + \dfrac{1}{1 + \cos \theta} = 2 \csc^2\theta$

(e) $\dfrac{\cos B}{\sin B} + \dfrac{\sin B}{1 + \cos B} = \csc B$

(f) $(1 - \sin x)(1 + \sin x) = \dfrac{1}{1 + \tan^2 x}$

(g) $(\sin A + \cos A)^2 = 1 + 2 \sin A \cos A$

(h) $(\tan x + 1)(\tan x - 1) = \sec^2 x - 2$

8. Verify each of the following identities by first expressing the left-hand side in terms of sines and cosines:

(a) $\dfrac{1}{\sec^2\theta} + \dfrac{1}{\csc^2\theta} = 1$

(b) $\dfrac{\sin y}{\csc y} + \dfrac{\cos y}{\sec y} = 1$

(c) $\sin^2 z \cot^2 z + \cos^2 z \tan^2 z = 1$

(d) $\dfrac{\sec\theta}{\sec\theta + \csc\theta} = \dfrac{\sin\theta}{\sin\theta + \cos\theta}$

(e) $\dfrac{\sin x + \tan x}{\cot x + \csc x} = \sin x \tan x$

(f) $\dfrac{\tan x + \cot x}{\sec x} = \csc x$

(g) $(\csc A - \cot A)^2 = \dfrac{1 - \cos A}{1 + \cos A}$

(h) $\dfrac{\sec y}{\tan y + \cot y} = \sin y$

9. Transform the left-hand side of each of the following identities into the right-hand side:

(a) $\sin^3\alpha = \sin\alpha - \cos^2\alpha \sin\alpha$

(b) $\cos^5 x = \cos x - 2\sin^2 x \cos x + \sin^4 x \cos x$

(c) $\tan^3\theta = \tan\theta \sec^2\theta - \tan\theta$

(d) $\tan^5\theta = \tan^3\theta \sec^2\theta - \tan\theta \sec^2\theta + \tan\theta$

(e) $\tan^6\beta = \tan^4\beta \sec^2\beta - \tan^2\beta \sec^2\beta + \sec^2\beta - 1$

10. Transform each of the following algebraic expressions into a simple trigonometric expression by making the substitution $x = a \tan\theta$, $x = a \sin\theta$, or $x = a \sec\theta$:

(a) $\dfrac{1}{1 + x^2}$

(b) $\dfrac{1}{\sqrt{x^2 - 1}}$

(c) $\sqrt{4 - x^2}$

(d) $\dfrac{1}{9 - x^2}$

(e) $\dfrac{\sqrt{1 - x^2}}{x}$

(f) $x^2\sqrt{x^2 - 1}$

11. Verify each of the following identities:

(a) $(\cot x - \csc x)^2 = \dfrac{1 - \cos x}{1 + \cos x}$

(b) $\dfrac{\tan y - \cot y}{\tan y + \cot y} = 2\sin^2 y - 1$

(c) $\dfrac{\sec\theta + \csc\theta}{\tan\theta + \cot\theta} = \sin\theta + \cos\theta$

(d) $\dfrac{\tan x - \sin x}{\sin^3 x} = \dfrac{\sec x}{1 + \cos x}$

(e) $(\sin x)(1 - \sin x\cos x) + (\cos x)(1 - \sin x\cos x) = \sin^3 x + \cos^3 x$

(f) $\dfrac{\sin\theta\cos\theta}{\cos^2\theta - \sin^2\theta} = \dfrac{\tan\theta}{1 - \tan^2\theta}$

(g) $\sec^2 A - \csc^2 A = \tan^2 A - \cot^2 A$

(h) $\dfrac{\csc^2\theta - 1}{\csc^2\theta} = \cos^2\theta$

(i) $\sin^3 y\cos y + \cos^3 y\sin y = \sin y\cos y$

(j) $\dfrac{\tan x - 1}{1 - \cot x} = \tan x$

12. Verify each of the following identities:

(a) $\dfrac{1}{\sec\theta + \tan\theta} = \sec\theta - \tan\theta$

(b) $\dfrac{1}{1 - \sin A} + \dfrac{1}{1 + \sin A} = 2\sec^2 A$

(c) $\dfrac{1 - \sin x}{\cos x} = \dfrac{\cos x}{1 + \sin x}$

(d) $\dfrac{\sin^3 x + \cos^3 x}{\sin x + \cos x} = 1 - \sin x\cos x$

(e) $\sec^4 y - \tan^4 y = 1 + 2\tan^2 y$

(f) $\dfrac{2\sin^2\theta - 1}{\sin\theta\cos\theta} = \tan\theta - \cot\theta$

(g) $\dfrac{\tan x - \sec x + 1}{\tan x + \sec x - 1} = \dfrac{\cos x}{1 + \sin x}$

(h) $\dfrac{\sin y}{\csc y - \cot y} = 1 + \cos y$

(i) $\tan^4 s + 2\tan^2 s + 1 = \sec^4 s$

(j) $(\sec\theta + \csc\theta)^2 = (1 + 2\sin\theta\cos\theta)\sec^2\theta\csc^2\theta$

5-10 The Addition Formulas and Double and Half Angle Formulas

Some extremely useful identities are those which express the trigonometric functions of the sum or difference of two numbers, α and β

in terms of functions of α and β alone. They are called the *addition formulas.*

Let us see how we can derive a formula for $\cos(\alpha - \beta)$ in terms of the sine and cosine of α and β. Let α and β be any two real numbers. Then $Q(\cos(\alpha - \beta), \sin(\alpha - \beta))$, $R(\cos \beta, \sin \beta)$, and $S(\cos \alpha, \sin \alpha)$ will be points on the unit circle. (See Figure 5-40.) If P is the point $(1, 0)$,

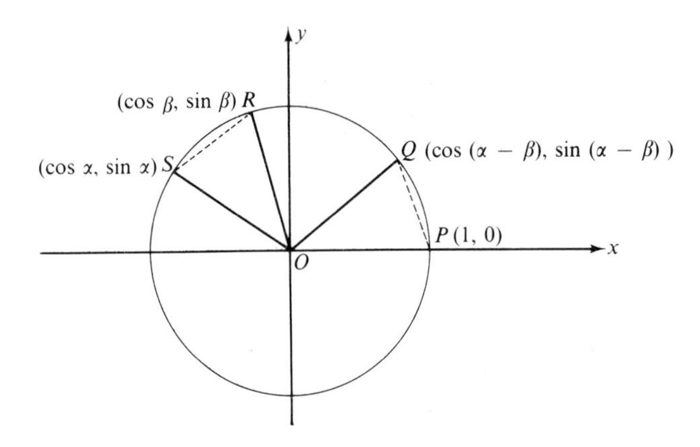

Figure 5-40

then one of the measures of angle QOP will be $\alpha - \beta$. Similarly one of the measures of angle SOR will be $\alpha - \beta$. Since these two angles have the same measure, chords SR and QP must have the same length. By the distance formula the distance between two points (x_1, y_1) and (x_2, y_2) in the plane is given by the formula

$$d = \sqrt{(x_1 - x_2)^2 + (y_1 - y_2)^2}$$

Thus

$$QP = \sqrt{(\cos(\alpha - \beta) - 1)^2 + (\sin(\alpha - \beta) - 0)^2}$$

and

$$SR = \sqrt{(\cos \alpha - \cos \beta)^2 + (\sin \alpha - \sin \beta)^2}$$

Equating these two and squaring, we get

$$(\cos(\alpha - \beta) - 1)^2 + \sin^2(\alpha - \beta) = (\cos \alpha - \cos \beta)^2 + (\sin \alpha - \sin \beta)^2$$

Multiplying out gives

$$\cos^2(\alpha - \beta) - 2 \cos(\alpha - \beta) + 1 + \sin^2(\alpha - \beta) = \cos^2\alpha - 2 \cos \alpha \cos \beta + \cos^2\beta + \sin^2\alpha - 2 \sin \alpha \sin \beta + \sin^2\beta$$

Since $\cos^2\theta + \sin^2\theta = 1$ by the first Pythagorean identity, this reduces to

$$2 - 2\cos(\alpha - \beta) = 2 - 2\cos\alpha\cos\beta - 2\sin\alpha\sin\beta$$

or

$$\cos(\alpha - \beta) = \cos\alpha\cos\beta + \sin\alpha\sin\beta \qquad \textbf{(5-1)}$$

Since (5-1) holds for any real numbers α and β, we can replace β by $-\beta$, and this gives us

$$\cos(\alpha - (-\beta)) = \cos\alpha\cos(-\beta) + \sin\alpha\sin(-\beta)$$

From Section 5-4 $\cos(-\theta) = \cos\theta$ and $\sin(-\theta) = -\sin\theta$; thus, we have

$$\cos(\alpha + \beta) = \cos\alpha\cos\beta - \sin\alpha\sin\beta \qquad \textbf{(5-2)}$$

We have already seen in Section 5-3 that for a real number θ, $0 \le \theta \le \pi/2$, $\sin\theta = \cos(\pi/2 - \theta)$ and $\cos\theta = \sin(\pi/2 - \theta)$. Now we can show that these identities hold for any real number θ whatever.

In (5-1) if we set $\alpha = \pi/2$ and $\beta = \theta$,

$$\begin{aligned}\cos(\pi/2 - \theta) &= \cos\pi/2\cos\theta + \sin\pi/2\sin\theta\\ &= 0\cdot\cos\theta + 1\cdot\sin\theta\\ &= \sin\theta\end{aligned}$$

Similarly if we set $\alpha = \pi/2$ and $\beta = \pi/2 - \theta$, then we have

$$\begin{aligned}\cos\theta &= \cos[\pi/2 - (\pi/2 - \theta)]\\ &= \cos\pi/2\cos(\pi/2 - \theta) + \sin\pi/2\sin(\pi/2 - \theta)\\ &= \sin(\pi/2 - \theta)\end{aligned}$$

Using these identities we can derive the formula for $\sin(\alpha - \beta)$ from (5-2).

For any real numbers α and β

$$\begin{aligned}\sin(\alpha - \beta) &= \cos[\pi/2 - (\alpha - \beta)]\\ &= \cos[(\pi/2 - \alpha) + \beta]\end{aligned}$$

and by (5-2) this is

$$= \cos(\pi/2 - \alpha)\cos\beta - \sin(\pi/2 - \alpha)\sin\beta$$

Since $\cos(\pi/2 - \alpha) = \sin\alpha$ and $\sin(\pi/2 - \alpha) = \cos\alpha$, this becomes

$$\sin(\alpha - \beta) = \sin\alpha\cos\beta - \cos\alpha\sin\beta \qquad \textbf{(5-3)}$$

Replacing β by $-\beta$, we get our fourth addition formula

$$\sin(\alpha + \beta) = \sin\alpha\cos\beta + \cos\alpha\sin\beta \qquad \textbf{(5-4)}$$

The formula for $\tan(\alpha - \beta)$ can be derived from formulas (5-1) and (5-3) and the first quotient identity.

$$\tan(\alpha - \beta) = \frac{\sin(\alpha - \beta)}{\cos(\alpha - \beta)}$$

$$= \frac{\sin \alpha \cos \beta - \cos \alpha \sin \beta}{\cos \alpha \cos \beta + \sin \alpha \sin \beta}$$

Dividing numerator and denominator by $\cos \alpha \cos \beta$, we get

$$= \frac{\dfrac{\sin \alpha}{\cos \alpha} - \dfrac{\sin \beta}{\cos \beta}}{1 + \dfrac{\sin \alpha}{\cos \alpha} \dfrac{\sin \beta}{\cos \beta}}$$

or

$$\tan(\alpha - \beta) = \frac{\tan \alpha - \tan \beta}{1 + \tan \alpha \tan \beta} \qquad \textbf{(5-5)}$$

Substituting $-\beta$ for β gives us

$$\tan(\alpha + \beta) = \frac{\tan \alpha + \tan \beta}{1 - \tan \alpha \tan \beta} \qquad \textbf{(5-6)}$$

One application of the addition formulas is the derivation of reduction formulas. By the use of reduction formulas, the trigonometric functions of any number can be reduced to a function of a number between zero and $\pi/2$.

Example 5-21. Show that $\cos(\pi/2 + \theta) = -\sin \theta$.

Solution: By formula (5-2)

$$\begin{aligned} \cos(\pi/2 + \theta) &= \cos \pi/2 \cos \theta - \sin \pi/2 \sin \theta \\ &= 0 \cdot \cos \theta - 1 \cdot \sin \theta \\ &= -\sin \theta \end{aligned}$$

Thus, for example, $\cos(2\pi/3) = \cos(\pi/2 + \pi/6) = -\sin \pi/6 = -1/2$.

Another application of these formulas arises in the process of solving certain differential equations. In these problems, it is desired to change an expression of the form $a \sin kx + b \cos kx$ to a sine or cosine function alone. The sum of a sine and cosine function having the same period is a sine or cosine function with the same period but usually a different amplitude. The graph of the new function will be shifted to the right or left by an amount α. To solve these problems, we need to observe that if x and y are any two numbers satisfying the equation $x^2 + y^2 = 1$, then (x, y) is a point on the unit circle, and $x = \cos\alpha$, $y = \sin\alpha$ for some number α.

Example 5-22. Express $f(x) = 2\sin x + 3\cos x$ in the form $f(x) = A\sin(x + \alpha)$.

Solution: First we multiply and divide by $\sqrt{2^2 + 3^2} = \sqrt{13}$. Then

$$f(x) = \sqrt{13}\left(\frac{2}{\sqrt{13}}\sin x + \frac{3}{\sqrt{13}}\cos x\right)$$

Now $(2/\sqrt{13})^2 + (3/\sqrt{13})^2 = 1$, hence $2/\sqrt{13} = \cos\alpha$ and $3/\sqrt{13} = \sin\alpha$ for some number α. Thus

$$f(x) = \sqrt{13}\,(\cos\alpha \sin x + \sin\alpha \cos x)$$

and by (5-4) this is

$$= \sqrt{13}\,\sin(x + \alpha)$$

where $\alpha = \arctan 3/2$. The method is exactly the same if x is replaced by kx.

By setting $\alpha = \beta = \theta$ in formulas (5-2), (5-4), and (5-6), we get the *double angle formulas,*

$$\sin 2\theta = 2\sin\theta\cos\theta \tag{5-7}$$

$$\cos 2\theta = \cos^2\theta - \sin^2\theta \tag{5-8}$$

$$\tan 2\theta = \frac{2\tan\theta}{1 - \tan^2\theta} \tag{5-9}$$

A few changes to formula (5-8) gives us the *half angle formulas.*

Since $\cos^2\theta = 1 - \sin^2\theta$ from the first Pythagorean identity, (5-8) becomes

$$\cos 2\theta = 1 - 2\sin^2\theta$$

Solving for $\sin\theta$ gives us

$$\sin^2\theta = \frac{1 - \cos 2\theta}{2} \tag{5-10}$$

or

$$\sin\theta = \pm\sqrt{\frac{1 - \cos 2\theta}{2}} \tag{5-11}$$

Similarly, replacing $\sin^2\theta$ by $1 - \cos^2\theta$ in (5-8) gives us

$$\cos 2\theta = 2 \cos^2\theta - 1$$

and solving for $\cos \theta$ we get

$$\cos^2\theta = \frac{\cos 2\theta + 1}{2} \tag{5-12}$$

or

$$\cos \theta = \pm \sqrt{\frac{\cos 2\theta + 1}{2}} \tag{5-13}$$

From (5-11), (5-13), and the quotient identity, we get

$$\tan \theta = \pm \sqrt{\frac{1 - \cos 2\theta}{1 + \cos 2\theta}} \tag{5-14}$$

where the sign chosen depends on the quadrant in which the angle falls.

Finally, replacing θ by $\phi/2$ in (5-11), (5-13), and (5-14) we arrive at the identities,

$$\sin \frac{\phi}{2} = \pm \sqrt{\frac{1 - \cos \phi}{2}} \tag{5-15}$$

$$\cos \frac{\phi}{2} = \pm \sqrt{\frac{1 + \cos \phi}{2}} \tag{5-16}$$

$$\tan \frac{\phi}{2} = \pm \sqrt{\frac{1 - \cos \phi}{1 + \cos \phi}} \tag{5-17}$$

We can put (5-17) in a more convenient form for computation purposes.

$$\begin{aligned}\tan \frac{\phi}{2} &= \pm \sqrt{\frac{1 - \cos \phi}{1 + \cos \phi}} \\ &= \pm \sqrt{\frac{(1 - \cos \phi)(1 - \cos \phi)}{(1 + \cos \phi)(1 - \cos \phi)}} \\ &= \pm \sqrt{\frac{(1 - \cos \phi)^2}{1 - \cos^2\phi}} \\ &= \pm \sqrt{\frac{(1 - \cos \phi)^2}{\sin^2\phi}}\end{aligned}$$

or

$$\tan \frac{\phi}{2} = \frac{1 - \cos \phi}{\sin \phi}$$

Here we have dropped the $\pm$ since $1 - \cos \phi$ is never negative and $\sin \phi$ and $\tan (\phi/2)$ always have the same sign.

An alternate form is

$$\tan \frac{\phi}{2} = \frac{\sin \phi}{1 + \cos \phi}$$

To summarize we have six addition formulas:

$$\sin(\alpha + \beta) = \sin \alpha \cos \beta + \cos \alpha \sin \beta \quad \textbf{(5-4)}$$

$$\cos(\alpha + \beta) = \cos \alpha \cos \beta - \sin \alpha \sin \beta \quad \textbf{(5-2)}$$

$$\tan(\alpha + \beta) = \frac{\tan \alpha + \tan \beta}{1 - \tan \alpha \tan \beta} \quad \textbf{(5-6)}$$

$$\sin(\alpha - \beta) = \sin \alpha \cos \beta - \cos \alpha \sin \beta \quad \textbf{(5-3)}$$

$$\cos(\alpha - \beta) = \cos \alpha \cos \beta + \sin \alpha \sin \beta \quad \textbf{(5-1)}$$

$$\tan(\alpha - \beta) = \frac{\tan \alpha - \tan \beta}{1 + \tan \alpha \tan \beta} \quad \textbf{(5-5)}$$

three double angle formulas

$$\sin 2\theta = 2 \sin \theta \cos \theta \quad \textbf{(5-7)}$$

$$\cos 2\theta = \cos^2\theta - \sin^2\theta \quad \textbf{(5-8)}$$

$$\tan 2\theta = \frac{2 \tan \theta}{1 - \tan^2\theta} \quad \textbf{(5-9)}$$

and three half angle formulas

$$\sin \frac{\phi}{2} = \pm \sqrt{\frac{1 - \cos \phi}{2}} \quad \textbf{(5-15)}$$

$$\cos \frac{\phi}{2} = \pm \sqrt{\frac{1 + \cos \phi}{2}} \quad \textbf{(5-16)}$$

$$\tan \frac{\phi}{2} = \pm \sqrt{\frac{1 - \cos \phi}{1 + \cos \phi}} = \frac{1 - \cos \phi}{\sin \phi} = \frac{\sin \phi}{1 + \cos \phi} \quad \textbf{(5-17)}$$

The student should not be dismayed at this formidable array of formulas. If the first three are memorized, most of the others can be written down almost immediately. The second three come from replacing β by $-\beta$. The next three are the result of setting $\alpha = \beta = \theta$, while the last three come from formula (5-8) and the first Pythagorean identity.

Formulas (5-10) and (5-12) are particularly useful forms in solving certain calculus problems in which it is necessary to replace a sine or cosine function raised to some exponent by a sum of trigonometric functions, each having exponent one.

Example 5-23. Replace $\sin^4 x$ by a sum of trigonometric functions each having exponent one.

Solution: By (5-10)

$$\sin^2 x = \frac{1}{2} - \frac{1}{2}\cos 2x$$

therefore

$$\sin^4 x = \left(\frac{1}{2} - \frac{1}{2}\cos 2x\right)^2$$
$$= \frac{1}{4} - \frac{1}{2}\cos 2x + \frac{1}{4}\cos^2 2x$$

By (5-12)

$$\cos^2 2x = \frac{1}{2} + \frac{1}{2}\cos 2(2x) = \frac{1}{2} + \frac{1}{2}\cos 4x$$

Thus,

$$\sin^4 x = \frac{1}{4} - \frac{1}{2}\cos 2x + \frac{1}{4}\left(\frac{1}{2} + \frac{1}{2}\cos 4x\right)$$
$$= \frac{3}{8} - \frac{1}{2}\cos 2x + \frac{1}{8}\cos 4x$$

Exercise 5-10

1. Use the addition formulas to derive the following reduction formulas:
   (a) $\sin(\pi/2 + \theta) = \cos\theta$ (b) $\sin(\pi + \theta) = -\sin\theta$
   (c) $\cos(\pi + \theta) = -\cos\theta$ (d) $\sin(\pi - \theta) = \sin\theta$
   (e) $\cos(\pi - \theta) = -\cos\theta$ (f) $\sin(3\pi/2 + \theta) = -\cos\theta$
   (g) $\cos(3\pi/2 + \theta) = \sin\theta$

2. Use the reduction formulas of problem 1 to express each of the following as a function of a number between 0 and $\pi/2$:
   (a) $\sin 7\pi/12$ (b) $\cos 9\pi/8$
   (c) $\sin 7\pi/4$ (d) $\cos 5\pi/6$
   (e) $\sin 4\pi/5$ (f) $\cos 13\pi/8$

3. (a) Derive a formula for $\cot(\alpha + \beta)$ in terms of $\cot\alpha$ and $\cot\beta$.
   (b) Derive a formula for $\cot(\alpha - \beta)$ from the formula of part (a).

4. Verify each of the following identities:
   (a) $\sin\alpha\cos\beta = 1/2\,[\sin(\alpha + \beta) + \sin(\alpha - \beta)]$

(b) $\sin \alpha \sin \beta = 1/2\,[\cos(\alpha - \beta) - \cos(\alpha + \beta)]$
(c) $\cos \alpha \cos \beta = 1/2\,[\cos(\alpha + \beta) + \cos(\alpha - \beta)]$

5. Use the identities of problem 4 to rewrite each of the following as a sum of sines or cosines:
   (a) sin 2x cos x (b) cos x cos 3x
   (c) sin 3x sin 2x (d) $\sin \theta \cos 4\theta$

6. Derive the following formulas by setting $\alpha + \beta = x$ and $\alpha - \beta = y$ in the identities of problem 4.
   (a) $\sin x + \sin y = 2 \sin \dfrac{x+y}{2} \cos \dfrac{x-y}{2}$
   (b) $\cos x + \cos y = 2 \cos \dfrac{x+y}{2} \cos \dfrac{x-y}{2}$
   (c) $\cos x - \cos y = -2 \sin \dfrac{x+y}{2} \sin \dfrac{x-y}{2}$

7. Express each of the following in the form $A \sin(x \pm \alpha)$. Be sure to specify α.
   (a) $4 \sin x + 3 \cos x$ (b) $\sin x + \cos x$
   (c) $\sin x + \sqrt{3} \cos x$ (d) $2 \sin x - 3 \cos x$
   (e) $\sqrt{3} \sin x - \cos x$ (f) $\sin x + 2 \cos x$

8. Find the exact values of the sine, cosine, and tangent of $\pi/12$ by observing that $\pi/12 = \pi/3 - \pi/4$ and using the addition formulas.

9. Find the exact values of the sine, cosine, and tangent of $5\pi/12$.

10. Find the exact values of the sine, cosine, and tangent of $7\pi/12$.

11. Find the exact values of the sine, cosine, and tangent of $\pi/8$ by using the half angle formulas.

12. Verify the statement in the text that $\sin \phi$ and $\tan (\phi/2)$ always have the same sign by considering the four cases in which the terminal side of the angle falls in each of the four quadrants.

13. Derive the alternate form for identity (5-17).

$$\tan \frac{\phi}{2} = \frac{\sin \phi}{1 + \cos \phi}$$

14. Replace $\cos^4 x$ by a sum of trigonometric functions, each having exponent one.

15. Verify each of the following identities:
   (a) $(\sin x + \cos x)^2 = 1 + \sin 2x$
   (b) $\cos 4\theta = 8 \cos^4\theta - 8 \cos^2\theta + 1$
   (c) $\csc 2A - \cot 2A = \tan A$
   (d) $\sin^2 \dfrac{\phi}{2} + \cos \phi = \cos^2 \dfrac{\phi}{2}$

(e) $\dfrac{\sin 2x}{\sin x} - \dfrac{\cos 2x}{\cos x} = \sec x$

(f) $\sin 4x = 4 \cos x (\sin x - 2 \sin^3 x)$

(g) $\cot \dfrac{\alpha}{2} = \dfrac{1}{\csc \alpha - \cot \alpha}$

(h) $\tan \dfrac{\alpha}{2} = \dfrac{1}{\csc \alpha + \cot \alpha}$

16. Verify each of the following identities:
(a) $\cos 2x = \cos^4 x - \sin^4 x$
(b) $\tan \theta \sin 2\theta = 2 \sin^2 \theta$
(c) $\cos 3\theta = 4 \cos^3 \theta - 3 \cos \theta$
(d) $\dfrac{1 + \cos 2x}{\sin 2x} = \cot x$
(e) $\dfrac{\sin 3x}{\sin x} - \dfrac{\cos 3x}{\cos x} = 2$
(f) $\dfrac{2 \tan x}{1 + \tan^2 x} = \sin 2x$
(g) $\csc 2A + \cot 2A = \cot A$
(h) $2 - 2 \tan \theta \cot 2\theta = \sec^2 \theta$

17. If $\sin \theta = 4/5$ and θ is in the first quadrant, find the values of the sine, cosine, and tangent of $\theta/2$, of 2θ.

18. If $\tan \theta = 1/2$ and θ is in the first quadrant, find the values of the sine, cosine, and tangent of $\theta/2$, of 2θ.

19. If $A + B + C = \pi$, show that:
(a) $\sin(B + C) = \sin A$
(b) $\sin(1/2)(B + C) = \cos(1/2)A$

5-11* De Moivre's Theorem

A complex number is a number of the form $x + iy$, where x and y are real numbers and $i^2 = -1$. Since there are two real numbers x and y involved, every complex number can be associated with a point (x, y) in the plane. Thus, for example, the complex number $2 - i$ can be assigned the point $(2, -1)$. Since a real number can be thought of as a complex number of the form $x + 0i$, real numbers are assigned points $(x, 0)$ on the x-axis.

We have already seen that every point in the plane (except the origin) determines an angle in standard position. If θ is some measure of this angle, then $\cos \theta = x/r$ and $\sin \theta = y/r$, where $r = \sqrt{x^2 + y^2}$. Thus, $x = r \cos \theta$, $y = r \sin \theta$, and the complex number $x + iy$ can be written in the form $r \cos \theta + ir \sin \theta = r(\cos \theta + i \sin \theta)$. This is called the *trigonometric form* of the complex number.

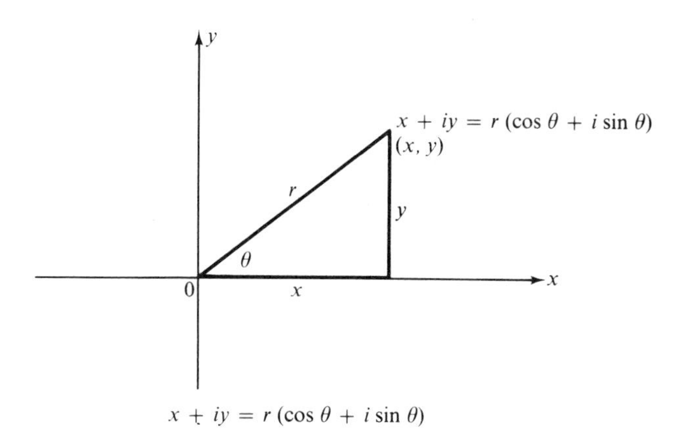

$x + iy = r(\cos \theta + i \sin \theta)$

Figure 5-41

Example 5-24. Express $1 + i$ in trigonometric form.

Solution: Here $x = 1$, $y = 1$, $r = \sqrt{1^2 + 1^2} = \sqrt{2}$. Since the point $(1, 1)$ is on the terminal side of the angle, we can see that one value for θ is $\pi/4$. Thus, $1 + i = \sqrt{2}(\cos \pi/4 + i \sin \pi/4)$, or more generally, $\sqrt{2}[\cos(\pi/4 + 2\pi k) + i \sin(\pi/4 + 2\pi k)]$ for any integer k.

Example 5-25. Express 3 in trigonometric form.

Solution: In this case, $x = 3$, $y = 0$, and $r = 3$. Plotting the point $(3, 0)$, we see that one value for θ is zero and $3 = 3(\cos 0 + i \sin 0)$. More generally, we write $3 = 3(\cos 2\pi k + i \sin 2\pi k)$ for any integer k.

De Moivre's Theorem. If n is a positive integer, then

$$(\cos \theta + i \sin \theta)^n = \cos n\theta + i \sin n\theta$$

We will not prove this theorem (the interested reader is referred to Dickson, *New First Course in the Theory of Equations*, page 3), however we can verify it for $n = 2$ and $n = 3$.

If $n = 2$, then we want to show that $(\cos \theta + i \sin \theta)^2 = \cos 2\theta + i \sin 2\theta$. Multiplying out the left-hand side of this equation gives us

$$\cos^2\theta + i(2 \sin \theta \cos \theta) + i^2 \sin^2\theta$$

Since $i^2 = -1$, this becomes

$$(\cos^2\theta - \sin^2\theta) + i(2 \sin \theta \cos \theta)$$

By double angle formulas (Section 5-10), $\cos^2\theta - \sin^2\theta = \cos 2\theta$, and $2 \sin \theta \cos \theta = \sin 2\theta$; thus, the left-hand side is equal to $\cos 2\theta + i \sin 2\theta$.

If $n = 3$, then we want to show that $(\cos \theta + i \sin \theta)^3 = \cos 3\theta + i \sin 3\theta$. Now

$$(\cos \theta + i \sin \theta)^3 = (\cos \theta + i \sin \theta)^2(\cos \theta + i \sin \theta)$$

and by the proof for $n = 2$, this becomes

$$= (\cos 2\theta + i \sin 2\theta)(\cos \theta + i \sin \theta)$$

Multiplying this out gives us

$$\cos 2\theta \cos \theta + i(\cos 2\theta \sin \theta + \sin 2\theta \cos \theta) + i^2 \sin 2\theta \sin \theta$$

Replacing i^2 by -1, we have

$$(\cos 2\theta \cos \theta - \sin 2\theta \sin \theta) + i(\cos 2\theta \sin \theta + \sin 2\theta \cos \theta)$$

By the addition formulas the term in the first parentheses is $\cos(2\theta + \theta) = \cos 3\theta$ and that in the second parentheses is $\sin(2\theta + \theta) = \sin 3\theta$. Thus, this is equal to $\cos 3\theta + i \sin 3\theta$.

One of the most useful applications of De Moivre's Theorem is in finding all of the nth roots of a number a, which may be real or complex. We have already noted in Section 4-1 that every number a has n different nth roots which are the solutions to the polynomial equation $x^n = a$. Of these roots at most two will be real and the rest will be complex.

To find these roots, we must first write a in trigonometric form,

$$a = r(\cos \theta + i \sin \theta)$$

Now an nth root of a will be a number x which satisfies the equation $x^n = r(\cos \theta + i \sin \theta)$. By De Moivre's Theorem $\sqrt[n]{r}\,[\cos(\theta/n) + i \sin(\theta/n)]$ will be such a number, since

$$\left[\sqrt[n]{r}\left(\cos\frac{\theta}{n} + i \sin \frac{\theta}{n}\right)\right]^n = r(\cos \theta + i \sin \theta)$$

Here $\sqrt[n]{r}$ is the positive real root of r. (See Section 4-1.)

To find *all* of the n nth roots, we must write a in a more general form, $a = r[\cos(\theta + 2\pi k) + i \sin(\theta + 2\pi k)]$ for any integer k. Then for any integer k,

$$\sqrt[n]{r}\left(\cos\frac{\theta + 2\pi k}{n} + i \sin \frac{\theta + 2\pi k}{n}\right)$$

will be a root of a. We will get only n different roots, however, corresponding to $k = 0, 1, 2, \ldots, n - 1$.

Example 5-26. Find the three cube roots of 8.

Solution: First we write 8 in trigonometric form

$$8 = 8(\cos 2\pi k + i \sin 2\pi k)$$

The three cube roots will be

$$\sqrt[3]{8}\left(\cos \frac{2\pi k}{3} + i \sin \frac{2\pi k}{3}\right) \quad \text{for } k = 0, 1, 2$$

If $k = 0$,

$$r_1 = 2(\cos 0 + i \sin 0) = 2$$

If $k = 1$,

$$r_2 = 2\left(\cos \frac{2\pi}{3} + i \sin \frac{2\pi}{3}\right) = 2(-1/2 + i\sqrt{3}/2) = -1 + i\sqrt{3}$$

If $k = 2$,

$$r_3 = 2\left(\cos \frac{4\pi}{3} + i \sin \frac{4\pi}{3}\right) = 2(-1/2 - i\sqrt{3}/2) = -1 - i\sqrt{3}$$

If we take $k = 3$, we get $2(\cos 2\pi + i \sin 2\pi)$, and this is r_1. In the same way $k = 4$ gives r_2, $k = 5$ gives r_3, and so on.

If we plot the points corresponding to these three roots on the plane we find that they are evenly spaced on the circumference of a circle of radius 2. (See Figure 5-42.)

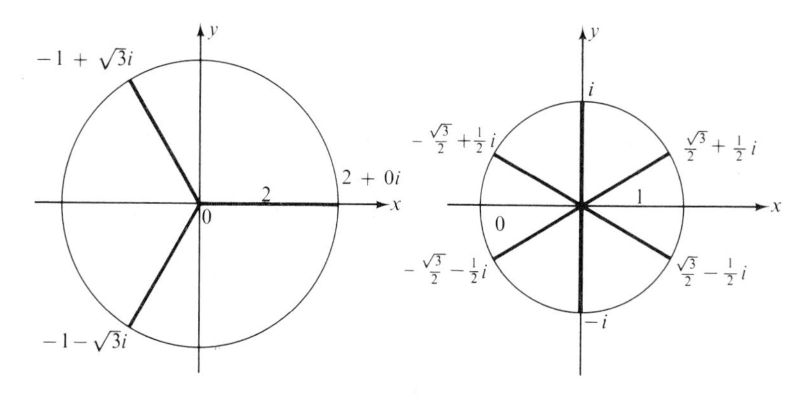

Figure 5-42

Example 5-27. Find the six sixth roots of -1.

Solution: In trigonometric form

$$-1 = 1[\cos(\pi + 2\pi k) + i \sin(\pi + 2\pi k)]$$

and its six sixth roots will be

$$\cos \frac{\pi + 2\pi k}{6} + i \sin \frac{\pi + 2\pi k}{6} \quad \text{for } k = 0, 1, 2, 3, 4, 5$$

$$k = 0: \quad r_1 = \cos \frac{\pi}{6} + i \sin \frac{\pi}{6} = \frac{\sqrt{3}}{2} + \frac{1}{2} i$$

$$k = 1: \quad r_2 = \cos \frac{\pi}{2} + i \sin \frac{\pi}{2} = i$$

$$k = 2: \quad r_3 = \cos \frac{5\pi}{6} + i \sin \frac{5\pi}{6} = \frac{-\sqrt{3}}{2} + \frac{1}{2} i$$

$$k = 3: \quad r_4 = \cos \frac{7\pi}{6} + i \sin \frac{7\pi}{6} = \frac{-\sqrt{3}}{2} - \frac{1}{2} i$$

$$k = 4: \quad r_5 = \cos \frac{3\pi}{2} + i \sin \frac{3\pi}{2} = -i$$

$$k = 5: \quad r_6 = \cos \frac{11\pi}{6} + i \sin \frac{11\pi}{6} = \frac{\sqrt{3}}{2} - \frac{1}{2} i$$

Points corresponding to these roots are evenly spaced on a circle of radius 1. (See Figure 5-42.)

Exercise 5-11

1. Express each of the following in general trigonometric form:
   (a) 5 (b) -3
   (c) $-1 + i$ (d) $1 + \sqrt{3}i$
   (e) i (f) $-2i$
2. Find the three cube roots of 27. Plot the points corresponding to the roots.
3. Find the four fourth roots of 81. Plot the points corresponding to the roots.
4. Find the three cube roots of -1. Plot the points corresponding to the roots.
5. Find the three cube roots of -1 algebraically by factoring the left-hand side of the equation $x^3 + 1 = 0$ into $(x + 1)(x^2 - x + 1)$ and using the

quadratic formula to find the roots of this equation. Compare your answer with the results of problem 4.

6. Find the three cube roots of 1
   (a) by De Moivre's Theorem.
   (b) by finding the roots of $x^3 - 1 = 0$ by factoring the left-hand side.
7. Find the two square roots of i. Check your answer by multiplication.
8. Find all of the solutions to the equation $x^4 + 1 = 0$.
9. Verify De Moivre's Theorem for $n = 4$. (Hint: Use the fact, proved in the text, that the theorem holds for $n = 3$.)
10. Use De Moivre's Theorem to find $(1 + i)^{10}$.
11. Use De Moivre's Theorem to find $[(1/2) + (\sqrt{3}/2)i]^{15}$.

5-12 * Trigonometric Equations

A trigonometric equation differs from a trigonometric identity in that it is true only for certain values of the variable, while the identity holds for all values for which the expression is defined. A trigonometric equation is sometimes called a *conditional* equation and we are usually interested in finding those values of the variable for which it is true. We might compare solving trigonometric equations to finding the roots of a polynomial equation. There are certain important differences, however. A polynomial equation has a finite number of roots, while a trigonometric equation will usually have infinitely many solutions. Unfortunately there are no general methods of solving these equations. To find the solutions we must rely on our knowledge of the trigonometric functions.

Example 5-27. Solve $\sin x = 0$.

Solution: The sine function is zero when $x = 0$ or $x = \pi$. Since the period of the sine function is 2π, then $0 + 2\pi k$ and $\pi + 2\pi k$ are also solutions for any integer k. A more compact statement of the solution would be $x = \pi k$, $k = 0, 1, 2, \ldots$.

Factoring, a technique that was useful in solving polynomial equations, can often be used to solve trigonometric equations.

Example 5-28. Find the values of x, $0 \leq x < 2\pi$, for which $2 \sin x \cos x - \sin x = 0$.

Solution: Factoring gives

$$\sin x(2 \cos x - 1) = 0$$

The left-hand side of this equation will be zero when $\sin x = 0$ or $2 \cos x - 1 = 0$. Sin x will be zero, as we have seen in Example 5-27, when $x = \pi k$, $k = 0, \pm 1, \pm 2, \ldots$; $2 \cos x - 1$ will be zero when $\cos x = 1/2$, and this occurs when $x = \pi/3 + 2\pi k$ or $x = 5\pi/3 + 2\pi k$. The solutions desired are, thus, 0, π, $\pi/3$, and $5\pi/3$.

Sometimes trigonometric identities may be used to put the equation in a different form which is easier to solve.

Example 5-29. Solve $\sin^2 x - \cos^2 x = 0$.

Solution: Replacing $\sin^2 x$ by $1 - \cos^2 x$ gives

$$1 - 2 \cos^2 x = 0$$

or

$$\cos x = \pm \frac{1}{\sqrt{2}}$$

If $\cos x = 1/\sqrt{2}$, then $x = \pi/4 + 2\pi k$ or $x = 7\pi/4 + 2\pi k$. If $\cos x = -1/\sqrt{2}$, then $x = 3\pi/4 + 2\pi k$ or $x = 5\pi/4 + 2\pi k$. A compact form of the general solution would be $x = (\pi/4)(2k + 1)$; $k = 0, \pm 1, \pm 2, \ldots$.

We can multiply both sides of an equation by the same trigonometric function or square both sides if this is helpful. However, if we do either of these we must check our solutions in the original equation, because such operations may introduce extraneous solutions.

Example 5-30. Solve $\sin x + \csc x = 2$.

Solution: Multiplying both sides of this equation by $\sin x$ gives us

$$\sin^2 x + 1 = 2 \sin x$$

or

$$\sin^2 x - 2 \sin x + 1 = 0$$

Factoring, $(\sin x - 1)^2 = 0$. Thus, $\sin x = 1$ and $x = \pi/2 + 2\pi k$.

$$\sin(\pi/2 + 2\pi k) + \csc(\pi/2 + 2\pi k) = \sin \pi/2 + \csc \pi/2 = 1 + 1 = 2$$

Example 5-31. Find the values of x, $0 \leq x < 2\pi$, which are solutions for $\sin x + \cos x = 1$.

Solution: First we write $\sin x = 1 - \cos x$ and then square both sides.

$$\sin^2 x = 1 - 2\cos x + \cos^2 x$$
$$1 - \cos^2 x = 1 - 2\cos x + \cos^2 x$$

or

$$2\cos^2 x - 2\cos x = 0$$

Factoring, we have

$$2\cos x(\cos x - 1) = 0$$

If $\cos x = 0$, then $x = \pi/2, 3\pi/2$. If $\cos x = 1$, then $x = 0$.

$$\text{for } x = 0 \qquad \sin 0 + \cos 0 = 0 + 1 = 1$$

$$\text{for } x = \frac{\pi}{2} \qquad \sin\frac{\pi}{2} + \cos\frac{\pi}{2} = 1 + 0 = 1$$

$$\text{for } x = \frac{3\pi}{2} \qquad \sin\frac{3\pi}{2} + \cos\frac{3\pi}{2} = -1 + 0 \neq 1$$

The only solutions, then, are 0 and $\pi/2$.

Equations involving trigonometric functions of $2x$, $x/2$, or any other multiple of x may be solved by similar methods, provided all of the functions contain the same multiple of x.

Example 5-32. Solve the equation $\tan 3x = 1$.

Solution: From our knowledge of the tangent function, we see that

$$3x = \frac{\pi}{4} + k\pi, \ k = 0, \pm 1, \pm 2, \ldots$$

Thus,

$$x = \frac{\pi}{12} + \frac{k\pi}{3}, \ k = 0, \pm 1, \pm 2, \ldots$$

To find the solutions in the interval $0 \leq x \leq 2\pi$, we assign the values 0, 1, 2, 3, 4, and 5 to k getting the solution set $\{\pi/12, 5\pi/12, 3\pi/4, 13\pi/12, 17\pi/12, 7\pi/4\}$.

If different multiples of x are involved, the equation can be solved by using the identities of Section 5-10 to change them all to the same multiple of x.

Example 5-33. Solve $\cos 2x - \sin x = 0$.

Solution: Using the identity $\cos 2x = \cos^2 x - \sin^2 x = 1 - 2\sin^2 x$, this equation becomes

$$1 - 2\sin^2 x - \sin x = 0$$

or

$$2\sin^2 x + \sin x - 1 = 0$$

Using the quadratic formula gives

$$\sin x = \frac{-1 \pm \sqrt{1+8}}{4}$$

or

$$\sin x = -1 \quad \frac{1}{2}$$

If $\sin x = -1$, then $x = 3\pi/2 + 2\pi k$, and if $\sin x = 1/2$, $x = \pi/6 + 2\pi k$ or $x = 5\pi/6 + 2\pi k$, $k = 0, \pm 1, \pm 2, \ldots$. For $0 \leq x \leq 2\pi$ we have the solution set $\{3\pi/2, \pi/6, 5\pi/6\}$.

Exercise 5-12

Find all of the solutions of the following:

1. $\cos x = -1$
2. $\tan x = 0$
3. $2\sin x - 1 = 0$
4. $\sin^2 x - 3\sin x = 0$
5. $\tan^2 x - 1 = 0$

Find the values of x, $0 \leq x < 2\pi$, which are solutions of the following:

6. $\cos^2 x - 4 = 0$
7. $\cos 2x + \sin^2 x = 0$
8. $2\tan^2 x + \sec^2 x = 2$
9. $4\sin^2 x - 3 = 0$
10. $\sin^2 x + \sin x - 2 = 0$
11. $\cos x - \sqrt{3}\sin x = 1$
12. $\cos 2x + \cos x + 1 = 0$
13. $\sin 2x + \cos x = 0$
14. $\sin 3x = 1/\sqrt{2}$

15. $\cos (x/2) = 1/2$

16. $\cos 2x + 1 = 0$

17. $\cos 3x = \sqrt{3}/2$

18. $\sin^2(x/2) = 1/2$

19. $\sin 4x = -1$

20. $\sin^2 2x + \sin 2x = 0$

21. $\cos^2(x/2) + \cos (x/2) - 2 = 0$

22. $\sin 2x + \sin x = 0$

23. $\sin^2(x/2) + \cos x = 0$

24. $\tan 2x + \tan x = 0$

25. $\sin 2x + 2 \sin^2(x/2) = 1$

26. $\sin 2x - 4 \sin x - \cos x + 2 = 0$

5-13* The Hyperbolic Functions

A group of functions of some interest in applications are those called the *hyperbolic functions*. These functions can be defined in terms of the exponential function e^x. The hyperbolic sine function, written sinh, is defined as follows:

Definition 5-9.
$$\sinh x = \frac{e^x - e^{-x}}{2}$$

While the hyperbolic cosine function, cosh, is defined

Definition 5-10.
$$\cosh x = \frac{e^x + e^{-x}}{2}$$

The hyperbolic tangent function, tanh, is defined to be the ratio of the two, thus,

Definition 5-11.
$$\tanh x = \frac{e^x - e^{-x}}{e^x + e^{-x}}$$

In addition, the hyperbolic cosecant (cosech), the hyperbolic secant (sech), and the hyperbolic cotangent (coth) can be defined as the reciprocals of these functions.

The hyperbolic functions can also be defined in terms of the coordinates of points on the hyperbola $x^2 - y^2 = 1$, just as the trigonometric functions can be defined by points on the unit circle, $x^2 + y^2 = 1$.

To show this analogy between the circular functions and the hyperbolic functions, we need to look at the trigonometric functions from a slightly different point of view. In a circle whose radius is r, the area of a sector determined by an angle whose radian measure is θ is $\pi r^2 (\theta/2\pi)$. In the unit circle, the area of this sector is $\theta/2$. We could therefore define the sine and cosine functions as follows:

Let θ be any real number, $-2\pi \leq \theta \leq 2\pi$, and let AOB be an angle in standard position which determines a sector of area $\theta/2$. Then the x-coordinate of A is $\cos\theta$ and the y-coordinate of A is $\sin\theta$. If θ is negative, then we will consider the sector to be swept out in a clockwise direction. (See Figure 5-43.)

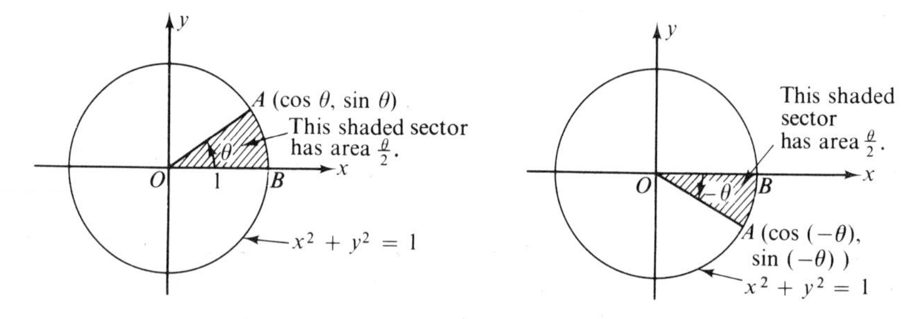

Figure 5-43

Now let us see how we can define the hyperbolic sine and the hyperbolic cosine in a similar manner. Let u be a positive real number, and let A be a point in the first quadrant on the hyperbola $x^2 - y^2 = 1$ such that the area of the region AOB is $u/2$. (See Figure 5-44.) Then the x-coordinate of A is $\cosh u$ and the y-coordinate of A is $\sinh u$. If u is negative, then we take the point A to be in the fourth quadrant.

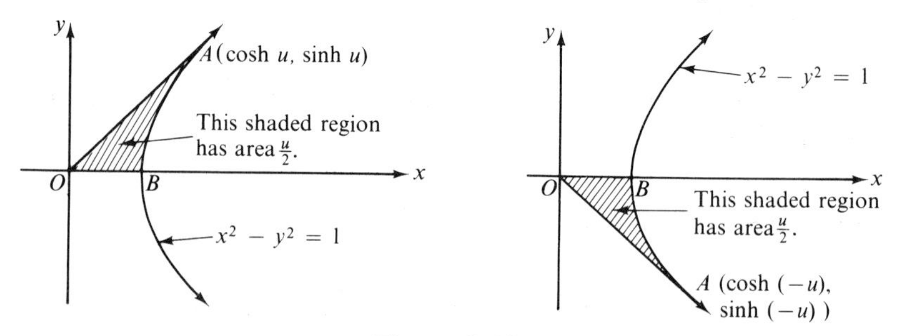

Figure 5-44

These two definitions of the hyperbolic functions are equivalent, that is, for example, Definition 5-9 and the definition above in terms of the y-coordinate of a point on the hyperbola give exactly the same set of ordered pairs for the hyperbolic sine function.

The graph of $y = \cosh x$ is called a *catenary*. It is the shape assumed by a flexible cable of uniform density which hangs freely from two points of support.

Exercise 5-13

1. Use (i) Definitions 5-9 and 5-10 and (ii) the definition of sinh and cosh as coordinates of points on the hyperbola $x^2 - y^2 = 1$ to verify each of the following:
   (a) $\sinh 0 = 0$ (b) $\cosh 0 = 1$
   (c) $\cosh^2 u - \sinh^2 u = 1$ (d) $\cosh(-u) = \cosh u$
   (e) $\sinh(-u) = -\sinh u$ (f) $\cosh u \geq 1$

2. Graph $y = \cosh x$. Give its domain and range. Is this function periodic? Is it one-to-one?

3. Graph $y = \sinh x$. Give its domain and range. Is this function periodic? Is it one-to-one?

4. Graph $y = \tanh x$. Give its domain and range. Is this function periodic? Is it one-to-one?

5. Verify each of the following identities:
   (a) $\cosh x + \sinh x = e^x$
   (b) $\cosh x - \sinh x = e^{-x}$
   (c) $\sinh 2x = 2 \cosh x \sinh x$
   (d) $\cosh 2x = \cosh^2 x + \sinh^2 x$
   (e) $\tanh 2x = \dfrac{2 \tanh x}{1 + \tanh^2 x}$

6. Find the inverse of the function $y = \cosh x$, $x \geq 0$, by setting $u = e^x$ in $y = [(e^x + e^{-x})/2]$ and solving for u. Then set $x = \ln u$.

7. Find the inverse of the function $y = \sinh x$.

Review Exercise

1. (a) Change to radian measure: 240°; 15°; 180°; 310°; −60°.
   (b) Change to degree measure: $5\pi/6$; $\pi/3$; $-\pi$; $3\pi/4$; 2.

2. If $\sin \theta = -3/4$, in what quadrants can the terminal side of the angle fall? Find the values of $\cos \theta$ and $\tan \theta$ for each case.

3. If $\tan \theta = 3$ and $\cos \theta < 0$, find $\sin \theta$ and $\cos \theta$.

4. Find the exact value of each of the following:
   (a) $\sin \pi$ (b) $\cos 0$
   (c) $\tan 5\pi/3$ (d) $\sec(-\pi/2)$
   (e) $\sin(-11\pi/4)$ (f) $\cos 13\pi$
   (g) $\tan 8\pi/3$ (h) $\cot(-2\pi)$

5. Express each of the following as a function of a number between 0 and $\pi/4$:
   (a) $\sin 12\pi/5$ (b) $\cos(-2\pi/5)$
   (c) $\tan 24\pi/7$ (d) $\csc(-35\pi/8)$

6. Complete the following table:

function	domain	range	period	cofunction	reciprocal function
sine					
cosine					
tangent					

7. Give the amplitude (where it applies), the period, and the phase shift of each of the following functions:
   (a) $y = 2\tan(3x - \pi)$ (b) $y = \cos(2x + 1)$
   (c) $y = (1/2)\sin(2x)$ (d) $y = \tan(2x - \pi/2)$
   (e) $y = 3\cos[(2/3)x - \pi/2]$ (f) $y = \sin(.5x + \pi/3)$

8. Graph each of the following functions:
   (a) $y = 2\sin 3x$ (b) $y = (1/2)\cos(4x - \pi)$
   (c) $y = \tan 2x$ (d) $y = 2\csc x$

*9. Graph $y = \sin x + \cos 2x$. What is the period of this function?

10. Give the domain and range of (a) $y = \arcsin x$; (b) $y = \arccos x$; (c) $y = \arctan x$.

11. Find the following:
    (a) $\arcsin(-1)$ (b) $\arccos(1/\sqrt{2})$
    (c) $\arctan(-1/\sqrt{3})$ (d) $\arcsin 1/2$
    (e) $\arccos(-\sqrt{3}/2)$ (f) $\arctan \sqrt{3}$

12. Find the following:
    (a) $\sin(\arccos 2/3)$ (b) $\tan(\arctan(-3))$
    (c) $\tan(\arcsin(-1/3))$ (d) $\sec(\arccos 1/2)$

13. Verify each of the following identities:
    (a) $\dfrac{\tan x - 1}{1 - \cot x} = \tan x$
    (b) $(\sin\theta + \cos\theta)^2 = 1 + 2\sin\theta\cos\theta$
    (c) $\sin^3 x = \sin x - \cos^2 x \sin x$
    (d) $\dfrac{1}{1 - \sin A} + \dfrac{1}{1 + \sin A} = 2\sec^2 A$

14. Given the formula $\sin(\alpha + \beta) = \sin \alpha \cos \beta + \cos \alpha \sin \beta$, derive the formulas for $\sin(\alpha - \beta)$ and $\sin 2\theta$.

15. Express $2 \sin x + 2 \cos x$ in the form $A \sin(x + \alpha)$. Be sure to specify α.

16. If $\sin \theta = 2/3$ and θ is in the first quadrant, find the values of the sine, cosine, and tangent of $\theta/2$; of 2θ.

17. Verify each of the following identities:
    (a) $\sin^4 x = 3/8 - 1/2 \cos 2x + 1/8 \cos 4x$
    (b) $\cos 2\theta = \cos^4\theta - \sin^4\theta$
    (c) $\cos \theta = \cos^2(\theta/2) - \sin^2(\theta/2)$

*18. Find the three cube roots of -1. Plot the points corresponding to these roots.

*19. Find all of the values of x, $0 \leq x < 2\pi$, which are solutions of $2 \sin^2 x - \sin x - 1 = 0$.

*20. (a) Define $\sinh x$, $\cosh x$, $\tanh x$.
    (b) Verify that $\cosh^2 x - \sinh^2 x = 1$.

Bibliography

Cooley, Hollis, David Gans, Morris Kline, Howard Wahlert, *Introduction to Mathematics,* Ch. 6. Cambridge: Houghton Mifflin Company, 1937.

Hooper, Alfred, *Makers of Mathematics,* Ch. 4. New York: Random House, 1948.

Kline, Morris, *Mathematics, a Cultural Approach,* Ch. 24. Reading, Massachusetts: Addison-Wesley Publishing Company, Inc., 1962.

———, *Mathematics in Western Culture,* Ch. 19. New York: Oxford University Press, 1953.

Whitehead, A. N., *An Introduction to Mathematics.* New York: Oxford University Press, 1948.

6

Applications of Trigonometry

6-1 Introduction

Trigonometry is more than just a tool for use in the calculus. The beginnings of trigonometry are found in the work of the great Greek mathematician-astronomers or even further back with the Egyptian surveyor-builders. The early uses of trigonometry were in astronomy, navigation, surveying, and engineering, and it is still used in these fields today. Our study of trigonometry would be incomplete if we did not investigate its work-a-day side.

In these applications we are usually concerned with finding the length of a certain line segment or the measure of a certain angle. Since these line segments and angles can be thought of as parts of a triangle, our problem resolves itself into finding the measure of some part of a triangle when the measures of certain other parts are known. The solution of this sort of problem is accurately described as "trigonometry," since this word means "triangle measurement."

The applications we are studying are in fields where the degree is the customary unit of measure for the angle; consequently, in this chapter angles will be measured in degrees. The notation sin 90° for example will be considered to be equivalent to $\sin \pi/2$, thus we have $\sin 90° = 1$.

Since many of the figures in the problems we will be solving will be measurements, it is important here to make some comments about accuracy. A measurement is never exact. If a measurement is given to be 12 feet, this means that the measurement is taken to the nearest foot. The true length is between 11.5 and 12.5 feet. Similarly a measurement given as 113.5 miles is taken to the nearest tenth of a mile and the actual length is between 113.45 and 113.55 miles. A measure of 12 feet is said to have two significant digits; a measure of 113.5 miles has four significant digits.

In a computation involving measured quantities, the answer should have no more significant digits than the *least accurate* measurement. Thus, if one of the measurements is 12 feet, the answer should not have more than two significant digits. Table 6-1 gives a rule of thumb for the equivalence between the number of significant digits in the measurement of the sides of a triangle and the measurement of the angles.

Table 6-1

Number of Significant Digits in the Measurement of the Sides of a Triangle	Measurement of Angles to the Nearest
2	degree
3	10 minutes
4	minute
5	10 seconds

For example, if the sides of a triangle are given to two significant digits, then the measurement of the angles should be calculated only to the nearest degree.

Although the numbers given in the examples and exercises are all approximations, we will use the equal sign with the understanding that we really mean "approximately equal to."

6-2 The Use of Tables and Reduction Techniques

Values of the trigonometric functions have been calculated with great accuracy and are available to us in tables. We have already seen that the trigonometric functions are periodic, hence it is sufficient to tabulate functional values for $0 \leq \theta < 360°$. Actually we shall see that a much shorter table is adequate. Any trigonometric function of θ, where θ is the degree measure of some angle, can be reduced to a function of a number R, where $0 \leq R \leq 90°$, with perhaps a change of sign. The sign chosen will depend on the quadrant in which the terminal side of the angle falls.

Let us see how we can find this number R. First, consider an angle whose measure is θ and whose terminal side lies in the second quadrant. [See Figure 6-1(a).] If we drop a perpendicular to the x-axis we

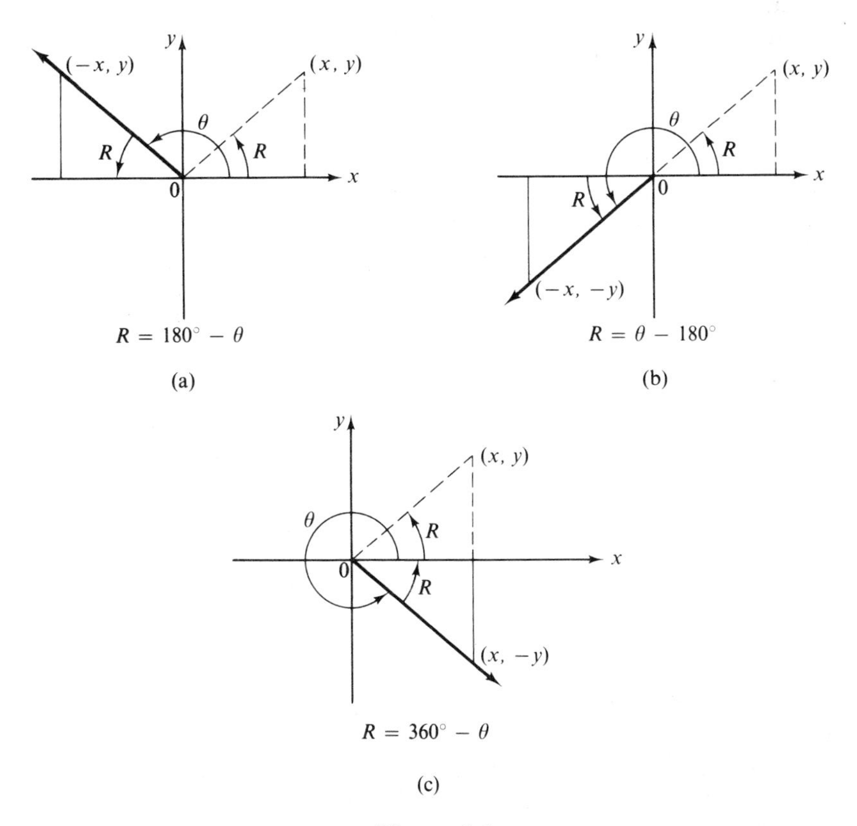

Figure 6-1

have a right triangle one of whose acute angles has measure $R = 180° - \theta$. Let us call this acute angle whose sides are the x-axis and the terminal side the *reference angle*. If we construct a triangle in Quadrant I congruent to this triangle then corresponding to the point (x, y) on the terminal side of the angle whose measure is R there is a point $(-x, y)$ on the terminal side of the angle whose measure is θ. Thus,

$$\sin \theta = \frac{y}{r} = \sin R$$

$$\cos \theta = \frac{-x}{r} = -\cos R$$

$$\tan \theta = \frac{-y}{x} = -\tan R$$

If the terminal side of the angle falls in Quadrant III [Figure 6-1(b)], then the reference angle has measure $R = \theta - 180°$ and

$$\sin \theta = \frac{-y}{r} = -\sin R$$

$$\cos \theta = \frac{-x}{r} = -\cos R$$

$$\tan \theta = \frac{-y}{-x} = \frac{y}{x} = \tan R$$

If the terminal side lies in the fourth quadrant, $R = 360° - \theta$ [Figure 6-1(c)] and

$$\sin \theta = \frac{-y}{r} = -\sin R$$

$$\cos \theta = \frac{x}{r} = \cos R$$

$$\tan \theta = \frac{-y}{x} = -\tan R$$

Note that in every case the reference angle is an acute angle whose sides are the x-axis and the terminal side of the angle in question. In summary, to reduce a function of θ to a function of a number R, $0 \leq R \leq 90°$:

1. Sketch the angle whose measure is θ in standard position.
2. Determine from the sketch the measure of the reference angle R.
3. Find the sign of the function in the given quadrant.

Example 6-1. Express the trigonometric functions of 192° as functions of R, where $0 \leq R \leq 90°$.

Solution: The terminal side of this angle falls in Quadrant III [Figure 6-2(a)] and $R = 192° - 180° = 12°$. Since the sine and cosine functions are negative and the tangent function is positive in Quadrant III, we have

$$\sin 192° = -\sin 12° \qquad \csc 192° = -\csc 12°$$
$$\cos 192° = -\cos 12° \qquad \sec 192° = -\sec 12°$$
$$\tan 192° = \tan 12° \qquad \cot 192° = \cot 12°$$

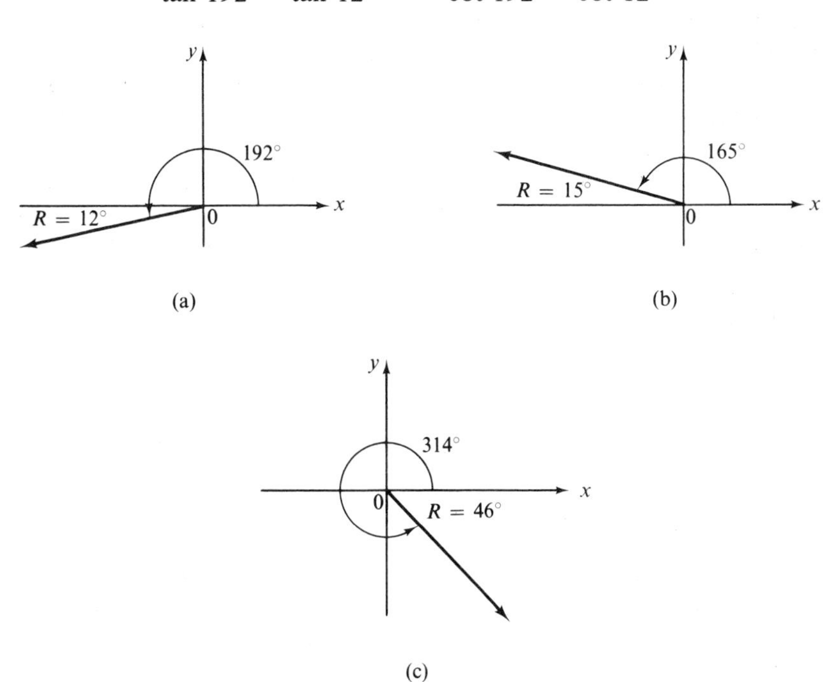

Figure 6-2

Example 6-2. Express tan 165° as a function of R, $0 \leq R \leq 90°$.

Solution: The terminal side of this angle lies in Quadrant II, and $R = 180° - 165° = 15°$. [See Figure 6-2(b).] Since the tangent function is negative in Quadrant II, $\tan 165° = -\tan 15°$.

Example 6-3. Express cos 674° as a function of R, $0 \leq R \leq 90°$.

Solution: Since $674° = 360° + 314°$ and the cosine function is periodic with period $360°$, $\cos 674° = \cos 314°$. The terminal side of this angle lies in Quadrant IV. [See Figure 6-2(c).] The reference angle has measure $R = 360° - 314° = 46°$ and the cosine function is positive in the fourth quadrant, so $\cos 674° = \cos 314° = \cos 46°$.

We can go even further than this in our reduction. Since any function of R is equal to the cofunction of $90° - R$, the function can be reduced still further to a function of a number α, where $0 \leq \alpha \leq 45°$.

In our tables, then, it is sufficient to list values of the trigonometric functions for angles whose measure is between 0° and 45°, inclusive. (See Table 2.) Values of θ, $0 \leq \theta \leq 45°$ are listed in the left-hand column and the values of $90° - \theta$ are listed in the column on the far right. The name of the cofunction of each function listed across the top of the table is listed across the bottom of the table. Thus, for $0 \leq \theta \leq 45°$ we use the left-hand column and find the name of the function at the top of the table, while for $45° \leq \theta \leq 90°$ we use the right-hand column and find the name of the function at the bottom of the table.

Example 6-4.

θ	$\sin\theta$	$\cdots$	$\cot\theta$	
9°00′	.1564	$\cdots$	6.314	81°00′
·	·		·	·
·	·		·	·
·	·		·	·
40°00′	.6428	$\cdots$	1.192	50°00′
	$\cos\theta$		$\tan\theta$	θ

$$\sin 9° = \cos 81° = .1564; \; \tan 50° = \cot 40° = 1.192$$

The student should realize that the values in Table 2 are, in general, approximations. Thus, if the table gives the value .1564 for sin 9°, we take this to mean that sin 9° is a number between .15635 and .15645. Tables are available which give these values to many more digits.

Table 2 lists values of θ at intervals of 10 minutes, so to find values between these we must use linear interpolation. (See Section 4-5.)

Example 6-5. Find $\sin 26°16'$.

Solution: From the table we find

$$10\left\{\begin{array}{l} 6\left\{\begin{array}{l}\sin 26°10' = .4410\\ \sin 26°16' = \end{array}\right\}d \\ \\ \sin 26°20' = .4436\end{array}\right\}.0026$$

Computing differences, we find that $6/10 = d/.0026$ or $d = .00156$. Thus, $\sin 26°16' = .4410 + .00156 = .4426$, rounded off to four digits.

Example 6-6. Find $\cos 56°13'$.

Solution: From the table we have

$$10\left\{\begin{array}{l} 3\left\{\begin{array}{l}\cos 56°10' = .5568\\ \cos 56°13' = \end{array}\right\}d \\ \\ \cos 56°20' = .5544\end{array}\right\}-.0024$$

Note that the cosine function is decreasing between 0° and 90°, hence the difference is taken to be negative. Equating ratios of differences we have $3/10 = d/(-.0024)$ or $d = -.00072$. Thus, $\cos 56°13' = .5568 - .00072 = .5561$ to four digits.

Example 6-7. Find θ if $0 < \theta < 90°$ and $\tan\theta = .7731$.

Solution: From the table we find that

$$10'\left\{\begin{array}{l} d\left\{\begin{array}{l}\tan 37°40' = .7720\\ \tan\theta \quad\quad = .7731\end{array}\right\}.0011 \\ \\ \tan 37°50' = .7766\end{array}\right\}.0046$$

Then $d/10' = 11/46$ and $d = (11/46)(10') \simeq 2'$. Thus, $\theta = 37°40' + 2' = 37°42'$.

Example 6-8. Find θ if $0 \le \theta \le 180°$ and $\cos\theta = -.2447$.

Solution: Since the cosine is negative, we must have $90° \le \theta \le 180°$. Using the identity $\cos(180° - \theta) = -\cos\theta$, we find that $\cos(180° - \theta) = .2447$ and $180° - \theta = 75°50'$. Thus, $\theta = 104°10'$.

Exercise 6-2

1. Express each of the following as a function of R, $0 \le R \le 90°$:
   (a) $\sin 400°$ (b) $\cos 475°$

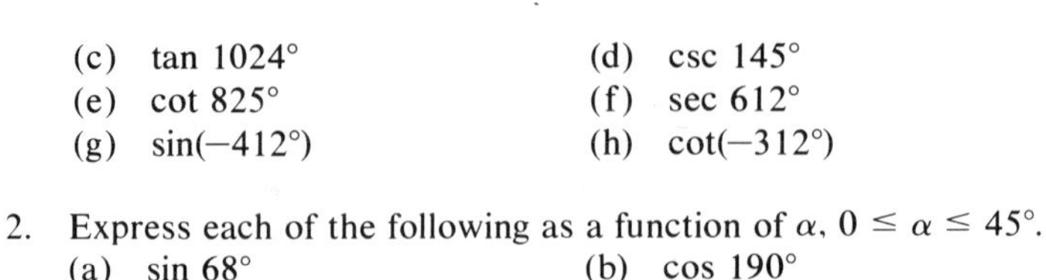

(c) $\tan 1024°$ (d) $\csc 145°$
(e) $\cot 825°$ (f) $\sec 612°$
(g) $\sin(-412°)$ (h) $\cot(-312°)$

2. Express each of the following as a function of α, $0 \leq \alpha \leq 45°$.
   (a) $\sin 68°$ (b) $\cos 190°$
   (c) $\tan 273°$ (d) $\sec 245°$
   (e) $\csc 432°$ (f) $\cot 101°$

3. Use reduction techniques and Table 2 to find the values of the trigonometric functions for $\theta = 294°$.

4. Use reduction techniques and Table 2 to find the values of the trigonometric functions for $\theta = 222°$.

5. Use reduction techniques and Table 2 to find the values of each of the following:
   (a) $\sin 127°$ (b) $\cos 1{,}000°$
   (c) $\tan 416°$ (d) $\cot 114°$
   (e) $\sec 514°$ (f) $\sin(-104°)$
   (g) $\sec(-92°)$ (h) $\tan(-125°)$

6. Use linear interpolation to find:
   (a) $\tan 54°46'$ (b) $\cos 46°42'$
   (c) $\tan 10°54'$ (d) $\cot 48°15'$
   (e) $\sin 127°14'$ (f) $\cos 135°26'$
   (g) $\sec 35°18'$ (h) $\csc 78°25'$

7. Use linear interpolation to find θ, $0 < \theta < 90°$, if:
   (a) $\sin \theta = .8988$ (b) $\cos \theta = .9899$
   (c) $\tan \theta = .8541$ (d) $\sec \theta = 1.037$
   (e) $\sin \theta = .3409$ (f) $\cos \theta = .8151$
   (g) $\tan \theta = 1.211$ (h) $\cot \theta = 1.033$

8. Tell whether each of the following statements is true or false. If a statement is false, change it making it true.
   (a) $\sin(-45°) < \sin(-10°)$
   (b) $\cos 30° < \cos 120°$
   (c) $\sin 37° \cdot \csc(-37°) = -1$
   (d) $\cos^2\theta + \sin^2(-\theta) = 1$
   (e) $\tan 60° = 2 \tan 30°$
   (f) $\cos(-36°) = \sin 54°$
   (g) $\sin 30° + \sin 10° = \sin 40°$

9. Find θ, $0 \leq \theta \leq 180°$, if:
   (a) $\cos \theta = -.9848$ (b) $\tan \theta = -1.664$
   (c) $\sec \theta = -2.508$ (d) $\cot \theta = -1.085$

6-3 Right Triangles

There are six principal measurements associated with a triangle—the lengths of its three sides and the measures of its three angles. If ABC is a triangle, then we will use the letters a, b, and c to stand for the lengths of the sides opposite the angles at A, B, and C respectively. The Greek letters α, β, and γ will denote the measures of angles A, B, and C respectively. (See Figure 6-3.)

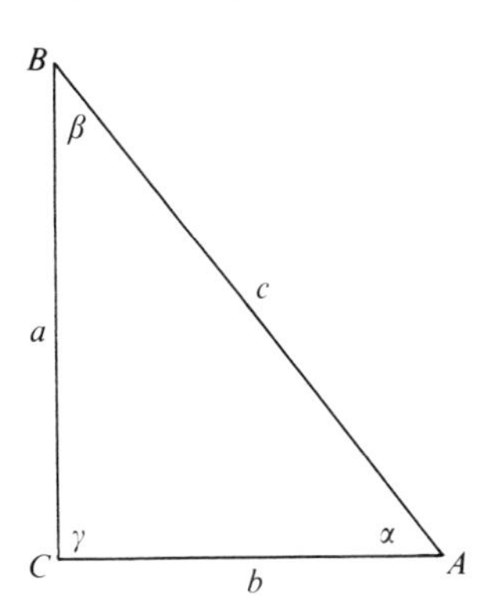

Figure 6-3

If ABC is a right triangle, then we will take angle C to be the right angle. If we know the lengths of two sides of a right triangle, or the length of one side and the measure of one of the acute angles, then we can find the three unknown measurements. Finding all of the unknown measurements of a triangle is called "solving the triangle." Solving a right triangle is particularly simple since we know so many different relationships between its parts.

The Pythagorean Theorem, for example, tells us that

$$c^2 = a^2 + b^2 \qquad \textbf{(6-1)}$$

and since $\gamma = 90°$, we know that

$$\alpha + \beta = 90° \qquad \textbf{(6-2)}$$

In addition to these two equations, we have the trigonometric functions, which tell us that

$$\begin{array}{ll} \sin \alpha = a/c & \sin \beta = b/c \\ \cos \alpha = b/c & \cos \beta = a/c \\ \tan \alpha = a/b & \tan \beta = b/a \end{array}$$

and so on. (See Definition 5-2.)

Let us see how we can solve a right triangle given the measures of two of its sides.

Example 6-9. Solve the right triangle ABC if $a = 43$ and $b = 24$. (See Figure 6-4.)

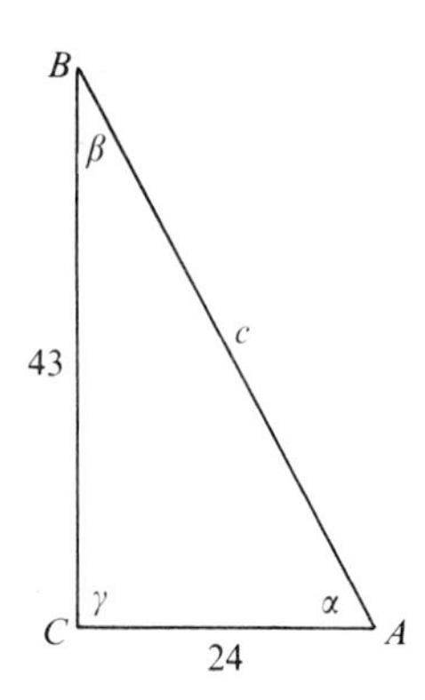

Figure 6-4

Solution: We need to find α, β, and c. First we find α.

$$\tan \alpha = \frac{43}{24} = 1.792$$

From the table we find α to the nearest degree (see Table 6-1) to be 61°. From Equation (6-2) we get $\beta = 90° - 61° = 29°$.

To find c we can either use Equation (6-1) or the fact that $\sec \alpha = c/b$. Since $b = 24$ and $\sec 61° = 2.063$, we have

$$c = (24)(2.063) = 49.512$$

which we round off to 50.

We can also solve a right triangle given the measure of one acute angle and one side.

Example 6-10. Solve the right triangle ABC if $\alpha = 35°10'$ and $c = 72.5$.

Solution: First we find that $\beta = 90° - \alpha = 54°50'$.

We use the trigonometric equation $\sin \alpha = a/c$ to find a. Since $\sin 35°10' = .5760$ and $c = 72.5$, then $a = (.5760)(72.5) = 41.76$. We round this off to three significant digits getting $a = 41.8$.

To find b we can use the equation $\cos \alpha = b/c$, or $b = (72.5)(.8175) = 59.3$ to three significant digits.

Exercise 6-3

Solve each of the following right triangles ABC, where $\gamma = 90°$. Be sure to give your answers only to the number of significant digits warranted. (See Section 6-1.) (Note: Your answers may be slightly different from those given in the back of the book due to rounding off errors.)

1. $a = 25.4$; $b = 38.2$
2. $a = 16$; $b = 27$
3. $a = 506.2$; $c = 984.8$
4. $\alpha = 35°$; $c = 11$
5. $\beta = 48°40'$; $c = 225$
6. $\alpha = 32°10'$; $a = 75.4$
7. Find the area of a parallelogram if two of its adjacent sides are 10 and 12 inches long and the measure of the included angle is 41°.
8. In Figure 6-5 the line BD crosses a pond. A surveyor needs to locate a point on this line, so he measures an angle of $51°10'$ at B and proceeds 151 feet to point C where he lays off an angle of 90°. How far must he proceed from C to find a point A on the line BD?

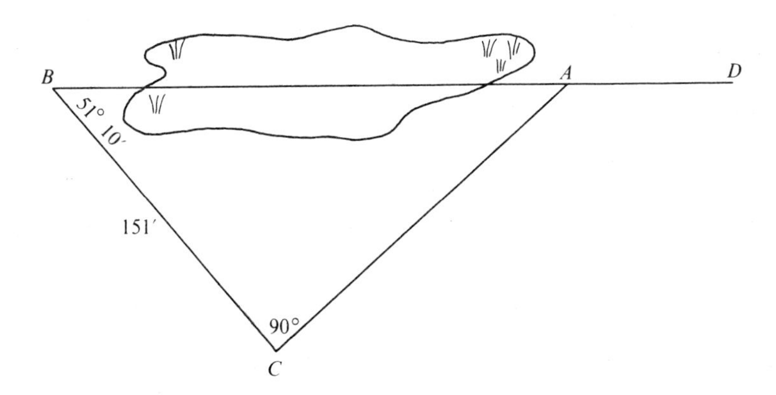

Figure 6-5

9. A 24 foot ladder leans against a wall with the foot of the ladder 7.5 feet from the bottom of the wall. What angle does the ladder make with the ground?

10. Find the height of an isosceles triangle if the base angles measure $36°10'$ and the opposite sides are 12.5 inches long. Find the length of the base.

11. The hypotenuse and one leg of a right triangle have lengths 37 inches and 24 inches respectively. Find the length of the third side by (a) the Pythagorean Theorem; (b) trigonometry.

12. A balloon, directly above a town, is seen by an observer 12 miles away. The angle of elevation of the balloon as seen by the observer is 4°. How high is the balloon above the town?

13. A road rises 12.5 feet in a horizontal distance of 235 feet. Find the angle of elevation of the road.

14. A tower stands on level ground and from a certain point x feet away from its base the angle of elevation of the top of the tower is 55°. From a point 1,000 feet further away the angle of elevation is 40°. Find the height of the tower. (Hint: Let h be the height of the tower. Then $\cot 55° = x/h$ and $\cot 40° = (1{,}000 + x)/h$. Solve these two equations for h.)

15. A regular pentagon has sides of length 5 inches. Find the *apothem,* the perpendicular distance from the center to a side. What is the area of the pentagon?

6-4 The Law of Sines and the Law of Cosines

Given certain information it is easy to solve a right triangle. All we need is the definitions of the trigonometric functions and a table. To solve a triangle which is not a right triangle is more difficult. We need formulas relating the measures of the sides and angles of *any* triangle. Two such formulas are the Law of Sines and the Law of Cosines.

As before, let ABC be a triangle and a, b, and c the lengths of the sides opposite vertices A, B, and C respectively. Let α, β, and γ be the measures of the angles at A, B, and C.

Let CD be the altitude drawn from vertex C to the opposite side, and let h be its length. Then CD is perpendicular to AB. [See Figure 6-6 (a).] Since ADC is a right triangle, $\sin \alpha = h/b$, or $h = b \sin \alpha$. Moreover DBC is a right triangle, so $\sin \beta = h/a$, and $h = a \sin \beta$. Equating these two gives $b \sin \alpha = a \sin \beta$, or

$$\frac{a}{\sin \alpha} = \frac{b}{\sin \beta}$$

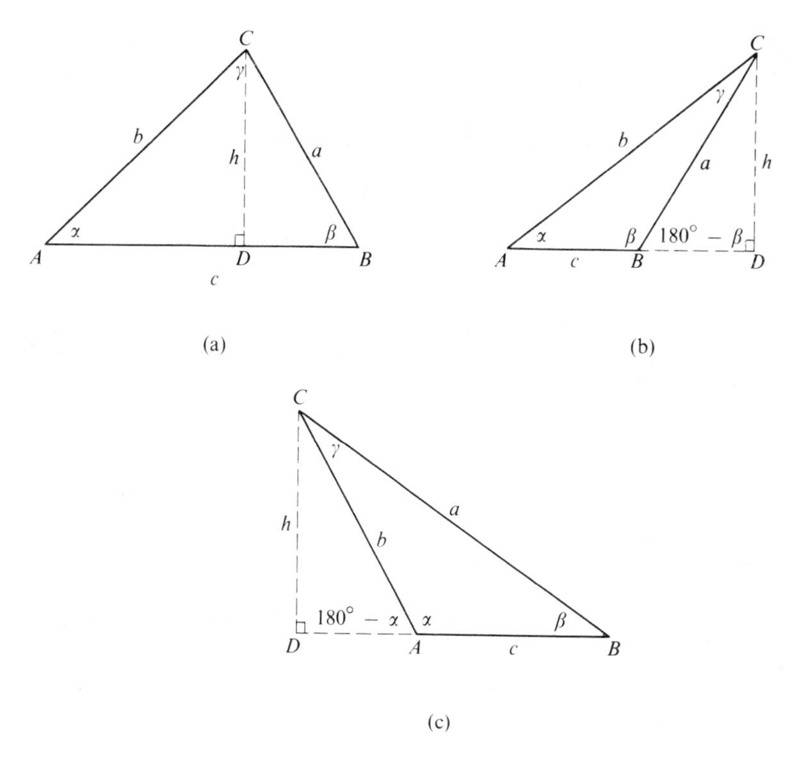

Figure 6-6

If the altitude CD lies in the exterior of the triangle [Figure 6-6 (b) or (c)] then we have (b) $\sin \alpha = h/b$ and $\sin(180° - \beta) = h/a$; or (c) $\sin(180° - \alpha) = h/b$ and $\sin \beta = h/a$. But $\sin(180° - \theta) = \sin \theta$; thus, we have $\sin \alpha = h/b$ and $\sin \beta = h/a$ as before. In a similar manner we can show that

$$\frac{a}{\sin \alpha} = \frac{c}{\sin \gamma}$$

We have thus derived:

Theorem 6-1. *The Law of Sines.* Let ABC be a triangle with a, b, and c the lengths of the sides opposite vertices A, B, and C respectively. Let α, β, and γ be the measures of the angles at A, B, and C. Then

$$\frac{a}{\sin \alpha} = \frac{b}{\sin \beta} = \frac{c}{\sin \gamma}$$

We can use the same figure to derive the Law of Cosines. Let AD and DB stand for the lengths of line segments AD and DB. From triangle ADC in Figure 6-6 (a) we see that $\cos \alpha = AD/b$ or $AD = b \cos \alpha$, and $\sin \alpha = h/b$ or $h = b \sin \alpha$. Moreover, since $AD + DB = c$, $DB = c - AD = c - b \cos \alpha$. Since DBC is a right triangle, by the Pythagorean Theorem $h^2 + DB^2 = a^2$. Substituting the values of h and DB into this equation gives

$$(b \sin \alpha)^2 + (c - b \cos \alpha)^2 = a^2$$
$$b^2\sin^2\alpha + c^2 - 2bc \cos \alpha + b^2\cos^2\alpha = a^2$$
$$b^2(\sin^2\alpha + \cos^2\alpha) + c^2 - 2bc \cos \alpha = a^2$$

Since $\sin^2\alpha + \cos^2\alpha = 1$, this becomes

$$a^2 = b^2 + c^2 - 2bc \cos \alpha$$

If the altitude CD lies outside the triangle [Figure 6-6 (b)] then $DB = AD - c = b \cos \alpha - c = -(c - b \cos \alpha)$. Since this quantity is squared in the derivation, we arrive at the same result. In the third case [Figure 6-6 (c)], $\cos \alpha = -\cos(180° - \alpha) = -(DA/b)$. Thus, $DA = -b \cos \alpha$. Since in this case $DB = c + DA = c - b \cos \alpha$, we have the same result as before.

In a similar manner we could derive the other two forms of the Law of Cosines:

$$b^2 = a^2 + c^2 - 2ac \cos \beta$$
$$c^2 = a^2 + b^2 - 2ab \cos \gamma$$

Theorem 6-2. *The Law of Cosines.* Let ABC be a triangle with a, b, and c the lengths of the sides opposite vertices A, B, and C, and α, β, and γ the measures of the angles at A, B, and C respectively. Then

$$a^2 = b^2 + c^2 - 2bc \cos \alpha$$
$$b^2 = a^2 + c^2 - 2ac \cos \beta$$
$$c^2 = a^2 + b^2 - 2ab \cos \gamma$$

Now let us see how we can use these formulas to solve the general triangle. From elementary geometry, we recall that in order to determine a unique triangle we must know at least three of the six principal measurements, and even this may not be sufficient in some cases.

Case 1. SSS. Given the lengths of three sides of a triangle we can always find a unique solution provided the sum of the measures of any two of the sides is greater than the third. There is no triangle whose sides measure 1, 1, and 2 units, for example. Provided this criterion (called the *triangle inequality*) is met, we can solve the triangle by using the Law of Cosines.

Example 6-11. Solve the triangle ABC if $a = 3$, $b = 5$, and $c = 6$.

Solution: Since the sum of any two of these numbers is greater than the third, this problem has a solution. By the Law of Cosines $a^2 = b^2 + c^2 - 2bc \cos \alpha$. Thus,

$$9 = 25 + 36 - 60 \cos \alpha$$

Solving for $\cos \alpha$,

$$\cos \alpha = \frac{52}{60} = .8667$$

so to the nearest degree,

$$\alpha = 30°$$

We can now use either the Law of Cosines or the Law of Sines to find β. By the Law of Cosines,

$$b^2 = a^2 + c^2 - 2ac \cos \beta$$
$$25 = 9 + 36 - 36 \cos \beta$$

or

$$\cos \beta = \frac{20}{36} = .5556$$
$$\beta = 56°$$

or, by the Law of Sines,

$$\frac{\sin \alpha}{a} = \frac{\sin \beta}{b}$$
$$\sin \beta = \frac{b \sin \alpha}{a} = \frac{(5)(.5)}{3} = .8333$$
$$\beta = 56°$$

Then $\gamma = 180° - (30° + 56°) = 94°$.

Case 2. SAS. Given the lengths of two sides and the measure of the included angle, we can always find a unique solution. Again we use the Law of Cosines.

Example 6-12. Solve triangle ABC if $a = 21$, $b = 13$, and $\gamma = 51°$.

Solution: By the Law of Cosines $c^2 = a^2 + b^2 - 2ab \cos \gamma$

$$c^2 = 441 + 169 - (546)(.6293)$$
$$= 266$$

To two digits, then

$$c = 16$$

Again we can use either the Law of Cosines or the Law of Sines to find either α or β. By the Law of Cosines we find

$$441 = 169 + 266 - 416 \cos \alpha$$

or

$$\cos \alpha = -\frac{6}{416} = -.0144$$

Since $\cos \alpha$ is negative, α must be greater than 90°. Using the identity $\cos (180° - \alpha) = -\cos \alpha$, we find that $\cos (180° - \alpha) = .0144$ and $180° - \alpha = 89°$. Thus, $\alpha = 91°$ and $\beta = 180° - (91° + 51°) = 38°$.

> *Case 3. AAS.* If we know the length of one side and the measures of any two angles, we can always find a unique solution provided the sum of the measures of the two angles is less than 180°. In this case, we use the Law of Sines.

Example 6-13. Solve the triangle ABC if $\alpha = 51°$, $\beta = 16°$, $a = 21$.

Solution: Since $\alpha + \beta + \gamma = 180°$, $\gamma = 180° - (51° + 16°) = 113°$. To find b, we use the Law of Sines.

$$\frac{b}{\sin \beta} = \frac{a}{\sin \alpha}$$
$$b = \frac{(21)(\sin 16°)}{\sin 51°}$$
$$= \frac{(21)(.2756)}{.7771}$$
$$= 7.4$$

Similarly,

$$\frac{c}{\sin \gamma} = \frac{a}{\sin \alpha}$$
$$c = \frac{(21)(.9205)}{.7771}$$
$$= 25$$

Case 4. SSA. The case in which we know the lengths of two sides and the measure of an angle opposite one of them is called the *ambiguous case* since it may have no solution at all, one unique solution, or two different solutions. These three possibilities are illustrated in Figure 6-7.

Given α, b, and a, if a is smaller than the perpendicular distance h from C to the opposite side ($h = b \sin \alpha$), then there is no solution. [See Figure 6-7 (a).]

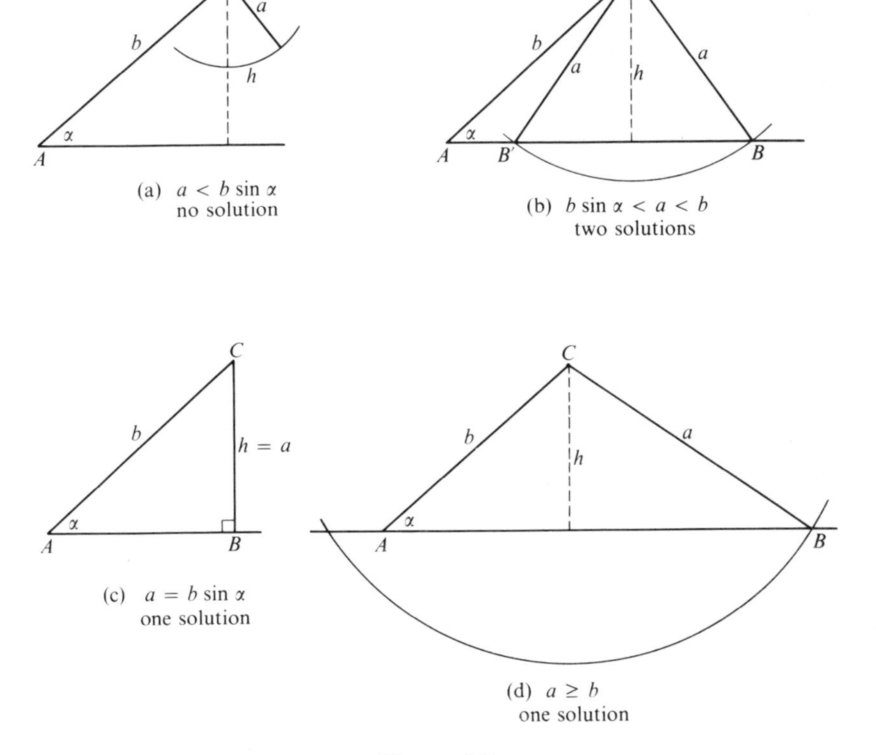

Figure 6-7

If a is larger than h, but smaller than b, then there are two solutions. [See Figure 6-7 (b).]

If a is equal to h, or if a is greater than or equal to b, there is one unique solution. [See Figure 6-7 (c) and (d).]

The solution(s), if any, are found by using the Law of Sines.

Example 6-14. Determine whether there is one solution, two solutions, or no solution at all for the triangle ABC if $\alpha = 42°$, $b = 12$, and $a = 7$.

Solution: We compute $h = b \sin \alpha = (12)(.6691) \simeq 8$. Since $a < h$, we conclude that there is no solution.

If we failed to notice this and proceeded to try to find β using the Law of Sines, we would find

$$\begin{aligned} \sin \beta &= \frac{(\sin \alpha)(b)}{a} \\ &= \frac{(12)(.6691)}{7} \simeq \frac{8}{7} \end{aligned}$$

which is clearly impossible, since the sine function is never greater than one.

Example 6-15. Find all of the solutions (if any) for triangle ABC if $\alpha = 42°$, $b = 12$, $a = 10$.

Solution: Again we compute $h = b \sin \alpha \simeq 8$. Since $h < a < b$, we have two solutions. To find these, we use

$$\begin{aligned} \sin \beta &= \frac{(\sin \alpha)(b)}{a} \\ &= \frac{(.6691)(12)}{10} = .8029 \end{aligned}$$

From Table 2 we find that, to the nearest degree, $\beta = 53°$. However, from Figure 6-7 we know that there are two values for β, one less than 90° and one greater than 90°. Since $\sin(180° - \beta) = \sin \beta$, we conclude that the other value for β must be $\beta' = 180° - 53° = 127°$. Thus, $\gamma = 85°$, $\gamma' = 11°$. We can find c for each of these cases by using the Law of Sines again.

$$c = \frac{(10)(.9962)}{(.6691)} = 15 \qquad c' = \frac{(10)(.1908)}{(.6691)} = 2.9$$

Example 6-16. Find all of the solutions (if any) if $\alpha = 42°$, $b = 12$, $a = 14$.

Solution: Since $a > b$, we will have one unique solution.

$$\begin{aligned} \sin \beta &= \frac{(\sin \alpha)(b)}{a} = \frac{(.6691)(12)}{14} = .5735 \\ \beta &= 35° \end{aligned}$$

$$\gamma = 180^\circ - (42^\circ + 35^\circ) = 103^\circ$$

and from the Law of Sines $c = (14)(.9744)/(.6691) = 20$.

If we had not noticed that a was greater than b and set about finding a second solution by taking $\beta' = 180^\circ - 35^\circ = 145^\circ$, then we would find that $\alpha + \beta' = 42^\circ + 145^\circ = 187^\circ$, which is impossible since the sum of the measures of the angles of a triangle is 180°.

In using both the Law of Sines and the Law of Cosines, we had to have the measure of at least one side. Knowing the measures of the three angles of a triangle (AAA) is not sufficient to determine a unique triangle.

Exercise 6-4

1. Show that the Pythagorean Theorem is a special case of the Law of Cosines for $\gamma = 90^\circ$.

2. Show that the area of any triangle ABC is given by the following formulas:

$$\text{area} = \left(\frac{1}{2}\right) bc \sin \alpha$$

$$\text{area} = \left(\frac{1}{2}\right) ac \sin \beta$$

$$\text{area} = \left(\frac{1}{2}\right) ab \sin \gamma$$

3. Give an alternate derivation of the Law of Sines by equating these area formulas (problem 2) and simplifying.

4. Derive the Law of Cosines from Figure 6-8 by first showing that the coordinates of C are $x = b \cos \alpha$; $y = b \sin \alpha$, and then using the distance formula and setting a equal to the distance from C to B.

5. Solve triangle ABC if:
   (a) $a = 10$, $b = 21$, $c = 12$
   (b) $a = 14$, $b = 9$, $c = 4$
   (c) $\alpha = 32^\circ$, $\beta = 68^\circ$, $c = 16$
   (d) $a = 11$, $b = 15$, $\gamma = 32^\circ$
   (e) $b = 101$, $c = 270$, $\alpha = 64^\circ 10'$
   (f) $\alpha = 14^\circ$, $\beta = 93^\circ$, $c = 11$
   (g) $\alpha = 126^\circ$, $\beta = 22^\circ$, $b = 22$
   (h) $a = 5$, $b = 10$, $c = 14$

6. Determine whether there is no solution, one solution, or two solutions for triangle ABC in each of the following cases:
   (a) $\alpha = 16^\circ$, $b = 24$, $a = 30$
   (b) $\alpha = 16^\circ$, $b = 24$, $a = 6$
   (c) $\alpha = 16^\circ$, $b = 24$, $a = 10$
   (d) $\alpha = 35^\circ$, $b = 25$, $a = 12$
   (e) $\alpha = 35^\circ$, $b = 25$, $a = 30$
   (f) $\alpha = 35^\circ$, $b = 25$, $a = 19$

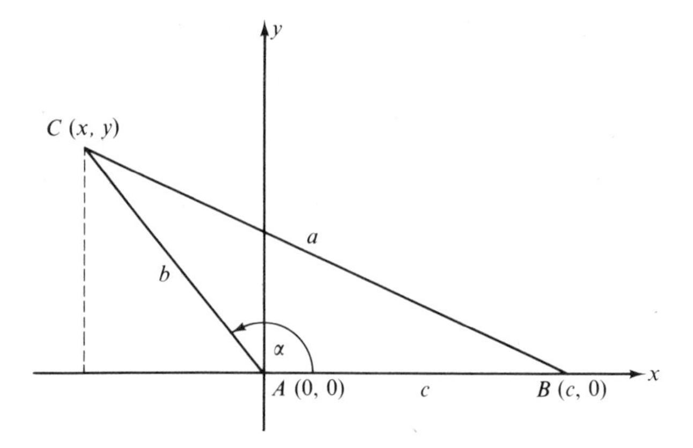

Figure 6-8

7. Find all of the solutions, if any, for the triangles of problem 6.

8. A triangle has sides of length 12, 18, and 24. Find the measure of the largest angle. (Hint: The largest angle is opposite the longest side.)

9. A triangle has sides of length 5, 10, and 12. Find the measure of the smallest angle.

10. Two angles of a triangle have measures 42° and 75°. The length of the longest side is 16 inches. Find the length of the shortest side.

11. Find the area of a triangle if two sides have lengths 62.5 inches and 21.4 inches and the included angle has measure 24°10′. (Hint: See Problem 2.)

12. A rhombus measures 12 inches on each side. If one of the angles has measure 52°, find the lengths of the two diagonals.

13. Use the Law of Cosines to show that if $\gamma > 90°$, then $c^2 > a^2 + b^2$ and if $\gamma < 90°$, then $c^2 < a^2 + b^2$.

Review Exercise

1. State the Law of Sines and the Law of Cosines.

2. Which law would you use in solving a triangle in which the following combinations of measures of sides and angles are given?
   (a) three sides
   (b) two angles and any side
   (c) two sides and the included angle
   (d) two sides and the angle opposite one of them

3. One angle of a right triangle has measure 34° and the side opposite it has length 15 inches. Find the area of the triangle.

4. Express each of the following as a function of R, $0 \leq R \leq 90°$:
   (a) sin 526°
   (b) cos 197°
   (c) tan 343°

5. (a) In triangle ABC if $a = 40.1$, $b = 16.2$, and $\gamma = 37°10'$, find c.
   (b) In triangle ABC if $a = 12$, $b = 18$, $c = 24$, find α.
   (c) In triangle ABC if $\alpha = 27°$, $\beta = 43°$, and $c = 14$, find b.

6. In triangle ABC, $b = 18$, $\alpha = 32°$, and $a = 12$. Are there two, one, or no solutions to this problem?

7. Is it possible to find the measure of the angles of a triangle if the sides are given to be 6.0, 8.0, and 15?

8. Would it be possible to find the distance to the moon from measurements made on the earth? What measurements would you require to carry out such a calculation?

Bibliography

Cooley, Hollis, David Gans, Morris Kline, Howard Wahlert, *Introduction to Mathematics,* Ch. 6. Cambridge: Houghton Mifflin Company, 1937.

Kline, Morris, *Mathematics in Western Culture,* Ch. 5. New York: Oxford University Press, 1953.

Whitehead, A. N., *An Introduction to Mathematics.* New York: Oxford University Press, 1948.

7

Analytic Geometry in the Plane

7-1 Introduction

In this chapter we will study the relationship between certain curves in the plane and equations in x and y. If we have an equation in x and y, then its solution set is the set of all ordered pairs of real numbers (x, y) which satisfy the equation. On the other hand, a curve in the plane is a set of points which can also be described as a set of ordered pairs of real numbers (x, y). If these two sets of ordered pairs are the same, then we say that the equation is *an equation for the curve* and the curve is *the graph of the equation.* This correspondence between curves and equations is not one-to-one since two (or more) equations may have the same graph. For example, $x + y = 1$ and $2x + 2y = 2$ have the same solution set, hence they have the same graph. If two equations have the same solution set then we say that they are *equivalent.* On the other hand, some equations have no graph since their solution set is the empty set. For example, there are no real number ordered pairs which satisfy the equation $x^2 + y^2 = -1$, consequently this equation has no graph.

In plane analytic geometry we study two types of problems: (1) given an equation in x and y, to draw its graph, and (2) given a curve in the plane to find an equation for it. The student has probably solved many problems of the first type, however, problems of the second type may be new to him. We will not attempt to solve this problem for all curves in the plane, but will restrict ourselves to some special curves—the line, the circle, the parabola, the ellipse, and the hyperbola.

7-2 Inclination and Slope

If a line intersects the x-axis in a single point, the angle of inclination of the line is the angle whose initial side is that part of the x-axis to the right of the point of intersection and whose terminal side is that part of the line that lies in the upper half plane. (See Figure 7-1.)

The measure of this angle is called the *inclination* of the line. The inclination of a horizontal line is defined to be zero. From this definition we can see that the inclination of a line that is not parallel to the x-axis will be some number between zero and π.

If θ is the inclination of a line, then the *slope* of this line, which we will designate by the letter m, is defined to be $\tan \theta$.

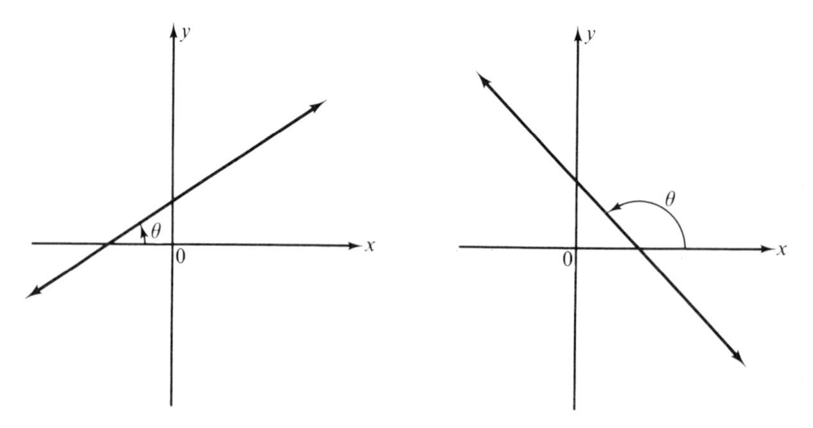

Figure 7-1

$$m = \tan \theta$$

For example, a line whose inclination is $\pi/4$ has slope $\tan \pi/4 = 1$. From this definition we at once draw several conclusions:

1. Since $\tan 0 = 0$, the slope of a horizontal line is zero.
2. Since $\tan \pi/2$ is undefined, the slope of a vertical line is undefined.
3. If the angle of inclination is acute, the slope of the line is positive and the line rises from left to right.
4. If the angle of inclination is obtuse, the slope of the line is negative and the line falls from left to right. (See Figure 7-2.)

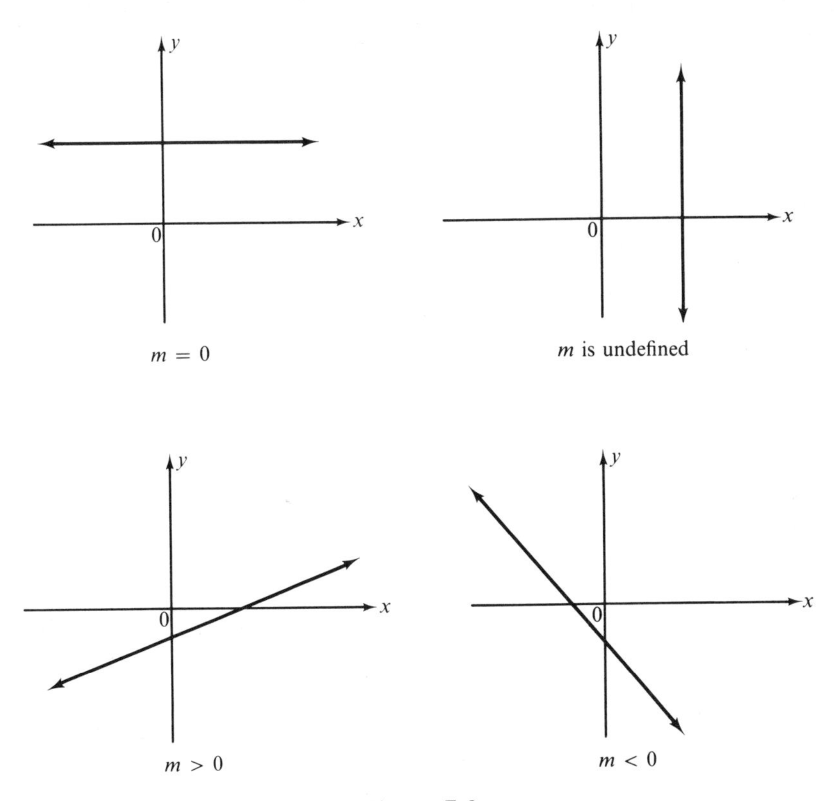

Figure 7-2

There are easier ways of finding the slope of a line than by measuring the angle of inclination and finding its tangent. If we know the coordinates of any two points on the line, we can compute the slope by using the following theorem:

Theorem 7-1. Let $P_1(x_1, y_1)$ and $P_2(x_2, y_2)$ be any two points on a nonvertical line, and let m be the slope of the line. Then

$$m = \frac{y_2 - y_1}{x_2 - x_1}$$

Proof: We will consider two cases, according to whether the angle of inclination is acute or obtuse.

Case 1. If $0 < \theta < \pi/2$, [Figure 7-3 (a)] then triangle P_1MP_2 is a right angle and the angle at P_1 has measure θ since the line P_1M is parallel to the x-axis. Thus, $\tan\theta = P_2M/P_1M = (y_2 - y_1)/(x_2 - x_1)$.

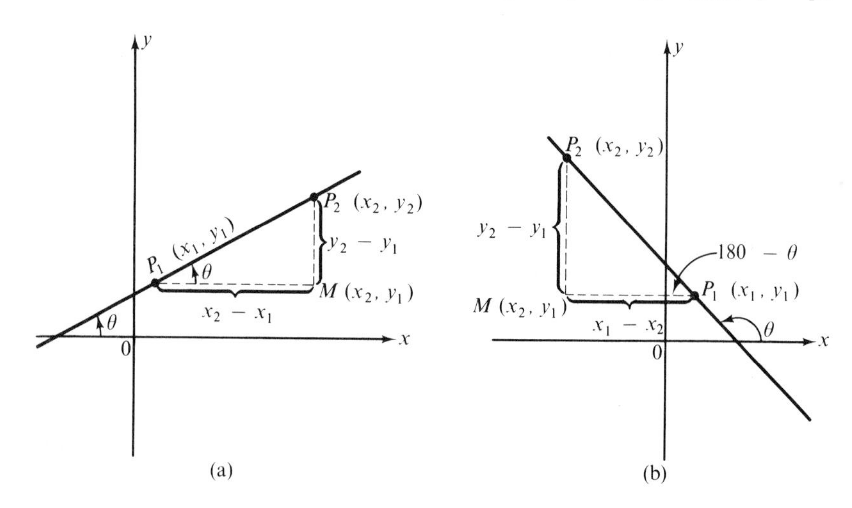

Figure 7-3

Case 2. If $\pi/2 < \theta < \pi$ [Figure 7-3 (b)], then the angle at P_1 has measure $180° - \theta$ and $\tan(180° - \theta) = -\tan\theta = (y_2 - y_1)/(x_1 - x_2)$. Thus, $\tan\theta = -(y_2 - y_1)/(x_1 - x_2) = (y_2 - y_1)/(x_2 - x_1)$.

If the line is horizontal, then $y_1 = y_2$ and $m = 0$. The slope of a vertical line is undefined. ■

Note that if we interchange the positions of P_1 and P_2 then $\tan\theta = (y_1 - y_2)/(x_1 - x_2) = -(y_2 - y_1)/-(x_2 - x_1) = (y_2 - y_1)/(x_2 - x_1)$ as before. Thus, the slope will be the same no matter which two points on the line we choose or in what order we choose them.

Example 7-1. Find the slope of the line which passes through the points $P_1(3, 1)$ and $P_2(0, -2)$.

Solution: $m = \dfrac{(y_2 - y_1)}{(x_2 - x_1)} = \dfrac{(-2 - 1)}{(0 - 3)} = \dfrac{-3}{-3} = 1$

Example 7-2. Sketch the line through the point (1, 0) whose slope is 2/3.

Solution: Since the slope is the ratio of the change in the y-coordinate to the change in the x-coordinate, this line increases 2 units vertically for every 3 units it increases horizontally. Starting at the point (1, 0) we move 2 units vertically and 3 units horizontally to find a second point (4, 2) on the line. (See Figure 7-4.)

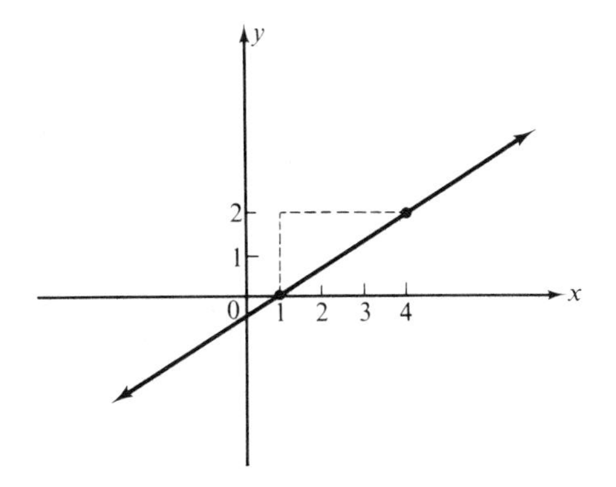

Figure 7-4

If we know the slope of two lines we can determine if the lines are parallel or perpendicular. This is the subject of the next two theorems.

Theorem 7-2. Two distinct lines L_1 and L_2 having slopes m_1 and m_2 respectively are parallel if and only if $m_1 = m_2$.

Proof: First we consider the case in which L_1 and L_2 are horizontal lines. Then $m_1 = m_2 = 0$. Conversely, if $m_1 = m_2 = 0$, the lines are both horizontal, hence parallel.

Suppose that L_1 and L_2 are two parallel lines which are neither horizontal nor vertical. (See Figure 7-5.) Then the x-axis is a transversal cutting these two parallel lines and from elementary geometry we know that their angles of inclination, being corresponding angles, must be congruent. Thus, $\theta_1 = \theta_2$, hence $\tan \theta_1 = \tan \theta_2$ and $m_1 = m_2$.

Conversely, if $m_1 = m_2$, then $\tan \theta_1 = \tan \theta_2$. Since the tangent function is one-to-one for $0 < \theta < \pi$ $(\theta \neq \pi/2)$ we must have $\theta_1 = \theta_2$ and the two lines are parallel. ■

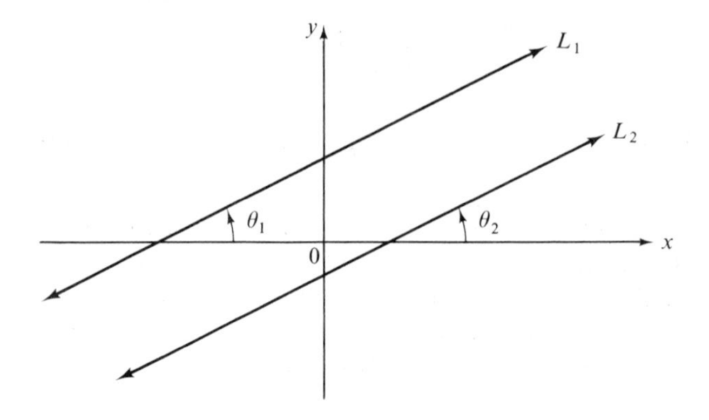

Figure 7-5

Theorem 7-3. Two lines L_1 and L_2, neither of which is vertical, whose slopes are m_1 and m_2 respectively are perpendicular if and only if $m_1 = -1/m_2$, or equivalently $m_1 m_2 = -1$.

Proof: Suppose that two lines L_1 and L_2, neither of which is vertical, are perpendicular. (See Figure 7-6.) Then the angle of

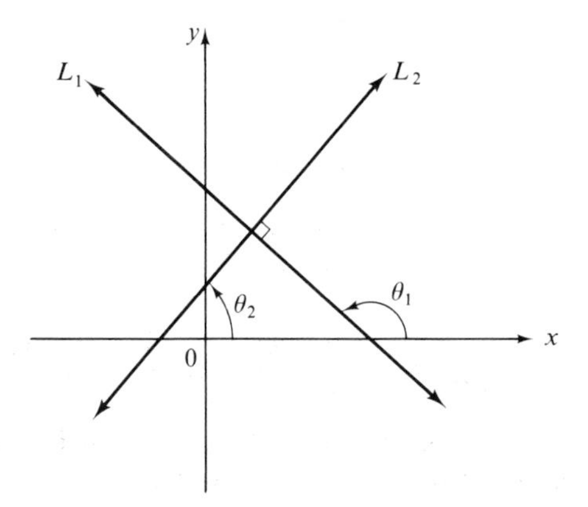

Figure 7-6

inclination of one is acute, while the other is obtuse. Let us assume that the angle of inclination of L_2 is acute. Then $\theta_1 = \theta_2 + \pi/2$ and

$$\begin{aligned}\tan\theta_1 &= \tan\left(\theta_2 + \frac{\pi}{2}\right)\\ &= \tan\left[\pi - \left(\frac{\pi}{2} - \theta_2\right)\right]\end{aligned}$$

$$= -\tan\left(\frac{\pi}{2} - \theta_2\right)$$
$$= -\cot\theta_2$$
$$= -\frac{1}{\tan\theta_2}$$

Thus, $m_1 = -1/m_2$.

Conversely, if $m_1 = -1/m_2$ then $\tan\theta_1 = -1/\tan\theta_2 = -\cot\theta_2 = \tan[\theta_2 + (\pi/2)]$, and $\theta_1 = \theta_2 + \pi/2$. Thus, the lines are perpendicular. ■

Example 7-3. The line through points (5, 2) and (6, −2) is parallel to the line through (1, 1) and (−1, 9) since the slope of both lines is −4.

Example 7-4. The line through points (5, 2) and (6, −2) is perpendicular to the line through (1, 1) and (5, 2) since the slope of the second line is 1/4.

Exercise 7-2

1. Find the slope of the line passing through each of the following pairs of points. Sketch each.
   (a) (1, 2), (5, 3)
   (b) (0, −2), (−1, 4)
   (c) (2, 3), (2, −3)
   (d) (7, −1), (1, −1)
   (e) (1, 1), (−4, 5)
   (f) (−4, −3), (1, −2)
   (g) (−1, 3), (−1, −4)
   (h) (2, −2), (3, 2)

2. If P_1 and P_2 are points on a line whose slope is m, find the missing coordinate.
   (a) $P_1(2, 1)$, $P_2(3, \_)$, $m = 2$
   (b) $P_1(-1, 4)$, $P_2(\_, 1)$, $m = -1$
   (c) $P_1(-3, \_)$, $P_2(0, 0)$, $m = 2/3$
   (d) $P_1(\_, 1)$, $P_2(5, 9)$, $m = -4/3$

3. For each of the following draw a line through the point given with the slope indicated. Give the coordinates of a second point on the line.
   (a) (1, 0), $m = 1/2$
   (b) (1, 1), $m = 3$
   (c) (−1, 2), $m = -1/2$
   (d) (0, 1), $m = -1$
   (e) (−1, −3), $m = -2/3$
   (f) (2, −5), $m = 3/2$

4. Use slopes to show that the points $P_1(0, 1)$, $P_2(2, -3)$, $P_3(3, -5)$ are collinear.

5. Use slopes to show that the points $P_1(-2, -3)$, $P_2(0, -2)$, $P_3(4, 0)$ are collinear.

6. Which of the following lines are parallel and which are perpendicular?
   (a) a line through $(4, 2)$ and $(5, -1)$
   (b) a line through $(4, -5)$ and $(1, 0)$
   (c) a line through $(-3, 7)$ and $(0, -2)$
   (d) a line through $(0, -2)$ and $(3, -1)$
   (e) a line through $(-1, 4)$ and $(1, 10)$

7. Find the missing coordinate in each of the following:
   (a) The line through $(2, 8)$ and $(-1, 4)$ is parallel to one through $(4, 1)$ and $(x, 5)$.
   (b) The line through $(5, 6)$ and $(2, 3)$ is perpendicular to one through $(0, y)$ and $(-1, 2)$.
   (c) The line through $(-1, 2)$ and $(4, 3)$ is parallel to one through $(2, 1)$ and $(3, y)$.
   (d) The line through $(-1, 2)$ and $(4, 3)$ is perpendicular to one through $(2, 1)$ and $(3, y)$.

8. Show that $P_1(2, 1)$, $P_2(3, 4)$ and $P_3(5, 0)$ are the vertices of a right triangle.

9. Use slopes to show that $P_1(-4, 3)$, $P_2(4, 1)$, $P_3(3, -2)$, and $P_4(-5, 0)$ are the vertices of a parallelogram.

7-3 Equations of Lines

An equation of the form $Ax + By + C = 0$, where A and B are not both zero, is called a *linear equation.* It is called linear because its graph is a line. In this section we will verify this statement and we will study several different forms of the linear equation. We will learn how to find an equation for a line given certain geometrical information such as two points on the line, or one point and the slope of the line.

If a line is parallel to the y-axis then every point on the line has the same x-coordinate and its equation can be written $x = h$, where h is a constant. Note that the variable y does not appear in this equation. It does not appear because no matter what value y may assume, x is still equal to the constant h. Thus, x does not depend on y. Similarly every point on a horizontal line has the same y-coordinate and its equation has the form $y = k$, where k is a constant.

Example 7-5. Find the equation of a vertical line which passes through the point $(2, 3)$.

Solution: Since the line is parallel to the y-axis [Figure 7-7 (a)] every point on the line has x-coordinate 2 and its equation is $x = 2$.

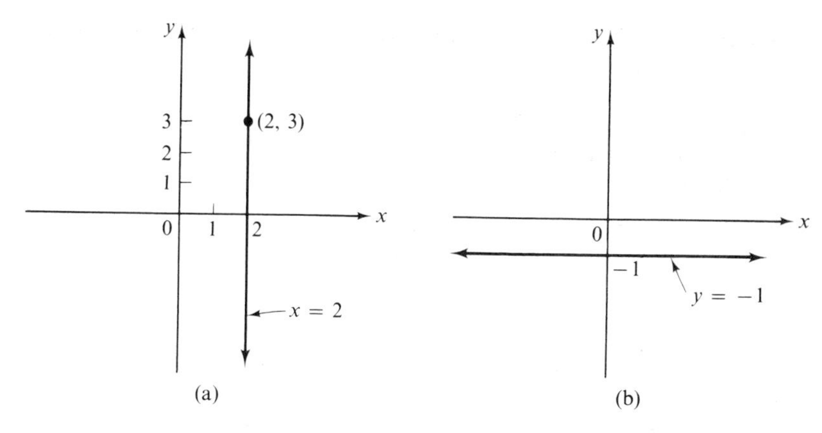

Figure 7-7

Example 7-6. Find the equation of a horizontal line one unit below the x-axis.

Solution: If this line lies one unit below the x-axis then the y-coordinate of every point on the line is -1 and its equation is $y=-1$. [See Figure 7-7 (b).]

If a line is not vertical, then it has some slope m. We have already seen in Example 7-2 that if we know the slope and one point on the line, we can sketch a unique line. To find an equation for this line we need the following theorem:

Theorem 7-4. Let L be a nonvertical line with slope m passing through the point $P_0(x_0, y_0)$. Then an equation for L is

$$y - y_0 = m(x - x_0) \tag{7-1}$$

Note: To show that this is an equation for L we must show two things. First, that every point on L satisfies this equation, and second, that any ordered pair (x, y) that satisfies this equation must be a point on L.

Proof: Let (x, y) be any point on L such that $x \neq x_0$. Then since the slope of the line is m and (x_0, y_o) and (x, y) are two points on the line, $m = (y - y_0)/(x - x_0)$, or $y - y_0 = m(x - x_0)$, and (x, y) satisfies Equation (7-1). If $x = x_0$, then since the line is not vertical, $y = y_0$, and the point must be P_0. Clearly P_0 satisfies Equation (7-1), since if we substitute $x = x_0$ and $y = y_0$ the equation becomes $0 = m(0)$. Thus, every point on L satisfies Equation (7-1).

Conversely, suppose (x, y) is some point which satisfies equation (7-1). If $x = x_0$, then $y - y_0 = 0$ and $y = y_0$ and (x, y) is the point P_0 which is on L. If $x \neq x_0$, then we can rewrite Equation (7-1) as $m = (y - y_0)/(x - x_0)$ and (x, y) must be a point on the line through (x_0, y_0) with slope m. Thus, (x, y) is on L.

We conclude that $y - y_0 = m(x - x_0)$ is an equation for L. ■

This is called the *point slope* form of the line.

Example 7-7. Find an equation for the line with slope 1/2 passing through the point $(-1, 2)$.

Solution: By Theorem 7-4 an equation for the line is

$$y - 2 = \left(\frac{1}{2}\right)(x + 1)$$

or equivalently,

$$x - 2y + 5 = 0$$

If we know two points (x_1, y_1) and (x_2, y_2) on the line, and $x_2 \neq x_1$, then the slope of the line is $m = (y_2 - y_1)/(x_2 - x_1)$ and we arrive at the following corollary:

Corollary 7-1. Let L be a nonvertical line passing through the points (x_1, y_1) and (x_2, y_2). Then an equation for L is

$$y - y_1 = \frac{y_2 - y_1}{x_2 - x_1}(x - x_1)$$

This is called the *two point* form of the line.

Example 7-8. Find an equation for the line passing through the points $(2, -1)$ and $(-1, 0)$.

Solution: Using the two point form we see that an equation for this line is

$$(y - 0) = \frac{0 + 1}{-1 - 2}(x + 1) \quad \text{or} \quad (y - 0) = \left(-\frac{1}{3}\right)(x + 1)$$

Equivalently we can write the equation $x + 3y + 1 = 0$.

Probably the most useful form of the linear equation is the *slope-y-intercept* form. If a line crosses the y-axis at the point $(0, b)$, then the number b is called the y-intercept of the line. By Theorem 7-4 a line with slope m passing through the point $(0, b)$ has the equation $(y - b) = m(x - 0)$, or $y = mx + b$.

Corollary 7-2. Let L be a nonvertical line with slope m and y-intercept b. Then an equation for L is

$$y = mx + b$$

The usefulness of this form lies in the fact that if any linear equation is rewritten in this form (i.e., "solved for" y) then the coefficient of x is the slope of the line and the constant term is the y-intercept.

Example 7-9. Find the slope and the y-intercept of the line whose equation is $3x - 2y + 7 = 0$.

Solution: Solving for y in this equation, we get the equivalent equation

$$y = \left(\frac{3}{2}\right)x + \frac{7}{2}$$

from which we read off the values $m = 3/2$, $b = 7/2$.

Another way of finding the slope of this line would be to choose any two points on it and use the formula $m = (y_2 - y_1)/(x_2 - x_1)$. If we choose the points $(1, 5)$ and $(3, 8)$ then we find that $m = (8 - 5)/(3 - 1) = 3/2$.

Every equation of the form $Ax + By + C = 0$ can be rewritten as $y = (-A/B)x - C/B$ if B is not zero, or as $x = -C/A$ if B is zero. Thus, $Ax + By + C = 0$ is equivalent to an equation of the form $y = mx + b$ or $x = h$, and we have shown that these equations are the equations of lines. We conclude that $Ax + By + C = 0$ is an equation for a line and we have finally justified the name *linear* applied to it.

Exercise 7-3

1. Find an equation of the form $Ax + By + C = 0$ for each of the lines described below.
   (a) through $(0, 2)$ with slope -2
   (b) through $(-1, 3)$ with slope $1/2$
   (c) through $(0, -2)$ with slope $3/2$
   (d) through $(3, -1)$ with slope 0
   (e) through $(3, -1)$ and $(2, 4)$

(f) through $(1, -1)$ and $(1, 4)$
(g) through $(2, -1)$ and $(-3, -1)$
(h) through $(1/2, 2)$ and $(-1/3, -2)$
(i) a vertical line three units to the left of the y-axis
(j) a horizontal line through $(2, -1)$
(k) a line parallel to the y-axis through $(4, -2)$
(l) a line with slope $-1/3$ and y-intercept 5

2. Find the slope and the y-intercept of each of the following. Sketch each.
(a) $3x - y = 2$
(b) $x + 2y - 4 = 0$
(c) $y + 4 = 0$
(d) $2x + 2y = 1$
(e) $x = 2$
(f) $3x + 4y - 12 = 0$

3. Which of the following lines are parallel? Which are perpendicular?
(a) $3x + y = 7$
(b) $x - 3y = -1$
(c) $x = 5$
(d) $2x - 6y - 4 = 0$
(e) $3x = y$
(f) $3x = 21$
(g) $y + 1 = 0$
(h) $3y = x$
(i) $3y + x - 10 = 0$
(j) $3x + 3y - 1 = 0$

4. Find an equation for the line which is
(a) parallel to $x + y - 7 = 0$ and passes through the point $(-1, 2)$.
(b) perpendicular to $x + y - 7 = 0$ and passes through the point $(-1, 2)$.
(c) parallel to $x = 2/3$ and passes through the point $(4, -2)$.
(d) perpendicular to $x = 2/3$ and passes through the point $(4, -2)$.
(e) parallel to $3x - 2y + 3 = 0$ and passes through the point $(1, 1)$.
(f) perpendicular to $3x - 2y + 3 = 0$ and passes through the point $(1, 1)$.

5. Determine the number k so that $2x - 4y = 7$ and $kx + 5y = -2$ will be (a) parallel; (b) perpendicular.

6. Determine the number k so that $x - 3y + 4 = 0$ and $2x + ky + 7 = 0$ will be (a) parallel; (b) perpendicular.

7. Let L be a line which is neither horizontal nor vertical and let $(a, 0)$ and $(0, b)$ be the points where L crosses the x- and y-axes respectively. If neither a nor b is zero, show that

$$\frac{x}{a} + \frac{y}{b} = 1$$

is an equation for L. This is called the *intercept* form of the linear equation and a is called the x-intercept.

8. Find the equation of a line
(a) with x-intercept 2 and y-intercept -1.
(b) with x-intercept $-1/2$ and y-intercept $2/3$.

9. If C is a circle with center at $O(1, 1)$ passing through the point $P(-2, 5)$, find the equation of the line tangent to the circle at P. (Hint: The tangent to the circle at P is perpendicular to the radius OP.)

10. Use Figure 7-8 and the converse of the Pythagorean Theorem (if $OP_1^2 + OP_2^2 = P_1P_2^2$, then triangle OP_1P_2 is a right triangle with right angle at O) to prove that if $m_1m_2 = -1$, then L_1 and L_2 are perpendicular.

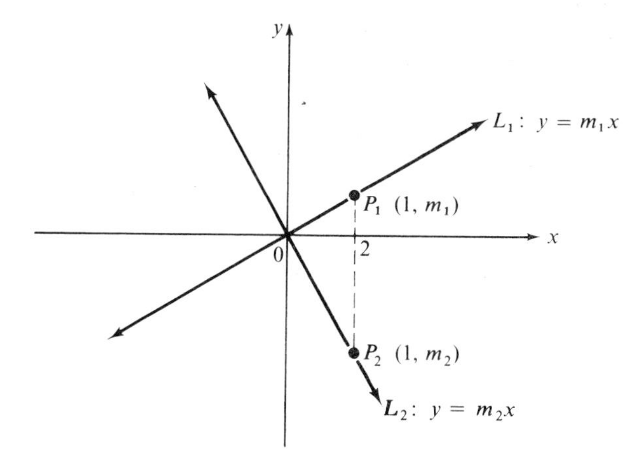

Figure 7-8

7-4 The Circle

A circle is the set of points in the plane whose distance from some fixed point O (the center) is a positive constant r (the radius). We have already seen in Chapter 1 that the distance between two points (x_1, y_1) and (x_2, y_2) in the plane is given by the formula

$$d = \sqrt{(x_1 - x_2)^2 + (y_1 - y_2)^2}$$

consequently we have the following theorem:

Theorem 7-5. Let C be a circle with center at (h, k) and radius r. Then an equation for C is

$$(x - h)^2 + (y - k)^2 = r^2 \qquad r > 0$$

Proof: Let (x, y) be a point on C. Then the distance from (x, y) to (h, k) is r, and by the distance formula

$$r = \sqrt{(x - h)^2 + (y - k)^2}$$

or

$$r^2 = (x - h)^2 + (y - k)^2$$

If $r > 0$, then these two equations are equivalent.

Conversely, suppose that (x, y) is a point which satisfies the equation

$$r^2 = (x - h)^2 + (y - k)^2$$

Then, since $r > 0$

$$r = \sqrt{(x - h)^2 + (y - k)^2}$$

and the distance from (x, y) to (h, k) is r. Therefore, (x, y) is on C. ■

Example 7-10. An equation for the circle with center $(2, -1)$ and radius $1/2$ is

$$(x - 2)^2 + (y + 1)^2 = \frac{1}{4}$$

or equivalently

$$4x^2 + 4y^2 - 16x + 8y + 19 = 0$$

We can see from this example that an equation for a circle could take the form $Ax^2 + Ay^2 + Bx + Cy + D = 0$, where the coefficients of x^2 and y^2 are the same. The question then arises, if the equation is in this form, how can we find its graph? The easiest way is to put it into the form $(x - h)^2 + (y - k)^2 = r^2$ (called *standard form*) so that the center (h, k) and the radius r can be read off from the equation. In order to do this we will need a technique known as "completing the square."

Given the two terms $x^2 + ax$, completing the square consists in finding a constant which when added to this expression will make it a perfect square. It is not hard to see that this constant must be $(a/2)^2$, since

$$x^2 + ax + \left(\frac{a}{2}\right)^2 = \left(x + \frac{a}{2}\right)^2$$

Example 7-11. Find the constant needed to complete the square for $x^2 + 7x$ and factor the resulting expression.

Solution: The constant needed is $(7/2)^2 = 49/4$, and $x^2 + 7x + 49/4 = (x + 7/2)^2$.

If the coefficient of x^2 is not one, then this coefficient should be factored out before completing the square.

Example 7-12. Find the constant needed to complete the square for $2x^2 - 3x$ and factor the resulting expression.

Solution: First we factor out the 2, getting

$$2\left(x^2 - \frac{3x}{2}\right)$$

To complete the square for $x^2 - 3x/2$ we must add $(-3/4)^2 = 9/16$ *inside the parentheses.* This is equivalent to adding $2(9/16) = 9/8$ since the expression in parentheses is multiplied by two. Thus, we have

$$2\left(x^2 - \frac{3x}{2} + \frac{9}{16}\right) = 2\left(x - \frac{3}{4}\right)^2$$

Now we are ready to change an equation from the form $Ax^2 + Ay^2 + Bx + Cy + D = 0$ to standard form.

Example 7-13. Change $x^2 + y^2 + 4x - 10y - 7 = 0$ to standard form and find its center and radius.

Solution: We must complete the square for $x^2 + 4x$ and $y^2 - 10y$ by adding $(4/2)^2 = 4$ to the first and $(-10/2)^2 = 25$ to the second. Since this is an equation, we must add these numbers to *both sides*. Thus, we have

$$x^2 + 4x + y^2 - 10y = 7$$
$$x^2 + 4x + 4 + y^2 - 10y + 25 = 7 + 4 + 25$$

and factoring, we get

$$(x + 2)^2 + (y - 5)^2 = 36$$

From this we see that the center of the circle is $(-2, 5)$ and the radius is $\sqrt{36} = 6$.

After completing the square we may find that the number on the right-hand side of the equation is zero or negative. Since this number represents the square of the radius, the graph in this case is not a circle at all but a single point or the empty set. These cases are called *degenerate forms* of the circle.

Example 7-14. Change $2x^2 + 2y^2 + 5x - y + 10 = 0$ to standard form and determine if its graph is a circle, a point, or the empty set.

Solution: First we divide through by 2 to make completing the square easier.

$$x^2 + \frac{5x}{2} + y^2 - \frac{y}{2} = -5$$

Adding $(5/4)^2$ and $(1/4)^2$ to both sides gives

$$x^2 + \frac{5x}{2} + \left(\frac{5}{4}\right)^2 + y^2 - \frac{y}{2} + \left(\frac{1}{4}\right)^2 = -5 + \frac{25}{16} + \frac{1}{16}$$

and factoring we get

$$\left(x + \frac{5}{4}\right)^2 + \left(y - \frac{1}{4}\right)^2 = -\frac{54}{16}$$

Since the constant on the right is negative while the left-hand side is never negative we conclude that there are no points (x, y) that satisfy this equation and its graph is the empty set.

Exercise 7-4

1. Find an equation of the circle
   (a) with center at $(4, -2)$ and radius 1.
   (b) with center at $(1, 0)$ and radius 1/2.
   (c) with center at $(1, 1)$ passing through the point $(4, -2)$.
   (d) with center at $(3, -1)$ passing through the point $(0, 1)$.
   (e) with center at $(3, -1)$ tangent to the x-axis.
   (f) with center at $(3, -1)$ tangent to the y-axis.
   (g) with a diameter whose end-points are $(2, 2)$ and $(2, -2)$.
   (h) which is tangent to both axes and passes through the point $(1, 2)$.

2. Find the constant needed to complete the square and factor the resulting expression.
   (a) $x^2 - 10x$ (b) $x^2 + 4x$
   (c) $x^2 + x$ (d) $3x^2 + 6x$
   (e) $2x^2 - 10x$ (f) $7x^2 + 2x$
   (g) $3x^2 - 4x$ (h) $2x^2 - x$

3. Change each of the following to standard form and determine if the graph is a circle, a point, or the empty set. If the graph is a circle, give its center and radius. Graph.
   (a) $x^2 + y^2 + 4x + 4y - 8 = 0$
   (b) $x^2 + y^2 - 6x + 2y + 10 = 0$
   (c) $x^2 + y^2 + 2x - 8y + 2 = 0$
   (d) $x^2 + y^2 + 3x - 2y + 1 = 0$
   (e) $x^2 + y^2 - 5x - 7y +.20 = 0$
   (f) $2x^2 + 2y^2 + 4x - 5 = 0$

(g) $3x^2 + 3y^2 - 7x - 2y + 4 = 0$
(h) $25x^2 + 25y^2 - 20x + 100y - 121 = 0$
(i) $5x^2 + 5y^2 + 2y - 1 = 0$

4. Find all values for k so that the graph of the equation

$$x^2 + y^2 + 2y + k = 0$$

will be (a) a circle; (b) a single point; (c) the empty set.

5. Show that the graph of $x^2 + y^2 + Bx + Cy + D = 0$ will be a circle if $D < (B^2 + C^2)/4$, a single point if $D = (B^2 + C^2)/4$, and the empty set if $D > (B^2 + C^2)/4$.

6. Use Figure 7-9 and the converse of the Pythagorean Theorem (if $BP^2 + AP^2 = AB^2$ then triangle ABP is a right triangle with right angle at P) to prove that if A and B are end-points of a diameter of a circle and if P is any point on the circle other than A or B, then APB is a right angle. (Hint: the equation of the circle is $x^2 + y^2 = r^2$.)

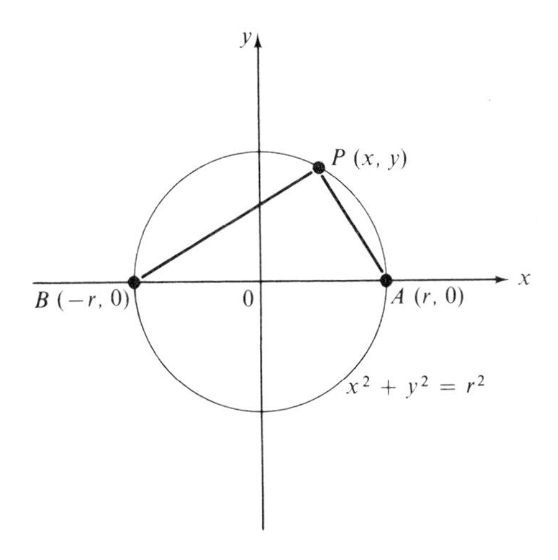

Figure 7-9

7-5 Conic Sections

The circle is an example of a family of curves called conic sections or simply conics. These curves were studied extensively by the Greeks, particularly Apollonius who lived in the third century B.C. Apollonius obtained these curves, the circle, the ellipse, the parabola, and the hyperbola, by taking the intersections of a plane and a right circular double cone.

This double cone can be generated by revolving a line which passes through the origin about the y-axis. (See Figure 7-10.)

Let α be the inclination of this generating line, $0 < \alpha < \pi/2$. Now suppose we take a plane which does not pass through the vertex of the double cone and whose inclination with the horizontal is θ.

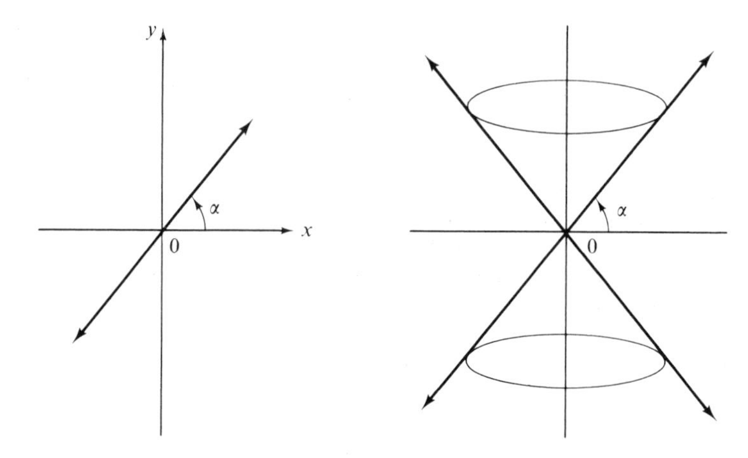

Figure 7-10

If $\theta = 0$, the plane is horizontal and its intersection with the cone is a *circle*. If $0 < \theta < \alpha$, the intersection is an *ellipse*. If $\theta = \alpha$, the plane is parallel to the generating line and the intersection is a *parabola*. If $\alpha < \theta \leq \pi/2$, the plane intersects both cones and the intersection is a *hyperbola*. (See Figure 7-11.)

If the plane passes through the vertex of the cone, the intersection will be either a single point (when $0 \leq \theta < \alpha$), a single line (when $\theta = \alpha$), or a pair of intersecting lines (when $\alpha < \theta \leq \pi/2$). These are called *degenerate forms* of the conics. Other degenerate forms are the empty set and a pair of parallel lines.

Our approach to the conics will be more algebraic than geometric. We will find that every conic has an equation of the form

$$Ax^2 + Bxy + Cy^2 + Dx + Ey + F = 0 \qquad \textbf{(7-2)}$$

where A, B, and C are not all three zero. This is called the general second degree equation.

Conversely we will see that every equation of this type has as its graph either one of the conics (a circle, an ellipse, a parabola, or a hyperbola) or some degenerate form. We have already observed that the equation of a circle takes the form of Equation (7-2) with $B = 0$ and $A = C$.

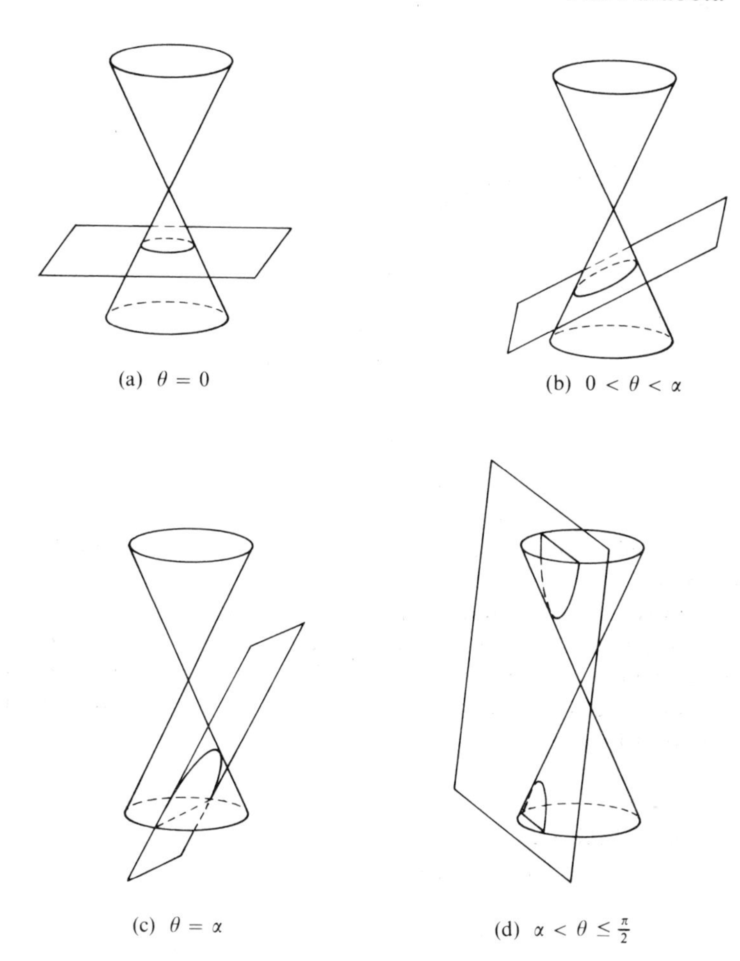

Figure 7-11

7-6 The Parabola

If D is a line and F is a point not on D, then a parabola is the set of all points P in the plane having the property that the distance from P to F is equal to the distance from P to D.

The point F is called the *focus* of the parabola and the line D is called the *directrix*. (See Figure 7-12.) The line through the focus perpendicular to the directrix is called the *axis* of the parabola and the point V where the parabola crosses the axis is called the *vertex* of the parabola.

If we choose the axis of the parabola to be the x-axis and the vertex

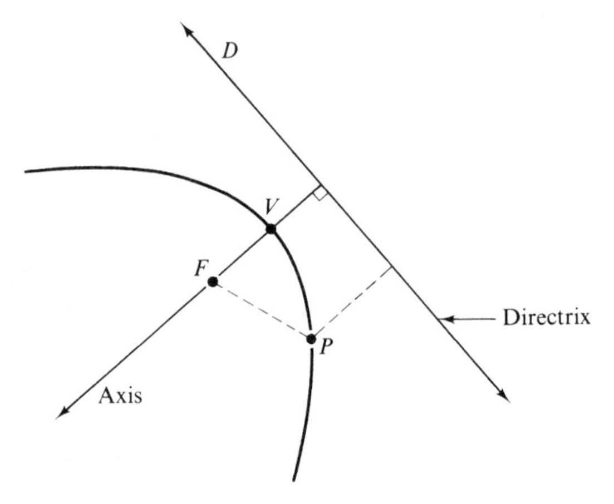

Figure 7-12

to be the origin, the equation for the parabola turns out to be particularly simple.

Let us take the focus to be the point $(a, 0)$ and the directrix the line $x = -a$, where $a > 0$. Then if (x, y) is any point on the parabola the distance from (x, y) to $(a, 0)$ is $\sqrt{(x-a)^2+y^2}$ by the distance formula. The distance from (x, y) to the line $x=-a$ is $x-(-a)=x+a$. Equating these two distances gives

$$\sqrt{(x-a)^2+y^2} = x+a$$

and squaring both sides we have

$$(x-a)^2 + y^2 = (x+a)^2$$
$$x^2 - 2ax + a^2 + y^2 = x^2 + 2ax + a^2$$

or

$$y^2 = 4ax \tag{7-3}$$

Every point on this parabola satisfies Equation (7-3). Conversely we could show by reversing our steps that every point (x, y) satisfying this equation is on the parabola.

This is the first form of the equation of the parabola. The axis of this parabola is the x-axis, the vertex is at the origin, and the parabola opens to the right. [Figure 7-13 (a).]

If we take the directrix to be $x = a$ and the focus to be $(-a, 0)$, then the equation will become

$$y^2 = -4ax \tag{7-4}$$

and the parabola will open to the left. [Figure 7-13 (b).]

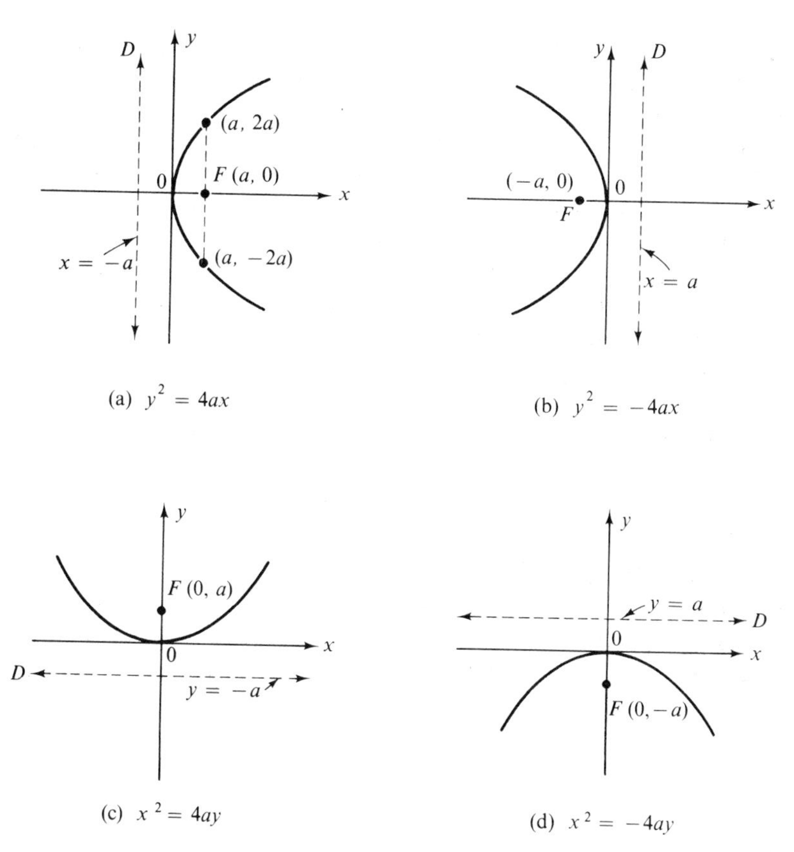

Figure 7-13

If we choose the directrix to be the line $y = -a$ and the focus the point $(0, a)$, then the axis of the parabola will be the y-axis, it opens upward, and its equation is

$$x^2 = 4ay \tag{7-5}$$

[See Figure 7-13 (c).]

If the directrix is $y = a$ and the focus $(0, -a)$ we get the fourth form of the equation for the parabola

$$x^2 = -4ay \tag{7-6}$$

This parabola opens downward. [See Figure 7-13 (d).]

Note that if the coefficient of the first degree term is positive when the equation is in one of these forms, then the parabola opens to the right or upward according to whether the first degree term is x or y.

Similarly, if the coefficient of the first degree term is negative, then the parabola opens to the left or downward according to whether the first degree term is x or y.

The line segment through F parallel to the directrix whose end-points are on the parabola is called the *latus rectum*. Its length is called the *focal width*. Interestingly enough, the focal width in each of the above cases turns out to be $4a$. We will show this for the case of Equation (7-3). In this equation, let $x = a$. Then $y^2 = 4a^2$ and $y = \pm 2a$. Thus, $(a, 2a)$ and $(a, -2a)$ are two points on the parabola which are end-points of the latus rectum and the distance between these two points is $4a$.

We can quickly graph the parabola if we know the vertex, the focus, and the focal width. Note that the parabola is symmetric with respect to its axis.

Example 7-15. Graph the parabola $y^2 = -8x$.

Solution: This equation has the form of Equation (7-4), so the vertex is at the origin, the axis of the parabola is the x-axis, and the parabola opens to the left. Since $4a = 8$, $a = 2$, and the focus is at $(-2, 0)$. The focal width is 8. [See Figure 7-14 (a).]

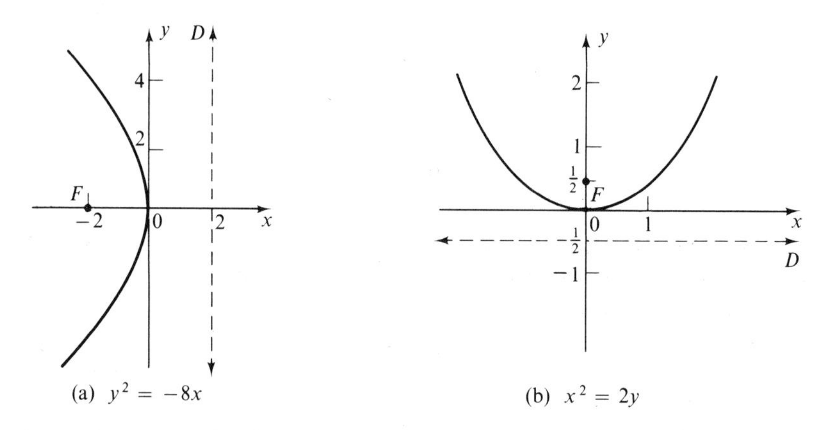

Figure 7-14

Example 7-16. Find the equation of the parabola with vertex at the origin and focus $(0, 1/2)$. Graph.

Solution: Since the vertex and focus are on the y-axis, this must be the axis of the parabola. The focus is above the vertex so the parabola opens upward. Its equation must have the form of Equation (7-5),

$x^2 = 4ay$, and $a = 1/2$. Thus, the equation is $x^2 = 2y$. [See Figure 7-14 (b).]

If we take the vertex of the parabola to be a point (h, k) in the plane and if the axis of the parabola is parallel to or lies on either the x- or the y-axis, then Equations (7-3), (7-4), (7-5), and (7-6) become

$$(y - k)^2 = 4a(x - h) \tag{7-7}$$

$$(y - k)^2 = -4a(x - h) \tag{7-8}$$

$$(x - h)^2 = 4a(y - k) \tag{7-9}$$

$$(x - h)^2 = -4a(y - k) \tag{7-10}$$

These are called standard forms of the equation of the parabola. The derivation of these forms is left for the student (Exercise 7-6, problems 12, 13, 14, and 15).

Example 7-17. Find the vertex, the focus, the equation of the directrix, the axis, and the focal width of $(x - 1)^2 = 8y$. Graph.

Solution: Since the first degree term is y and its coefficient is positive, this parabola opens upward. From the equation we see that $h = 1$, $k = 0$, so the vertex is $(1, 0)$. Since $4a = 8$, $a = 2$, and the focus must be two units above the vertex. Graphing this information we see that the coordinates of F must be $(1, 2)$ and the axis of the parabola is the line $x = 1$. The directrix is perpendicular to the axis and two units below the vertex, hence its equation is $y = -2$. The focal width is 8. [See Figure 7-15 (a).]

Example 7-18. Find the equation of the parabola with vertex $(2, -1)$ and focus $(0, -1)$. Graph.

Solution: Plotting these two points, we see that the focus is to the left of the vertex, thus the parabola must open to the left and its equation has the form of Equation (7-8). We have $h = 2$, $k = -1$, and a, the distance from F to V, is 2. Thus, the equation is

$$(y + 1)^2 = -8(x - 2)$$

[See Figure 7-15 (b).]

If we multiply out Equations (7-7), (7-8), (7-9), and (7-10), we can see that they will assume the form of the general second degree equation

$$Ax^2 + Bxy + Cy^2 + Dx + Ey + F = 0$$

where $B = 0$ and either A or C, but not both, is zero.

Conversely, any equation of this form must be the equation of a parabola (or perhaps of a degenerate form).

In order to graph a parabola whose equation is in this general form we will want to put it into one of the standard forms (7-7), (7-8), (7-9), or (7-10) so that we can read off the coordinates of the vertex and the focal width. To put the equation of the parabola in standard form we will again use the technique of completing the square.

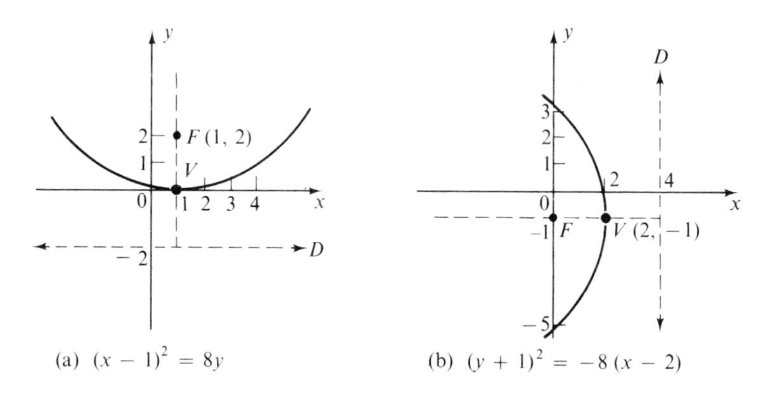

Figure 7-15

Example 7-19. Put $x^2 + 6x - 2y + 7 = 0$ in standard form, find the vertex, the focus, and the focal width.

Solution: We first rewrite the equation as

$$x^2 + 6x = 2y - 7$$

Completing the square gives

$$x^2 + 6x + 9 = 2y - 7 + 9$$

and factoring we have

$$(x + 3)^2 = 2(y + 1)$$

This has the form of Equation (7-9) and the parabola opens upward. The vertex is $(-3, -1)$, the focal width is 2. Since $4a = 2$, $a = 1/2$ and the focus is $(-3, -1/2)$.

In this section we have considered only the parabola whose axis is parallel to or coincident with either the x-axis or the y-axis. If the axis of the parabola is not parallel to one of these axes, then its equation is not so simple. Its equation will still be a general second degree equation, of course, however we will find that the equation has an xy term. We will take up this case in more detail in Section 7-10.

The parabola appears in many applications. If we rotate the parabola about its axis we generate a surface called a *paraboloid*. The "dish" antenna of a radar system, the reflecting mirror of a telescope, and the reflector of your automobile headlight are all portions of a paraboloid. This surface is used for these structures because of a very unique and interesting property. If rays parallel to the axis (these may be light rays, radio waves, sound waves) strike the reflecting inner surface of a paraboloid, they are reflected to one point, the focus (Figure 7-16). Thus, all rays striking the paraboloid are concentrated at one point, where a detection device can be placed. This also works in reverse. If a source of light is placed at the focus, then the light will be reflected so that it seems to come from the entire surface in a beam which is parallel to the axis of the paraboloid.

If air resistance is neglected, then the path of a projectile is parabolic.

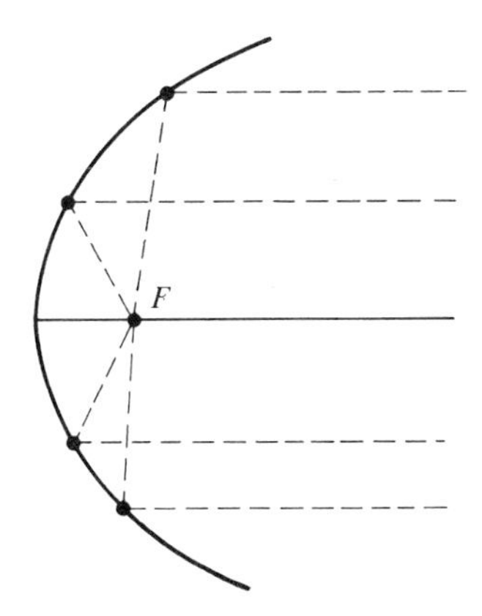

Figure 7-16

Exercise 7-6

1. Find an equation for each of the parabolas described below. Graph each.
   (a) vertex (0, 0), focus (3, 0)
   (b) vertex (0, 0), focus (0, -1)
   (c) vertex (0, 0), directrix $x = 1$
   (d) vertex (0, 0), directrix $y = -1/2$

(e) focus $(-8, 0)$, directrix $x = 8$
(f) vertex $(0, 3)$, focus $(0, 0)$
(g) vertex $(-3, 5)$, focus $(-1, 5)$
(h) focus $(0, 3)$, directrix $x = 2$
(i) focus $(2, 2)$, directrix $y = 6$
(j) vertex $(4, -2)$, focus $(4, 3)$

2. For each of the following find the vertex, the focus, the equation of the directrix, the axis, and the focal width. Sketch each.
   (a) $y^2 = -2x$
   (b) $x^2 = 3y$
   (c) $y^2 = (1/4)x$
   (d) $x^2 = -10y$
   (e) $y^2 = 2(x + 1)$
   (f) $(x - 1)^2 = -(y + 3)$
   (g) $(y + 2)^2 = -12(x - 3)$
   (h) $(x + 2)^2 = 16(y - 1)$
   (i) $(y - 1)^2 = x$
   (j) $x^2 = 3(y + 1)$

3. Put each of the following in standard form, find the vertex, the focus, the focal width, and graph.
   (a) $y^2 - 2x - 1 = 0$
   (b) $x^2 - 2x + 8y + 9 = 0$
   (c) $y^2 + 6y + 4x + 1 = 0$
   (d) $y^2 - 7y + 2x = 0$
   (e) $x^2 + 3x - 7y + 2 = 0$
   (f) $2x^2 + x - y = 0$
   (g) $4x^2 - 10x + 2y = 0$
   (h) $3y^2 - 2y - 4x + 1 = 0$

4. The graph of each of the following is a degenerate form of the parabola. Describe the graph of each.
   (a) $x^2 - 2x + 1 = 0$
   (b) $x^2 - 1 = 0$
   (c) $y^2 + 2y - 3 = 0$
   (d) $x^2 + 2x + 2 = 0$

5. Find the coordinates of the focus, the equation of the directrix, and the equation of the axis of the parabolas whose equations are (7-7), (7-8), (7-9), and (7-10).

6. Find the equations of all parabolas having vertex at $(1, -2)$ and focal width 3. Sketch each.

7. Find the equations of all parabolas having focus at $(-1, 1)$ and focal width 2. Sketch each.

8. Find the equations of all parabolas having directrix $x = 2$, axis $y = -1$ and focal width 1. Sketch each.

9. Find the equations of all parabolas having directrix $y = -3$, axis $x = 0$ and focal width 8. Sketch each.

10. Find the equation of a parabola with vertex at the origin, axis the y-axis, passing through the point $(1, 3)$.

11. A parabolic arch is 50 feet wide at the base and 25 feet high. Find the width of the arch at a point 5 feet above the base. (Hint: Place the parabola on a set of coordinate axes with vertex at the origin and axis the y-axis and find its equation.)

12. Derive Equation (7-7) by taking the vertex of the parabola to be (h, k), the focus a units to the right, and the directrix a units to the left of the vertex. (Hint: See the derivation of Equation (7-3).)

13. Derive Equation (7-8) by taking the vertex of the parabola to be (h, k), the focus a units to the left, and the directrix a units to the right of the vertex.

14. Derive Equation (7-9) by taking the vertex of the parabola to be (h, k), the focus a units above, and the directrix a units below the vertex.

15. Derive Equation (7-10) by taking the vertex of the parabola to be (h, k), the focus a units below, and the directrix a units above the vertex.

7-7 The Ellipse

If F_1 and F_2 are two distinct points, then an ellipse is the set of all points P in the plane having the property that the sum of the distances from P to F_1 and from P to F_2 is a constant.

The two points F_1 and F_2 are called the *foci* of the ellipse. The line segment with end-points on the ellipse passing through the foci is called the *major axis,* and the line segment with end-points on the ellipse and perpendicular to the major axis at its midpoint is called the *minor axis*. The intersection of these two axes is the *center* of the ellipse, and the end-points of the major axis are its *vertices.*

If we place our ellipse so that its center is at the origin and its major axis is on the x-axis, then we arrive at a particularly simple equation for the ellipse.

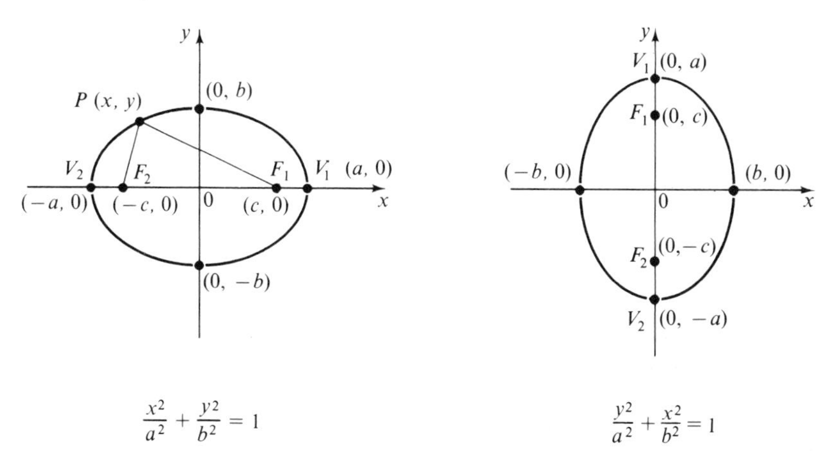

Figure 7-17

Let us take the foci to be the points $F_1(c, 0)$ and $F_2(-c, 0)$, and let $P(x, y)$ be any point on the ellipse. (See Figure 7-17.) Then the distance from P to F_1 is $\sqrt{(x-c)^2+y^2}$ and the distance from P to F_2 is $\sqrt{(x+c)^2+y^2}$. Since the sum of these distances is to be a constant, let us take this constant to be $2a$, where $2a > 2c$, because the sum of the lengths of two sides of a triangle is larger than the length of the third side. Thus, we have

$$\sqrt{(x-c)^2+y^2} + \sqrt{(x+c)^2+y^2} = 2a$$

Transposing and squaring both sides gives us

$$(x-c)^2 + y^2 = 4a^2 - 4a\sqrt{(x+c)^2+y^2} + (x+c)^2 + y^2$$

Simplifying gives

$$a\sqrt{(x+c)^2+y^2} = cx + a^2$$

and squaring again, we have

$$a^2[(x+c)^2 + y^2] = c^2x^2 + 2a^2cx + a^4$$

Simplifying again

$$(a^2 - c^2)x^2 + a^2y^2 = a^2(a^2 - c^2)$$

Dividing through by $a^2(a^2 - c^2)$ we arrive at

$$\frac{x^2}{a^2} + \frac{y^2}{a^2 - c^2} = 1$$

Since $a > c$, $a^2 - c^2$ is positive, so we set $b^2 = a^2 - c^2$ and the equation becomes

$$\frac{x^2}{a^2} + \frac{y^2}{b^2} = 1 \qquad \textbf{(7-11)}$$

There are several interesting points to be noted here.

1. Since we have defined $b^2 = a^2 - c^2$, then $a > b$.
2. When $y = 0$, $x = \pm a$, hence the coordinates of the vertices are $(a, 0)$ and $(-a, 0)$ and the length of the major axis is $2a$.
3. When $x = 0$, $y = \pm b$, hence the length of the minor axis is $2b$.

If we choose the major axis of the ellipse to be on the y-axis, then the equation of the ellipse will be

$$\frac{y^2}{a^2} + \frac{x^2}{b^2} = 1 \qquad \textbf{(7-12)}$$

(See Figure 7-17.)

In this case again $a > b$ and the lengths of the major and minor axes are $2a$ and $2b$ respectively. The coordinates of the vertices of this ellipse are $(0, a)$ and $(0, -a)$. Note that if the larger of the two numbers is in the denominator of the x^2 term, then the major axis lies on the x-axis, while if the larger of the two numbers is in the denominator of the y^2 term the major axis lies on the y-axis.

Example 7-20. Find an equation for the ellipse with center at the origin, one focus at (0, 3), with major axis of length 10. Graph the ellipse.

Solution: We have $c = 3$ and $2a = 10$, or $a = 5$. Since $b^2 = a^2 - c^2 = 25 - 9 = 16$, $b = 4$. Since the focus is on the y-axis, the major axis lies on the y-axis and the equation is

$$\frac{y^2}{25} + \frac{x^2}{16} = 1$$

(See Figure 7-18.)

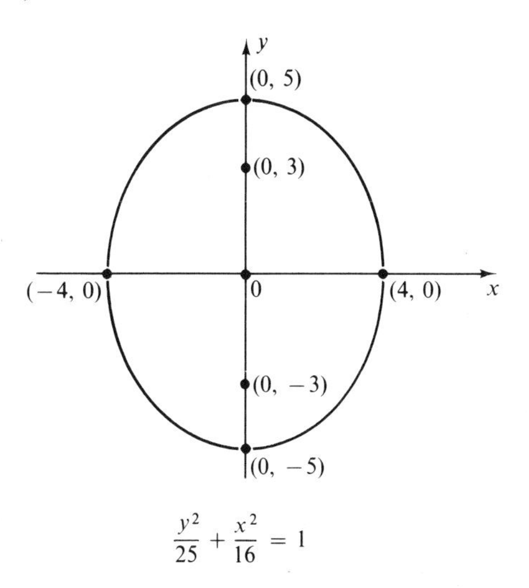

$$\frac{y^2}{25} + \frac{x^2}{16} = 1$$

Figure 7-18

If the center of the ellipse is at the point (h, k) and the major axis is parallel to or lies on either the x- or the y-axis, then Equations (7-11) and (7-12) become

$$\frac{(x-h)^2}{a^2} + \frac{(y-k)^2}{b^2} = 1 \qquad \textbf{(7-13)}$$

$$\frac{(y-k)^2}{a^2} + \frac{(x-h)^2}{b^2} = 1 \qquad \textbf{(7-14)}$$

These are called standard forms of the equation of the ellipse. Derivation of these forms is left to the student (Exercise 7-7, problems 12 and 13).

Example 7-21. Find an equation for the ellipse with center at $(1, -2)$, one focus at $(1, 1)$ and $a = 4$.

Solution: The major axis of the ellipse is on the line $x = 1$ and c must be 3, since this is the distance from the center $(1, -2)$ to the focus $(1, 1)$. Then $b^2 = 16 - 9 = 7$ and the equation must be

$$\frac{(y+2)^2}{16} + \frac{(x-1)^2}{7} = 1$$

Note that since the major axis is parallel to the y-axis the larger of the two numbers, a^2, is in the denominator of the y^2 term. [See Figure 7-19(a).]

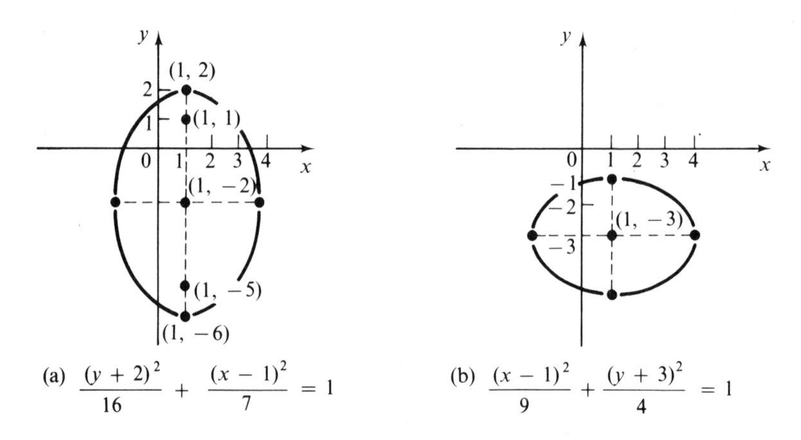

Figure 7-19

Example 7-22. Graph

$$\frac{(x-1)^2}{9} + \frac{(y+3)^2}{4} = 1$$

Solution: The center of this ellipse is at $(1, -3)$, its major axis is parallel to the x-axis and $a = 3$, $b = 2$. (We repeat that always $a > b$.) To find the vertices, we move three units to the right and to the left of the center. The ends of the minor axis are two units above and below the center. [See Figure 7-19(b).]

The *eccentricity* of an ellipse is defined to be the ratio of c to a.

$$e = \frac{c}{a}$$

Since $0 < c < a$, the eccentricity of an ellipse is a number between zero and one. The eccentricity is a measure of how "circular" the ellipse is. If e is very close to zero, then c is very small and since $b^2 = a^2 - c^2$, b and a are very nearly the same. Thus, for e close to zero the ellipse looks very much like a circle. Although it does not fit our definition, the circle is sometimes considered to be a special case of the ellipse with eccentricity equal to zero.

If e is close to one, then c and a are very nearly the same and b is close to zero. Thus, for eccentricities close to one, the ellipse is very elongated. (See Figure 7-20.)

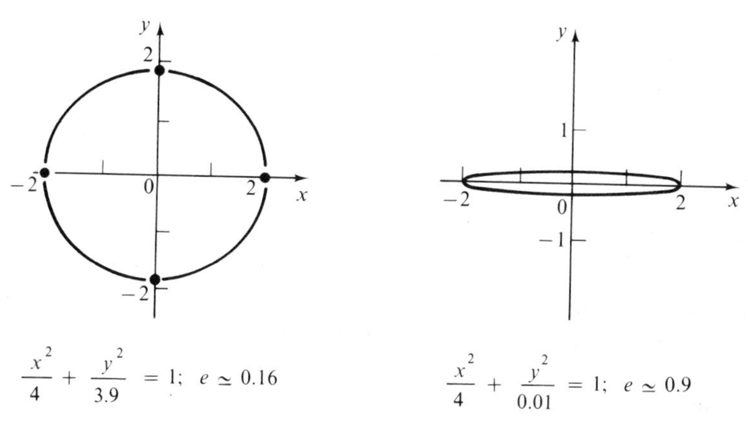

Figure 7-20

Example 7-23. Given the ellipse with equation $9x^2 + 25y^2 = 225$, find its major axis, its eccentricity, the coordinates of the foci and vertices, and graph the ellipse.

Solution: First we must put this equation in standard form by dividing through by 225. This gives

$$\frac{x^2}{25} + \frac{y^2}{9} = 1$$

Since $a > b$, $a = 5$, $b = 3$ and the major axis is along the x-axis. As $c^2 = a^2 - b^2$, $c^2 = 25 - 9 = 16$ and $c = 4$. Thus, $e = c/a = 4/5$. The foci are at (4, 0) and (−4, 0) and the vertices are at (5, 0) and (−5, 0). [See Figure 7-21(a).]

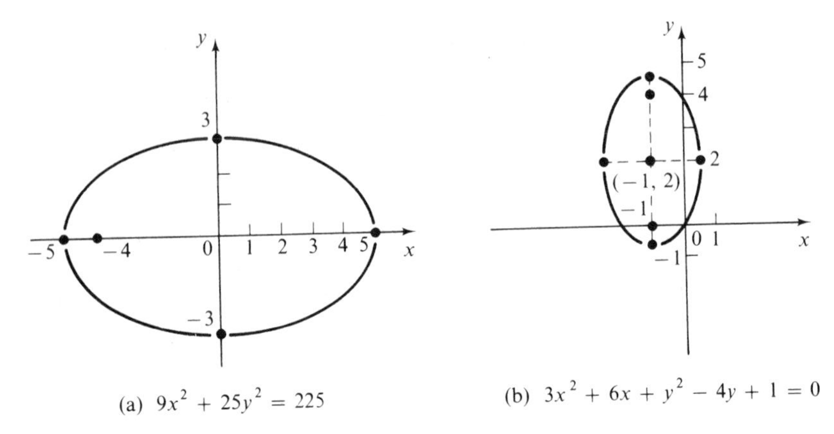

(a) $9x^2 + 25y^2 = 225$

(b) $3x^2 + 6x + y^2 - 4y + 1 = 0$

Figure 7-21

If we multiply out Equations (7-13) and (7-14) we can see that they will have the form of the general second degree equation

$$Ax^2 + Bxy + Cy^2 + Dx + Ey + F = 0$$

where $B = 0$ and A and C have the same sign, $A \neq C$. (If $A = C$ then the graph is a circle.)

Conversely, any equation of this form will be the equation of an ellipse (or perhaps of a degenerate form). If the equation is in this general form we can put it into one of the standard forms (7-13) or (7-14) by again using the technique of completing the square.

Example 7-24. Put $3x^2 + 6x + y^2 - 4y + 1 = 0$ into standard form, find its center, its foci and vertices, its eccentricity, and then graph it.

Solution: First we rewrite the equation, factoring a three out of the first two terms.

$$3(x^2 + 2x) + (y^2 - 4y) = -1$$

We complete the square on x by adding one inside the parentheses, which is equivalent to adding 3.

$$3(x^2 + 2x + 1) + (y^2 - 4y + 4) = -1 + 3 + 4$$

Factoring

$$3(x + 1)^2 + (y - 2)^2 = 6$$

Dividing through by 6, we get

$$\frac{(x + 1)^2}{2} + \frac{(y - 2)^2}{6} = 1$$

Since a^2, the larger number, is in the denominator of the y^2 term, the major axis is parallel to the y-axis. The center is at $(-1, 2)$, $a = \sqrt{6}$, $b = \sqrt{2}$, and $c = 2$. The foci are at $(-1, 4)$ and $(-1, 0)$; the vertices at $(-1, 2 + \sqrt{6})$ and $(-1, 2 - \sqrt{6})$. The eccentricity is $2/\sqrt{6}$. [See Figure 7-21(b).]

We have considered here only ellipses whose major axis is parallel to or lies on either the x- or the y-axis. If this is not the case then the equation will not have the form of Equation (7-13) or (7-14) but will be a general second degree equation with an xy term present. We will take up this case in Section 7-10.

Exercise 7-7

1. For each of the following ellipses, find the coordinates of the center, the foci, and the vertices; find the eccentricity, and graph.
   (a) $x^2 + y^2/9 = 1$
   (b) $x^2/9 + y^2 = 1$
   (c) $x^2/5 + y^2/4 = 1$
   (d) $(x - 1)^2/10 + y^2 = 1$
   (e) $(x + 2)^2/9 + (y + 3)^2/25 = 1$
   (f) $(x - 1)^2/16 + (y + 2)^2/7 = 1$
   (g) $(x + 1)^2/4 + y^2/16 = 1$
   (h) $x^2/25 + (y + 3)^2/169 = 1$

2. Find an equation for each ellipse described below. Graph.
   (a) $a = 6$, $b = 4$, center at the origin, major axis along the x-axis
   (b) $a = 5$, $b = 1$, center at the origin, major axis along the y-axis
   (c) one focus at $(-2, 0)$, center at the origin, $b = 3$
   (d) one focus at $(0, -1)$, center at the origin, $a = 2$
   (e) center at $(-2, 3)$, one focus at $(0, 3)$, eccentricity $2/3$
   (f) foci at $(2, 1)$ and $(2, 5)$, eccentricity $2/5$
   (g) vertices at $(8, 4)$ and $(-2, 4)$, end-points of the minor axis have coordinates $(3, 5)$ and $(3, 3)$
   (h) foci at $(-5, -3)$ and $(1, -3)$, $b = 1$
   (i) center at $(0, -4)$, eccentricity $1/2$, one vertex at $(4, -4)$
   (j) foci at $(-5, -7)$ and $(-5, -1)$, $a = 4$
   (k) one focus at $(2, 1)$, one vertex at $(3, 1)$, eccentricity $1/2$

3. Find the coordinates of the foci and the vertices for Equations (7-13) and (7-14).

4. Put the following in standard form and graph:
   (a) $3x^2 + 4y^2 = 12$
   (b) $16x^2 + 9y^2 = 144$
   (c) $2x^2 + y^2 + 4y = 0$
   (d) $x^2 + 9y^2 - 8x + 7 = 0$

(e) $9x^2 + 4y^2 - 18x + 16y = 11$
(f) $4x^2 + y^2 + 16x - 4y + 16 = 0$
(g) $4x^2 + 12y^2 - 4x - 24y + 1 = 0$
(h) $5x^2 + 9y^2 - 10x - 18y = 45$

5. Find the equations of all ellipses with one focus at the origin, major axis of length $2a$, minor axis of length $2b$, axes parallel to the coordinate axes.

6. If we define the focal width of an ellipse to be the length of a line segment through a focus with end-points on the ellipse, show that the focal width of the ellipse $(x^2/a^2) + (y^2/b^2) = 1$ is $2b^2/a$.

7. A rectangle is enscribed in the ellipse $(x^2/a^2) + (y^2/b^2) = 1$ so that two of its opposite sides pass through the foci and its sides are parallel to the axes of the ellipse. Find the area of the rectangle.

8. The graphs of each of the following are degenerate forms of the ellipse. Describe the graph of each.
(a) $2x^2 + y^2 = 0$
(b) $x^2 - 2x + 4y^2 + 5 = 0$
(c) $x^2 + 5y^2 + 10y + 5 = 0$

9. The earth moves around the sun in an elliptical path whose eccentricity is approximately .01673. Is this path nearly circular or is it elongated?

10. An ellipse has foci at $F_1(1, 0)$ and $F_2(0, 1)$, and $a = 2$. Use the definition of an ellipse to find an equation for this conic. Graph. (Hint: Let $P(x, y)$ be any point on the ellipse. Then $PF_1 + PF_2 = 2a$.)

11. Find the eccentricity of an ellipse if the two lines joining one end of the minor axis to the two foci are perpendicular.

12. Derive Equation (7-13) by taking the center of the ellipse to be (h, k) and the two foci to be c units to the right and to the left of the center. (Hint: See the derivation of (7-11).)

13. Derive Equation (7-14) by taking the center of the ellipse to be (h, k) and the two foci to be c units above and below the center.

7-8 The Hyperbola

If F_1 and F_2 are two distinct points, then a hyperbola is the set of all points P in the plane having the property that the difference of the distances from P to F_1 and from P to F_2 is a constant.

The two points F_1 and F_2 are called the *foci* of the hyperbola. From Figure 7-22 we see that the graph is in two parts, or *branches*. The points of intersection of the hyperbola with the line through the foci

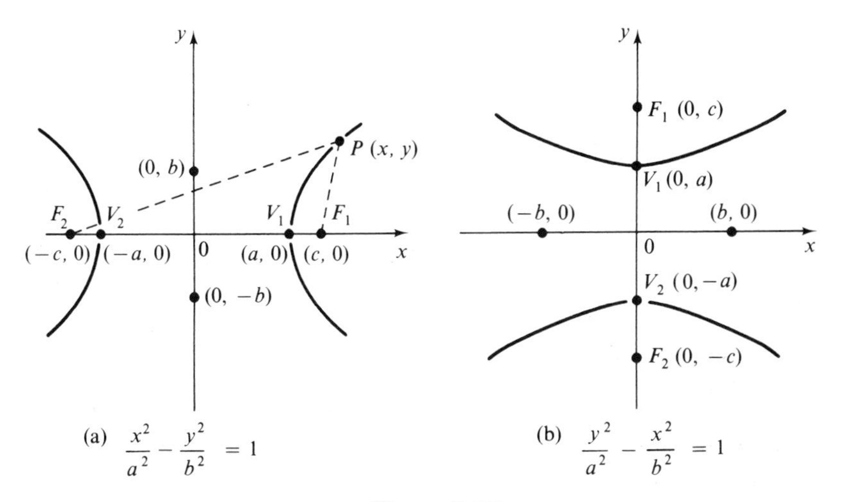

Figure 7-22

are called the *vertices* of the hyperbola. The line segment whose endpoints are the vertices is called the *transverse axis* and its midpoint is the *center* of the hyperbola.

The development of an equation for the hyperbola is very similar to that of the ellipse. Again we take the foci to have coordinates $(c, 0)$ and $(-c, 0)$ so that the center is at the origin and the transverse axis lies on the x-axis. If the difference of the distances is $2a$ and $P(x, y)$ is any point on the hyperbola, then

$$\left|\sqrt{(x-c)^2+y^2} - \sqrt{(x+c)^2+y^2}\right| = 2a$$

This time, however, $a < c$, since the difference of the lengths of two sides of a triangle is always smaller than the length of the third side.

Squaring (twice) and simplifying as before we get

$$\frac{x^2}{a^2} - \frac{y^2}{c^2 - a^2} = 1$$

Since $a < c$, $c^2 - a^2$ is a positive number and we set $b^2 = c^2 - a^2$. (Note that this is *not* the same as for the ellipse. For the ellipse we set $b^2 = a^2 - c^2$, since $a > c$ in that case.) Thus, we have

$$\frac{x^2}{a^2} - \frac{y^2}{b^2} = 1 \qquad \textbf{(7-15)}$$

If we choose the transverse axis to lie on the y-axis, the coordinates of the foci are $(0, c)$ and $(0, -c)$ and the equation becomes

$$\frac{y^2}{a^2} - \frac{x^2}{b^2} = 1 \qquad \textbf{(7-16)}$$

(See Figure 7-22.)

There are several interesting points to be noted here.

1. Since $b^2 = c^2 - a^2$, both b and a are less than c, but we have no information as to which of the two numbers, a or b, is the larger. Thus, in deciding whether the transverse axis is on the x- or the y-axis we look, not at the size of the denominators, but at the signs of the terms. If the x^2 term is positive, then the transverse axis lies on the x-axis; if the y^2 term is positive, the transverse axis lies on the y-axis.
2. When $y = 0$ in Equation (7-15), $x = \pm a$, so $(a, 0)$ and $(-a, 0)$ are the vertices of the hyperbola. Similarly in Equation (7-16) the vertices have coordinates $(0, a)$ and $(0, -a)$.
3. When $x = 0$ in Equation (7-15), $y^2 = -b^2$. These values for y are complex, thus the hyperbola does not cross the y-axis. Similarly in Equation (7-16) the hyperbola does not cross the x-axis.

The line segment whose end-points are $(0, b)$ and $(0, -b)$ in the graph of Equation (7-15) [and $(b, 0)$ and $(-b, 0)$ in the graph of Equation (7-16)] is called the *conjugate* axis of the hyperbola. Although these points are not on the graph of the hyperbola, they turn out to be extremely useful in graphing the hyperbola. If we draw a rectangle through the points $(\pm a, 0)$ and $(0, \pm b)$ whose sides are parallel to the x- and y-axes, the diagonals of this rectangle are asymptotes for the curve of Equation (7-15). The hyperbola never crosses these lines but approaches closer and closer to them as a point on the hyperbola moves away from the center. Thus, they serve as guide lines in drawing the hyperbola.

Although the parabola and one branch of the hyperbola may look very much alike, it is important to realize that they are not the same curve at all. The parabola, for instance, does not have any asymptotes. We could not fit even a small piece of a parabola onto a hyperbola any more than we could fit a line segment onto a circle.

Example 7-25. Graph

$$\frac{x^2}{4} - \frac{y^2}{9} = 1$$

and find the coordinates of its foci and vertices.

Solution: Here $a = 2$, $b = 3$ and the transverse axis lies along the x-axis. Thus, the vertices are (2, 0) and (−2, 0). Since $b^2 = c^2 - a^2$, $c^2 = a^2 + b^2 = 13$ and the foci are at $(\pm\sqrt{13}, 0)$. [See Figure 7-23(a).]

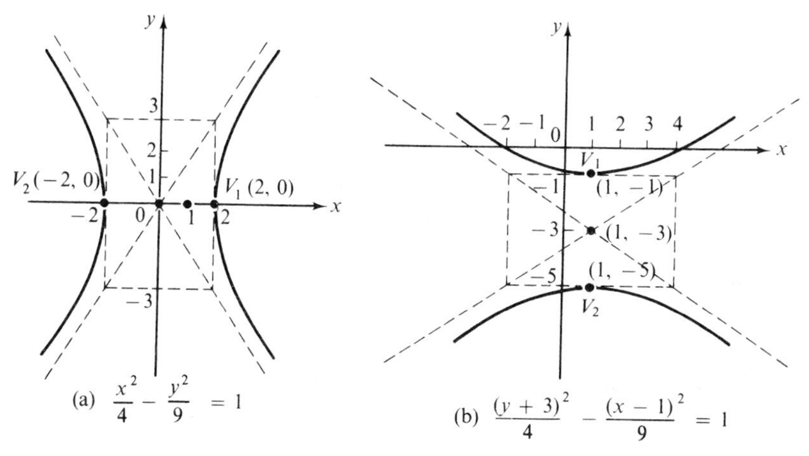

Figure 7-23

If the center of the hyperbola is at the point (h, k) and the transverse axis is parallel to or lies on either the x- or the y-axis, then Equations (7-15) and (7-16) become

$$\frac{(x-h)^2}{a^2} - \frac{(y-k)^2}{b^2} = 1 \qquad \textbf{(7-17)}$$

$$\frac{(y-k)^2}{a^2} - \frac{(x-h)^2}{b^2} = 1 \qquad \textbf{(7-18)}$$

These are the standard forms of the equation of the hyperbola. Derivation of these is left to the student (Exercise 7-8, problems 10 and 11).

Example 7-26. Find an equation for the hyperbola with center at (1, −2), one focus at (1, 1) and $a = 2$.

Solution: The transverse axis of the hyperbola is on the line $x = 1$, and $c = 3$ since this is the distance from center to focus. Since $b^2 = c^2 - a^2$, we find $b^2 = 9 - 4 = 5$ and the equation is

$$\frac{(y+2)^2}{4} - \frac{(x-1)^2}{5} = 1$$

Note that since the transverse axis is parallel to the y-axis, the sign of the y^2 term is positive, and a^2 is in the denominator of this term.

Example 7-27. Graph

$$\frac{(y+3)^2}{4} - \frac{(x-1)^2}{9} = 1$$

Solution: The center of this hyperbola is at $(1, -3)$, its transverse axis is parallel to the y-axis, and $a = 2$, $b = 3$. To graph this we sketch a rectangle with center at $(1, -3)$ and sides 3 units to the right and left of center and 2 units above and below center. [See Figure 7-23(b).]

The eccentricity of the hyperbola is defined to be c/a, as it is for the ellipse, however since $c > a$ in this case, the eccentricity of a hyperbola is always greater than one. (The eccentricity of a parabola is defined to be one.) If the eccentricity e is very close to one, then c and a are very nearly the same and b is very close to zero. (See Figure 7-24.) If e is very large, then a is relatively small compared to c, and b is large. If $a = b$, then the asymptotes are perpendicular and $e = \sqrt{2}$. This hyperbola is called an *equilateral hyperbola.*

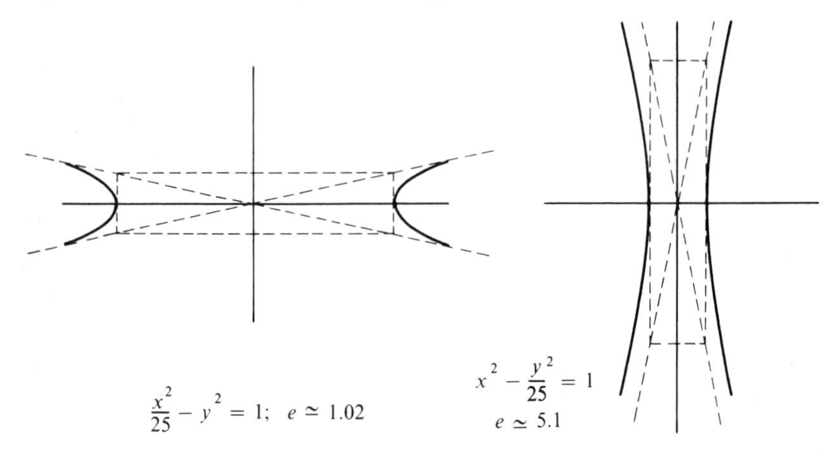

Figure 7-24

If we multiply out Equations (7-17) and (7-18) we will find that we have again the general second degree equation

$$Ax^2 + Bxy + Cy^2 + Dx + Ey + F = 0$$

where $B = 0$ and A and C have different signs. (If $A = -C$ the graph is an equilateral hyperbola.) Conversely, any equation of this form must be the equation of a hyperbola or of a degenerate form.

As in the case of the other conics, an equation in this general form can be put into standard form (7-17) or (7-18) by completing the square.

Example 7-28. Put $7x^2 - 4y^2 - 14x - 16y + 19 = 0$ into standard form.

Solution: First we rewrite the equation as

$$(7x^2 - 14x) + (-4y^2 - 16y) = -19$$

then factor a 7 out of the first two terms and a -4 out of the second two.

$$7(x^2 - 2x) - 4(y^2 + 4y) = -19$$

Completing the square we add a 1 and a 4 inside the parentheses, which is equivalent to adding 7 and -16.

$$7(x^2 - 2x + 1) - 4(y^2 + 4y + 4) = -19 + 7 - 16$$

Factoring

$$7(x - 1)^2 - 4(y + 2)^2 = -28$$

and dividing through by -28 gives

$$-\frac{(x-1)^2}{4} + \frac{(y+2)^2}{7} = 1$$

Table 7-1

	Ellipse	Hyperbola
I	$\frac{(x-h)^2}{a^2} + \frac{(y-k)^2}{b^2} = 1$ $\frac{(y-k)^2}{a^2} + \frac{(x-h)^2}{b^2} = 1$	$\frac{(x-h)^2}{a^2} - \frac{(y-k)^2}{b^2} = 1$ $\frac{(y-k)^2}{a^2} - \frac{(x-h)^2}{b^2} = 1$
II	$a > b$	either a or b may be the larger
III	The major axis is parallel to or lies on the x-axis if the larger number a^2 is in the denominator of the x^2 term when the equation is in standard form; the y-axis if the larger number is in the denominator of the y^2 term.	The transverse axis is parallel to or lies on the x-axis if the x^2 term is positive when the equation is in standard form; the y-axis if the y^2 term is positive.
IV	$c < a$ and $b^2 = a^2 - c^2$	$c > a$ and $b^2 = c^2 - a^2$
V	$e = c/a;\ 0 < e < 1$	$e = c/a;\ e > 1$

or

$$\frac{(y+2)^2}{7} - \frac{(x-1)^2}{4} = 1$$

As in the case of the parabola and the ellipse, if the transverse axis of the hyperbola neither lies on nor is parallel to the x- or y-axis, then the equation for the hyperbola will have an xy term.

The ellipse and the hyperbola have many similar features and students often confuse the two. As an aid, we offer a summary of the main differences between these two conics in Table 7-1.

The figures of the ellipse and the hyperbola occur frequently in astronomy. The student no doubt knows that the planets move around the sun in orbits which are elliptical with the sun at one focus of the ellipse. If an object enters the gravitational field of the sun its path will be an ellipse if it is captured; if it is not captured, its path will be a hyperbola.

Exercise 7-8

1. For each of the following hyperbolas, find the coordinates of the center, the foci, and the vertices; find the eccentricity, and graph.
   (a) $y^2/9 - x^2 = 1$
   (b) $x^2 - y^2/9 = 1$
   (c) $x^2/4 - y^2/5 = 1$
   (d) $(x-1)^2/8 - y^2 = 1$
   (e) $(y+2)^2/4 - (x+3)^2/4 = 1$
   (f) $(x-1)^2/4 - (y+2)^2/9 = 1$
   (g) $(y+2)^2 - (x+1)^2/8 = 1$
   (h) $y^2/4 - (x+3)^2/12 = 1$

2. Find an equation for each of the hyperbolas described below. Graph.
   (a) center at the origin, foci $(\pm 3, 0)$, $b = 2$
   (b) foci at $(0, \pm 4)$, eccentricity 2
   (c) vertices $(\pm 5, 0)$, eccentricity $\sqrt{2}$
   (d) center at $(1, 1)$, one vertex at $(1, -3)$, $b = 2$
   (e) foci at $(\pm 3, 1)$, eccentricity 1.5
   (f) foci at $(1, 0)$ and $(1, 6)$, $b = 2$
   (g) vertices at $(0, 0)$ and $(-10, 0)$, eccentricity 2
   (h) center at $(4, 1)$, one vertex at $(0, 1)$, one focus at $(10, 1)$

3. Put the following in standard form and graph:
   (a) $3x^2 - 4y^2 = 12$

(b) $5x^2 - 10y^2 + 100 = 0$
(c) $4y^2 - 8x^2 = 1$
(d) $2x^2 - 3y^2 = 3$
(e) $x^2 - 4y^2 + 4x = 0$
(f) $x^2 - y^2 - 2x - 4y - 4 = 0$
(g) $x^2 - 2y^2 + 4y = 0$
(h) $5x^2 - 4y^2 + 8y + 16 = 0$
(i) $4x^2 - 9y^2 - 8x - 54y - 113 = 0$
(j) $3x^2 - y^2 - 6x - 11 = 0$

4. Fill in the steps, omitted in this section, in going from

$$|\sqrt{(x-c)^2 + y^2} - \sqrt{(x+c)^2 + y^2}| = 2a$$

to

$$\frac{x^2}{a^2} - \frac{y^2}{c^2 - a^2} = 1$$

(Hint: If $|x - y| = k$, then $x - y = \pm k$.)

5. Two hyperbolas are said to be *conjugate hyperbolas* if the transverse axis of each is the conjugate axis of the other.
(a) Find the equation of the hyperbola which is conjugate to $x^2/9 - y^2/16 = 1$ and graph both on the same set of axes.
(b) Find the equation of the hyperbola which is conjugate to $x^2 - y^2 = 1$ and graph both on the same set of axes.

6. Show that the equations of the asymptotes of the hyperbola $x^2/a^2 - y^2/b^2 = 1$ are $bx - ay = 0$ and $bx + ay = 0$.

7. If we define the focal width of a hyperbola to be the length of a line segment through a focus with end-points on the hyperbola, show that the focal width of $x^2/a^2 - y^2/b^2 = 1$ is $2b^2/a$.

8. The graphs of each of the following are degenerate forms of the hyperbola. Describe the graph of each.
(a) $x^2 - y^2 = 0$
(b) $x^2 - y^2 - x + y = 0$

9. Use the definition to find the equation of the hyperbola with foci $F_1(\sqrt{2}, \sqrt{2})$ and $F_2(-\sqrt{2}, -\sqrt{2})$ if the difference of the distances is $2\sqrt{2}$. (Hint: Let $P(x, y)$ be any point on the hyperbola. Then $PF_1 - PF_2 = 2\sqrt{2}$.)

10. Derive Equation (7-17) by taking the center of the hyperbola to be (h, k) and the two foci to be c units to the right and to the left of the center. (Hint: See the derivation of (7-15).)

11. Derive Equation (7-18) by taking the center of the hyperbola to be (h, k) and the two foci to be c units above and below the center.

7-9 Translation of Axes

We have seen that the equation of a circle, a parabola, an ellipse, or a hyperbola is particularly simple if the vertex of the parabola or the center of the other conics is at the origin of the coordinate system. If the center of the conic (or the vertex of the parabola) is at some point (h, k) different from the origin, then the equation can be simplified by choosing a new set of coordinate axes, parallel to the original ones, so that the new origin is at the point (h, k). This is called a *translation of axes*.

Choosing a new set of axes will change the coordinates of points and thus will change the equations of curves, but it is easy to see that this translation will not change any of the intrinsic properties of the curves. A circle will still be a circle, and its radius will be the same. The values of a and b and the eccentricity of the ellipse and hyperbola will be unchanged. The focal width of the parabola will be the same.

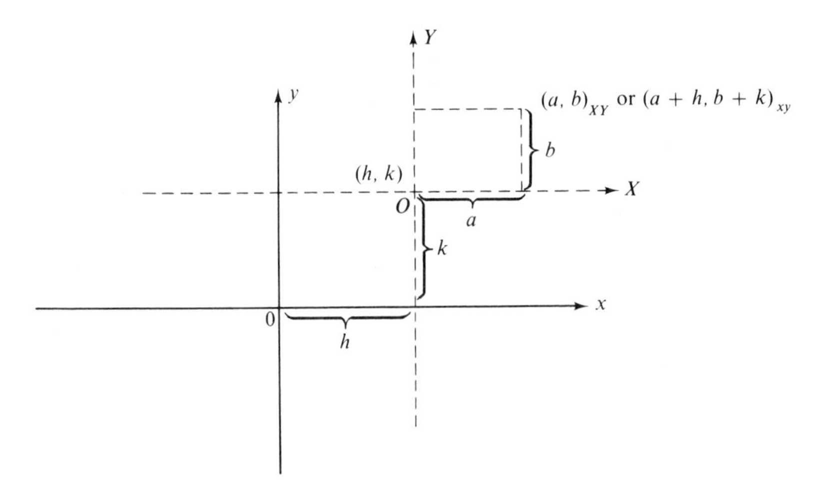

Figure 7-25

If the original axes are called the x- and y-axes, and a new set of coordinate axes, called the X- and Y-axes are chosen, parallel to the x- and y-axes respectively, so that their intersection is the point (h, k) (these coordinates being taken with respect to the x- and y-axes), then the relationship between x- and y-coordinates and X- and Y-coordinates are given by the equations

$$x - h = X$$
$$y - k = Y$$

or, equivalently

$$x = X + h$$
$$y = Y + k$$

For example, if a point has coordinates (a, b) with respect to the new X- and Y-axes, then its coordinates with respect to the old x- and y-axes will be $(a + h, b + k)$. (See Figure 7-25.)

Example 7-29. Simplify the equation $x^2 + y^2 + 2x - 4y = 4$ by a translation of axes.

Solution: Completing the square gives

$$(x + 1)^2 + (y - 2)^2 = 9$$

This is a circle with center at $(-1, 2)$. If we choose new axes X and Y so that $X = x + 1$ and $Y = y - 2$, then the equation of the circle with respect to these new coordinates will be

$$X^2 + Y^2 = 9$$

[See Figure 7-26 (a).]

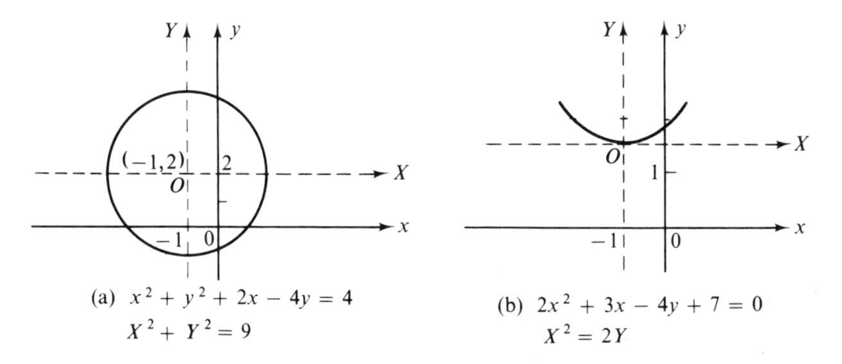

(a) $x^2 + y^2 + 2x - 4y = 4$
$X^2 + Y^2 = 9$

(b) $2x^2 + 3x - 4y + 7 = 0$
$X^2 = 2Y$

Figure 7-26

Example 7-30. Simplify the equation $2x^2 + 3x - 4y + 7 = 0$ by a translation of axes.

Solution: Dividing through by 2 and completing the square on x, we get

$$x^2 + \frac{3x}{2} + \left(\frac{3}{4}\right)^2 = 2y - \frac{7}{2} + \frac{9}{16}$$

or

$$\left(x + \frac{3}{4}\right)^2 = 2\left(y - \frac{47}{32}\right)$$

Setting

$$X = x + \frac{3}{4}$$

$$Y = y - \frac{47}{32}$$

the equation becomes

$$X^2 = 2Y$$

[See Figure 7-26 (b).]

As a result of this technique of translation of axes, any equation of the form

$$Ax^2 + By^2 + Cx + Dy + \mathrm{E} = 0$$

can be changed to an equation of the form

$$A'X^2 + B'Y^2 + E' = 0$$

if neither A nor B are zero, or to the form

$$A'X^2 + D'Y = 0$$

or

$$B'Y^2 + C'X = 0$$

if B or A is zero.

Exercise 7-9

Simplify each of the following equations by a translation of axes. Draw both sets of axes and sketch the graph.

1. $x^2 + y^2 + 2x - 4y - 1 = 0$
2. $2x^2 + 3y^2 - 4x + 6y + 2 = 0$
3. $2y^2 - 7y + 2x - 5 = 0$
4. $3x^2 + 2x - 6y + 4 = 0$
5. $x^2 - 2y^2 + 4x + 4y - 7 = 0$
6. $9x^2 - 16y^2 + 90x + 192y - 495 = 0$

7. $4x^2 + 9y^2 - 16x + 72y + 124 = 0$

8. $x^2 + 2y^2 - 6x + y = 0$

7-10 Rotation of Axes and the General Second Degree Equation

In the preceding sections we have seen that if the xy term is missing from the general second degree equation, then it is easy to identify the conic it represents. We summarize these results here.

> **Theorem 7-6.** If the graph of $Ax^2 + Cy^2 + Dx + Ey + F = 0$ (where A and C are not both zero) is not a degenerate form, then it is
>
> (1) a circle if $A = C$.
> (2) a parabola if either A or C is zero.
> (3) an ellipse if A and C have the same sign and $A \neq C$.
> (4) a hyperbola if A and C have different signs.

The degenerate forms of the circle and the ellipse are a single point and the empty set; the degenerate forms of the parabola are a single line or two parallel lines; and the degenerate form of a hyperbola is a pair of intersecting lines. (See Exercise 7-4, problem 5; Exercise 7-6, problem 4; Exercise 7-7, problem 8; Exercise 7-8, problem 8.)

In this section we will study the general second degree equation in which the xy term is *not* missing. In this case the graph of the conic will be a parabola, an ellipse, or a hyperbola (or perhaps a degenerate form) whose axis neither lies on nor is parallel to the x- or y-axis. (The equation of a circle will never have an xy term.)

In studying such a curve we will want to find its equation with respect to a new set of axes, an X-axis and a Y-axis, so chosen that the axis of the conic is either parallel to or lies on one of the new axes. This change of coordinate system will change the equation of the conic; however, it will not change any of the intrinsic properties of the conic. Its eccentricity, for example, will not change, nor will the distance from center to focus, although the coordinates of these points will be different. This change of coordinate system is called a *rotation of axes*. [See Figure 7-27 (a).]

In order to change the form of the equation from an equation in x and y to an equation in X and Y, we need to derive formulas relating the two sets of coordinates.

Let Ox and Oy be a pair of coordinate axes with intersection O, and

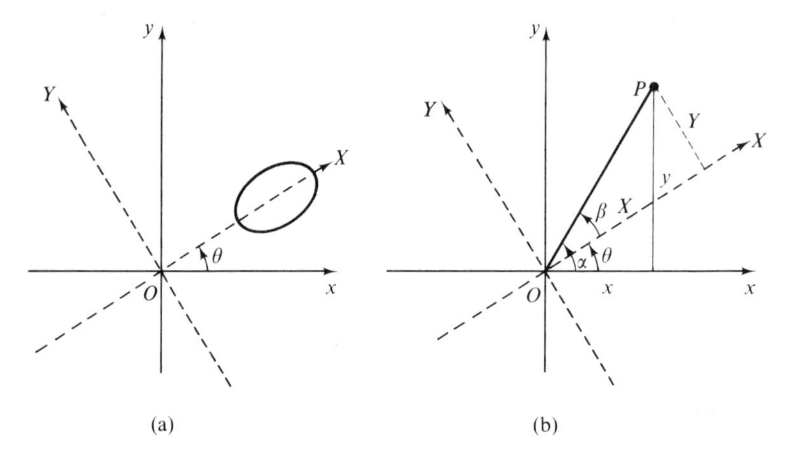

Figure 7-27

let OX and OY be a pair of mutually perpendicular lines also intersecting at O. Let θ be the measure of angle XOx, $0 < \theta < \pi/2$. Consider point P in the plane with coordinates (x, y) and let α be the inclination of line OP with the x-axis [Figure 7-27 (b)]. Then $\sin \alpha = y/OP$, $\cos \alpha = x/OP$, or

$$\begin{aligned} x &= OP \cos \alpha \\ y &= OP \sin \alpha \end{aligned} \tag{7-19}$$

Let β be the inclination of line OP with the X-axis. Then since the coordinates of P are (X, Y) with respect to these new axes, we have

$$\begin{aligned} \sin \beta &= \frac{Y}{OP} \\ \cos \beta &= \frac{X}{OP} \end{aligned} \tag{7-20}$$

where $\beta = \alpha - \theta$.

Substituting $\alpha = \beta + \theta$ in Equations (7-19) we get

$$x = OP \cos (\beta + \theta)$$
$$y = OP \sin (\beta + \theta)$$

or, by the addition formulas

$$x = OP(\cos \beta \cos \theta - \sin \beta \sin \theta)$$
$$y = OP(\sin \theta \cos \beta + \sin \beta \cos \theta)$$

Substituting Equations (7-20) into these equations gives us

$$x = X\cos\theta - Y\sin\theta$$
$$y = X\sin\theta + Y\cos\theta \tag{7-21}$$

These are called the *rotation formulas.*

Example 7-31. Find the new equation for $3x^2 - 2xy + 3y^2 - 2 = 0$ that results when the x- and y-axes are rotated through $\pi/4$. Graph the curve and both sets of axes.

Solution: Since $\theta = \pi/4$, $\sin\theta = \cos\theta = 1/\sqrt{2}$, and the rotation formulas become

$$x = \frac{X}{\sqrt{2}} - \frac{Y}{\sqrt{2}}$$
$$y = \frac{X}{\sqrt{2}} + \frac{Y}{\sqrt{2}}$$

Substituting these values into the equation we get

$$3\left(\frac{X}{\sqrt{2}} - \frac{Y}{\sqrt{2}}\right)^2 - 2\left(\frac{X}{\sqrt{2}} - \frac{Y}{\sqrt{2}}\right)\left(\frac{X}{\sqrt{2}} + \frac{Y}{\sqrt{2}}\right) + 3\left(\frac{X}{\sqrt{2}} + \frac{Y}{\sqrt{2}}\right)^2 - 2 = 0$$

and simplifying gives

$$2X^2 + 4Y^2 - 2 = 0$$

or, equivalently,

$$X^2 + 2Y^2 - 1 = 0$$

We recognize this as an ellipse, and putting it into standard form

$$X^2 + \frac{Y^2}{\left(\frac{1}{2}\right)} = 1$$

we see that its center is at the origin, its major axis lies on the X-axis, and $a = 1$, $b = 1/\sqrt{2}$. (See Figure 7-28.)

The problem now is, How can we find this angle θ that we need to eliminate the xy term from the general second degree equation?

Suppose that we start with the equation

$$Ax^2 + Bxy + Cy^2 + Dx + Ey + F = 0$$

where $B \neq 0$. We then make the substitutions

$$x = X\cos\theta - Y\sin\theta$$
$$y = X\sin\theta + Y\cos\theta$$

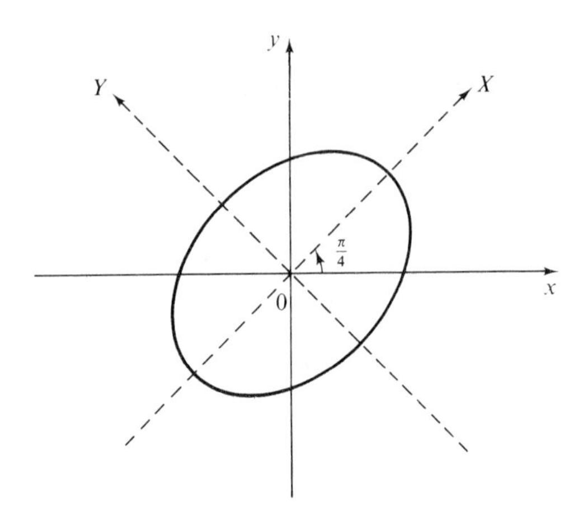

Figure 7-28

Multiplying out and collecting terms, we get an equation of the same form

$$A'X^2 + B'XY + C'Y^2 + D'X + E'Y + F' = 0$$

where

$$\begin{aligned} A' &= A \cos^2\theta + B \sin\theta \cos\theta + C \sin^2\theta \\ B' &= 2(C - A) \sin\theta \cos\theta + B(\cos^2\theta - \sin^2\theta) \\ C' &= A \sin^2\theta - B \sin\theta \cos\theta + C \cos^2\theta \\ D' &= D \cos\theta + E \sin\theta \\ E' &= E \cos\theta - D \sin\theta \\ F' &= F \end{aligned} \qquad \textbf{(7-22)}$$

We want to find a value for θ that will make B', the coefficient of the XY term, zero. Accordingly, we set

$$B' = 2(C - A) \sin\theta \cos\theta + B(\cos^2\theta - \sin^2\theta) = 0$$

Recalling the double angle formulas ($\sin 2\theta = 2 \sin\theta \cos\theta$, and $\cos 2\theta = \cos^2\theta - \sin^2\theta$) the equation above can be simplified to

$$(C - A) \sin 2\theta + B \cos 2\theta = 0$$

We shall see that we can always find a value for θ, $0 < \theta < \pi/2$, that will satisfy this equation. First, suppose $C = A$. Then the equation becomes

$$B \cos 2\theta = 0$$

Since $B \neq 0$, this equation is satisfied if $\theta = \pi/4$. Now suppose $A \neq C$. Then $\cos 2\theta \neq 0$, and dividing through by $\cos 2\theta\,(C - A)$ gives

$$\frac{\sin 2\theta}{\cos 2\theta} + \frac{B}{C - A} = 0$$

or

$$\tan 2\theta = \frac{B}{A - C}$$

Since for $0 < \theta < \pi/2$ $(\theta \neq \pi/4)$ the range of $\tan 2\theta$ is the set of all real numbers, this equation will always have a solution. Actually we do not need to find a value for θ. We only need the values of $\sin \theta$ and $\cos \theta$ to use in the rotation formulas (7-21). We can find these without ever finding θ by using the half angle formulas

$$\sin \theta = \sqrt{\frac{1 - \cos 2\theta}{2}}$$

$$\cos \theta = \sqrt{\frac{1 + \cos 2\theta}{2}}$$

Some examples will illustrate how this can be done.

Example 7-32. Simplify the equation $x^2 + 4xy + y^2 - 2 = 0$ by a rotation of axes to remove the xy term.

Solution: Since $A = C$, then $\theta = \pi/4$ and $\sin \theta = \cos \theta = 1/\sqrt{2}$. The rotation formulas are

$$x = \frac{X}{\sqrt{2}} - \frac{Y}{\sqrt{2}}$$

$$y = \frac{X}{\sqrt{2}} + \frac{Y}{\sqrt{2}}$$

Substituting these values into the equation and simplifying, we get

$$3X^2 - Y^2 - 2 = 0$$

or

$$\frac{X^2}{\left(\frac{2}{3}\right)} - \frac{Y^2}{2} = 1$$

which we recognize as a hyperbola with center at the origin and transverse axis along the X-axis.

Example 7-33. Simplify the equation $4x^2 + 24xy + 11y^2 - 100 = 0$ by a rotation of axes to remove the xy term. Sketch the graph.

Solution: If θ is the measure of the angle of rotation, then

$$\tan 2\theta = \frac{B}{A - C} = \frac{24}{(4 - 11)} = -\frac{24}{7}$$

If $0 < \theta < \pi/2$, then $0 < 2\theta < \pi$, and 2θ must be in the second quadrant, $y = 24$ and $x = -7$. (See Figure 7-29.) Thus, $r = 25$ and $\cos 2\theta = -7/25$.

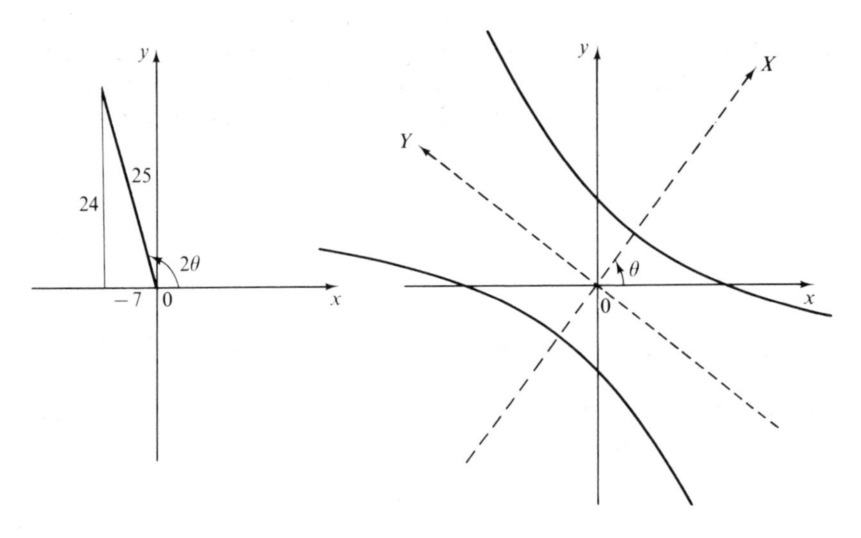

Figure 7-29

From the half angle formulas we get $\sin\theta = \sqrt{(1 + 7/25)/2} = 4/5$ and $\cos\theta = \sqrt{(1 - 7/25)/2} = 3/5$. The rotation formulas thus become

$$x = \frac{3X}{5} - \frac{4Y}{5}$$

$$y = \frac{4X}{5} + \frac{3Y}{5}$$

Substituting these into the original equation gives

$$4\left(\frac{3X}{5} - \frac{4Y}{5}\right)^2 + 24\left(\frac{3X}{5} - \frac{4Y}{5}\right)\left(\frac{4X}{5} + \frac{3Y}{5}\right) + 11\left(\frac{4X}{5} + \frac{3Y}{5}\right)^2 - 100 = 0$$

Simplifying, we get

$$20X^2 - 5Y^2 = 100$$

or

$$\frac{X^2}{5} - \frac{Y^2}{20} = 1$$

which we recognize as a hyperbola with center at the origin and major axis on the X-axis.

Now it must be admitted that this calculation is quite messy and, in fact, would have been a great deal worse if the coefficients had not been carefully chosen. The student will therefore welcome a method of determining what kind of conic the general equation represents without rotating the axes to find the new equation.

It can be shown (by another simple but messy calculation) that

$$B^2 - 4AC = (B')^2 - 4A'C'$$

where A, B, and C are the coefficients of x^2, xy, and y^2 in the general equation of the conic, and A', B', and C' are the new coefficients after the axes are rotated through an angle θ.

Now if θ is so chosen that $B' = 0$, then $B^2 - 4AC = -4A'C'$, and the product $A'C'$ tells us whether the conic is a parabola, an ellipse, or a hyperbola. If $A'C' > 0$, for example, then A' and C' have the same sign, and the conic is an ellipse. If $A'C' = 0$, then either A' or C' is zero, and the graph is a parabola. If $A'C' < 0$, then A' and C' have different signs and the curve is a hyperbola. Thus, the sign of $B^2 - 4AC$, which is called the *discriminant* of the general second degree equation will classify the conic.

Theorem 7-7. If the graph of $Ax^2 + Bxy + Cy^2 + Dx + Ey + F = 0$, $B \neq 0$, is not a degenerate form, then it is

(1) a parabola if $B^2 - 4AC = 0$.
(2) an ellipse if $B^2 - 4AC < 0$.
(3) a hyperbola if $B^2 - 4AC > 0$.

Exercise 7-10

1. Identify the graph of each of the following. You may assume that the graph is not degenerate.
   (a) $y^2 + x^2 + 4x + 12y + 4 = 0$
   (b) $y^2 - x^2 + 4x + 12y + 4 = 0$
   (c) $y^2 + 2x^2 + 4x + 12y + 4 = 0$
   (d) $y^2 - 2x^2 + 4x + 12y + 4 = 0$
   (e) $y^2 + 4x + 12y + 4 = 0$
   (f) $y + 2x^2 + 4x + 4 = 0$

2. Find the new equation for each of the following that results when the x- and y-axes are rotated through the angle indicated. Sketch the graph and both sets of axes.
   (a) $xy = 4$; $\theta = \pi/4$
   (b) $25x^2 + 6\sqrt{3}xy + 19y^2 = 16$; $\theta = \pi/6$
   (c) $3x^2 - 2\sqrt{3}xy + y^2 - 4x - 4\sqrt{3}y = 0$; $\theta = \pi/3$

3. Find the rotation formulas to change each of the following to an equation having no xy term:
   (a) $3x^2 + 6xy + 3y^2 = 10x$
   (b) $2x^2 - 3xy - 2y^2 + 2x - 3y - 7 = 0$
   (c) $3x^2 + 4xy + 10x - 7y + 15 = 0$
   (d) $xy = 2$
   (e) $7x^2 + 3xy + 11y^2 - 2y = 4$

4. By a rotation of axes, find a new equation having no xy term for each of the following. Put in standard form and graph, showing both sets of axes.
   (a) $17x^2 - 12xy + 8y^2 = 20$
   (b) $4x^2 - 6xy - 4y^2 + 3\sqrt{10}x + 9\sqrt{10}y - 40 = 0$
   (c) $x^2 + 2xy + y^2 + \sqrt{2}x = 0$
   (d) $4x^2 - 4xy + y^2 + \sqrt{5}x = 0$
   (e) $x^2 + 4xy + 4y^2 - 2\sqrt{5}x - \sqrt{5}y = 0$
   (f) $7x^2 + 8xy + y^2 + 81 = 0$
   (g) $xy + 4x + 3y + 4 = 0$

5. Use Theorem 7-7 to identify the graph of each of the following conics. You may assume that the graphs are not degenerate.
   (a) $x^2 + 2xy = 7$
   (b) $5x^2 - 4xy + 4y^2 = 12$
   (c) $4x^2 + 8xy + 3y^2 - 4 = 0$
   (d) $x^2 - 4xy + 4y^2 - 4x + 8 = 0$
   (e) $xy + x - y = 0$
   (f) $2x^2 - 4xy + 2y^2 - 7x + 10 = 0$

6. Use Theorem 7-7 to show that if either A or C is zero but $B \neq 0$ in the general second degree equation, then the graph is a hyperbola.

7. Use Theorem 7-7 to show that if $A = C = \pm B/2$ in the general second degree equation, then the graph is a parabola.

8. Use Theorem 7-7 to show that if A and C have different signs in the general second degree equation, then the graph is a hyperbola.

9. Use Equations (7-22) to show that $(B')^2 - 4A'C' = B^2 - 4AC$.

10. Show that if the graph of $Ax^2 + Bxy + Cy^2 + Dx + Ey + F = 0$ is a parabola with $B \neq 0$ and $A \neq C$, and if θ is the measure of the angle of rotation to remove the xy term, then $\cos 2\theta = (A - C)/(A + C)$.

7-11 Polar Coordinates

The rectangular coordinate system we are accustomed to use is not the only coordinate system for the plane. There are many others and one of the most useful of these is the system of polar coordinates.

In this system we start with a fixed point O, called the *pole*, and a fixed ray OA with end-point at O. [See Figure 7-30 (a).] This ray is called the *polar axis*. To locate a point P in the plane we can give the distance r (from P to O) and θ (the measure of the angle having the polar axis for its initial side and OP for its terminal side). The polar coordinates of P are (r, θ), where the distance r is listed first. We will take θ to be in radians; however, it could just as easily be chosen to be degrees.

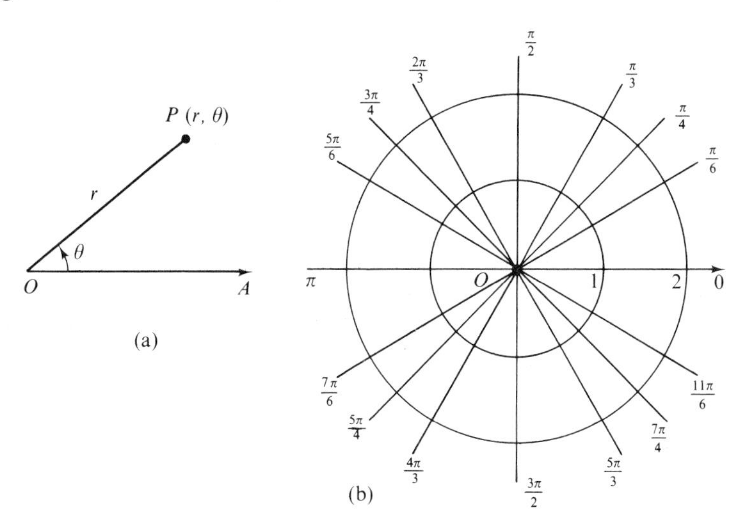

Figure 7-30

Thus, in polar coordinates points are located by finding the intersection of concentric circles (corresponding to values of r) with rays from the pole (corresponding to values of θ). As a matter of fact, polar coordinate graph paper is available which is printed with the kind of grid shown in Figure 7-30 (b).

As in trigonometry, θ can be taken to be positive or negative since the angle is measured counterclockwise or clockwise. This brings up a significant difference between polar and rectangular coordinates. A point has infinitely many representations (r, θ) in polar coordinates. For example, the point $(1, \pi/2)$ also has the coordinates $(1, -3\pi/2)$ and $(1, 5\pi/2)$, or, in general $(1, \pi/2 + 2\pi k)$ where k is any integer. We will also allow r to be negative. The point $(-r, \theta)$ will be taken to be the same as the point $(r, \theta + \pi)$. Thus, if r is negative, it is measured backward along the extension of the ray corresponding to θ. (See Figure 7-31.) It is true, however, that in the polar plane, as in the rectangular plane, to every ordered pair (r, θ) there corresponds exactly one point.

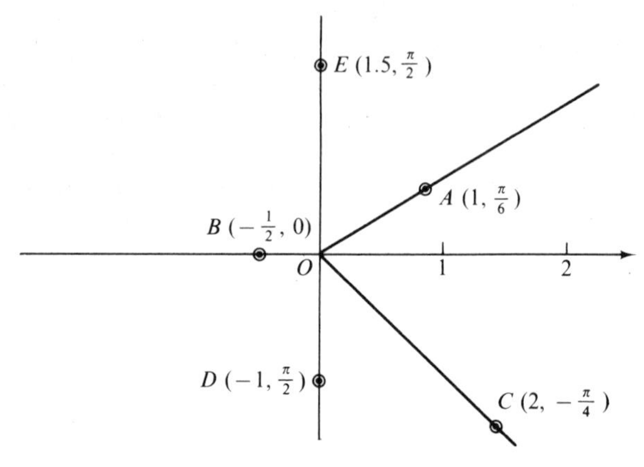

Figure 7-31

If we superimpose a rectangular coordinate system on a polar coordinate system so that the origin coincides with the pole and the positive x-axis lies on the polar axis, then there is a simple relationship between the two kinds of coordinates. (See Figure 7-32.)

Since $\cos\theta = x/r$ and $\sin\theta = y/r$ we have

$$\begin{aligned} x &= r\cos\theta \\ y &= r\sin\theta \end{aligned} \qquad \textbf{(7-23)}$$

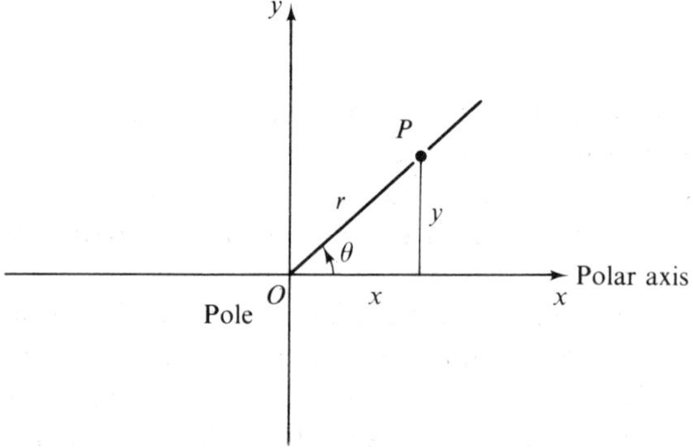

Figure 7-32

Solving Equations (7-23) for r and θ, we get

$$r = \sqrt{x^2 + y^2}$$

$$\theta = \arctan \frac{y}{x}$$

The equations $\cos \theta = x/\sqrt{x^2 + y^2}$ and $\sin \theta = y/\sqrt{x^2 + y^2}$ are also frequently useful.

Example 7-34. Change $x^2 + y^2 - 4x = 0$ to polar coordinates.

Solution: Since $x^2 + y^2 = r^2$ and $x = r \cos \theta$, this equation becomes

$$r^2 - 4r \cos \theta = 0$$

or

$$r(r - 4 \cos \theta) = 0$$

Since the solution set for $r - 4 \cos \theta = 0$ includes the pole, then we lose no points by omitting the factor r. Thus, the equation of this circle in polar coordinates is $r = 4 \cos \theta$.

Example 7-35. Change $r = 2 \cos \theta + 3 \sin \theta$ to rectangular coordinates.

Solution: Replacing $\cos \theta$ and $\sin \theta$ by x/r and y/r gives

$$r = \frac{2x}{r} + \frac{3y}{r}$$

or

$$r^2 = 2x + 3y$$

Since $r^2 = x^2 + y^2$, this becomes

$$x^2 + y^2 - 2x - 3y = 0$$

We recognize this as the equation of a circle.

Why should we bother to learn about a new kind of coordinate system when rectangular coordinates have always been sufficient for our needs? It turns out that some curves have equations that are particularly simple in polar coordinates, while their equations in rectangular coordinates are rather complicated. Consider, for example, the Spiral of Archimedes, whose equation in polar coordinates is $r = \theta$ (Figure 7-33). The equation of this curve in rectangular coordinates is $x^2 + y^2 = (\arctan y/x)^2$. Clearly it is convenient to have an alternate system to turn to if rectangular coordinates prove to be

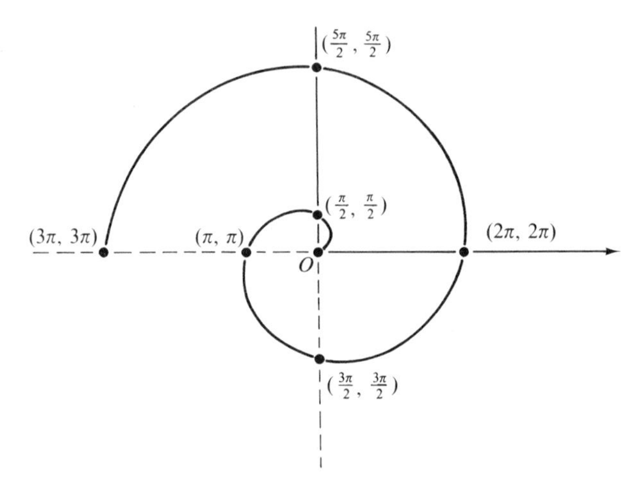

Figure 7-33

unsatisfactory for a particular problem. In deriving the equations for the orbits of the planets, polar coordinates are usually employed because the development is simpler this way.

In polar coordinates the equation of one type of curve can take many different forms. For example,

$$\theta = a$$

$$r \cos \theta = a$$

$$r = a \csc \theta$$

$$r(a \cos \theta + b \sin \theta) = 1$$

are all equations of lines. For this reason it is often difficult to identify a particular kind of curve by looking at its equation in polar coordinates. In graphing these curves, we can always try changing to rectangular coordinates. If this equation turns out to be more complicated than the original one, then we must make a table of values for r and θ, plot the points, then connect them with a smooth curve. Table 7-2 gives values of $\sin \theta$ and $\cos \theta$ for $0 < \theta < 2\pi$ which will be useful in graphing these curves.

Our work will be easier if we are aware of any symmetries the graph may have.

If we get an equivalent equation when (a) r is replaced by $-r$ or when (b) θ is replaced by $\theta + \pi$ then the graph is symmetric with respect to the pole.

If we get an equivalent equation when θ is replaced by $-\theta$ then the graph is symmetric with respect to the polar axis.

Table 7-2

θ	0	$\pi/6$	$\pi/4$	$\pi/3$	$\pi/2$	$2\pi/3$	$3\pi/4$	$5\pi/6$	
cos θ	1	.866	.707	.5	0	−.5	−.707	−.866	
sin θ	0	.5	.707	.866	1	.866	.707	.5	
θ	π	$7\pi/6$	$5\pi/4$	$4\pi/3$	$3\pi/2$	$5\pi/3$	$7\pi/4$	$11\pi/6$	2π
cos θ	−1	−.866	−.707	−.5	0	.5	.707	.866	1
sin θ	0	−.5	−.707	−.866	−1	−.866	−.707	−.5	0

If we get an equivalent equation when θ is replaced by $-\theta$ and r by $-r$ at the same time, then the graph is symmetric with respect to a line through $\theta = \pi/2$.

If any of these conditions hold, then the graph will have the symmetry indicated, however the converse is not true. It is possible for a graph to exhibit one of these symmetries although the condition may not hold.

Example 7-36. The graph of $r^2 = 2 \sin \theta$ is symmetric with respect to the pole since it is unchanged when r is replaced by $-r$.

The graph of $r = \sin 2\theta$ is also symmetric with respect to the pole since if θ is replaced by $\theta + \pi$ it becomes $r = \sin (2\theta + 2\pi) = \sin 2\theta$.

The graph of $r = 2 \cos \theta$ is symmetric with respect to the polar axis since $r = 2 \cos (-\theta) = 2 \cos \theta$.

The graph of $r = \sin \theta$ is symmetric with respect to the line through $\theta = \pi/2$ since if we replace r by $-r$ and θ by $-\theta$, the equation becomes $-r = \sin (-\theta) = -\sin \theta$, which is equivalent to $r = \sin \theta$.

Example 7-37. Graph $r = 1 + \cos \theta$.

Solution: We observe that $r = 1 + \cos (-\theta) = 1 + \cos \theta$, and conclude that the graph is symmetric with respect to the polar axis. Accordingly, although the period of the cosine function is 2π, we need make a table only of values for θ, $0 \leq \theta \leq \pi$ (See Table 7-3).

Table 7-3

θ	0	$\pi/6$	$\pi/4$	$\pi/3$	$\pi/2$	$2\pi/3$	$3\pi/4$	$5\pi/6$	π
cos θ	1	.866	.707	.5	0	−.5	−.707	−.866	−1
r (approx.)	2	1.9	1.7	1.5	1	.5	.3	.1	0

This curve is shown in Figure 7-34 and is called a *cardioid.*

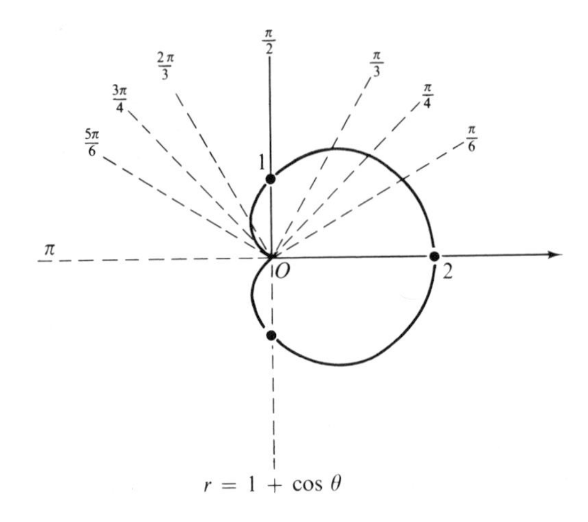

Figure 7-34

The cardioid has an equation of the form

$$r = a(1 \pm \cos\theta)$$

or

$$r = a(1 \pm \sin\theta)$$

The cardioid is one of a class of extremely beautiful curves whose equations are much simpler in polar coordinates than in rectangular coordinates. We will describe some of these curves below.

1. *The limaçon*

The equation of the limaçon is of the form

$$r = a \pm b\cos\theta$$

or

$$r = a \pm b\sin\theta$$

The cardioid is a special case of the limaçon in which $a = b$. If $b > a > 0$, then the curve has an inner loop. [See Figure 7-35 (a).] If $a > b > 0$, then r is never zero, and the curve is pictured in Figure 7-35 (b).

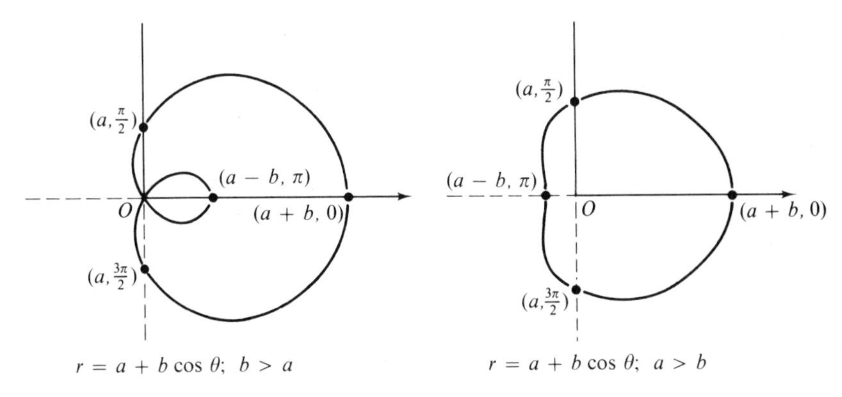

Figure 7-35

2. *The rose*

The equation of the rose has the form

$$r = a \sin n\theta$$

or

$$r = a \cos n\theta \quad \text{for } n = 2, 3, 4, \ldots$$

If n is odd, then the rose will have n petals. If n is even, it will have $2n$ petals. Thus, the graph of $r = a \cos 3\theta$ is a three-leaved rose, while that of $r = a \cos 2\theta$ is a four-leaved rose. [See Figure 7-36 (a).] In graphing the roses, it is helpful to find those values of θ between 0 and 2π that will make $\cos n\theta = \pm 1$ (or $\sin n\theta = \pm 1$). Each of these rays will be a ray of symmetry for one of the petals of the rose.

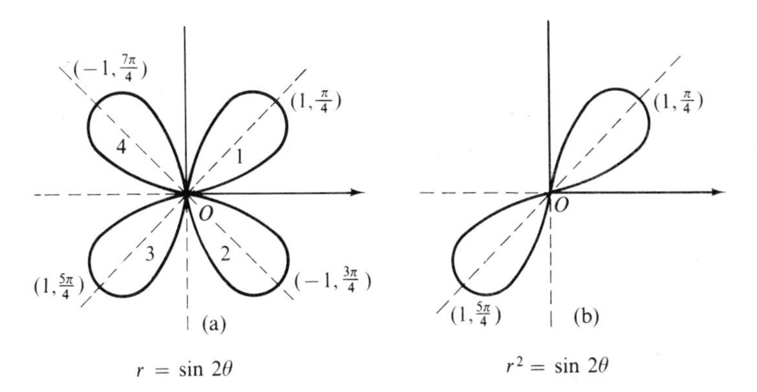

Figure 7-36

Example 7-38. Graph $r = \sin 2\theta$.

Solution: This is a four-leaved rose. Setting $\sin 2\theta = \pm 1$, we get $2\theta = \pi/2, 3\pi/2, \pi/2 + 2\pi, 3\pi/2 + 2\pi$, or $\theta = \pi/4, 3\pi/4, 5\pi/4$, and $7\pi/4$ in the range $0 \leq \theta < 2\pi$. These rays will serve as guide lines for drawing our petals. Since the graph is symmetric both with respect to the pole and with respect to the line through $\theta = \pi/2$, we need values only for $0 \leq \theta \leq \pi/2$ (See Table 7-4).

Table 7-4

θ	0	$\pi/8$	$\pi/4$	$3\pi/8$	$\pi/2$
2θ	0	$\pi/4$	$\pi/2$	$3\pi/4$	π
$\sin 2\theta$	0	.7	1	.7	0

It is interesting to note that for $\pi/2 < \theta < \pi$, r is negative; thus, these values for θ give us the petal in quadrant IV. For $\pi < \theta < 3\pi/2$, r is positive, and we get the petal in quadrant III; when $3\pi/2 < \theta < 2\pi$, r is negative again and we get the petal in quadrant II.

3. *The lemniscate*

The lemniscate has an equation of the form

$$r^2 = a^2 \cos 2\theta$$

or

$$r^2 = a^2 \sin 2\theta$$

This equation looks very much like that of the four-leaved rose, except that it has r^2 instead of r. This difference means that the values of θ that would make $\cos 2\theta$ (or $\sin 2\theta$) negative are excluded, since r^2 is always nonnegative. Thus, the graph of $r^2 = \sin 2\theta$ is similar to the graph of the rose of Example 7-38 with petals 2 and 4 omitted. [See Figure 7-36 (b).]

Exercise 7-11

1. Draw a grid as in Figure 7-30 and locate the points whose polar coordinates are given: $A(2, 0)$; $B(1, \pi/2)$; $C(-1, \pi)$; $D(2, 3\pi/2)$; $E(-2, \pi/6)$; $F(0, \pi/4)$; $G(1.5, -\pi/4)$; $H(-1, -\pi)$.
2. Draw the graph of $\theta = \pi/4$, of $\theta = -\pi/6$, of $\theta = -3\pi/2$, of $r = 1$, of $r = 2$, of $r = -1$.

3. Change each of the following equations to polar coordinates:
   (a) $x = 2$ (b) $y = -1$
   (c) $x^2 + y^2 = 4$ (d) $xy = 1$
   (e) $y^2 = 2x$ (f) $2x^2 + 3y^2 = 6$
   (g) $2x^2 + 2y^2 = x$ (h) $x^2 + y^2 = \arctan y/x$

4. Change each of the following equations to rectangular coordinates.
   (a) $r = 2 \sec \theta$ (b) $r = 2 \cos \theta$
   (c) $r = \sin \theta$ (d) $r(2\cos \theta + 3 \sin \theta) = 1$
   (e) $r = 2 \sin \theta + 5 \cos \theta$ (f) $r = \sin 2\theta$
   (g) $r^2 = \cos 2\theta$ (h) $r = 1 + \cos \theta$
   (i) $r = 2 + \sin \theta$ (j) $r = 2/(1 - \cos \theta)$

5. Show that each of the following is an equation for a line by changing each to rectangular coordinates:
   (a) $\theta = a$ (b) $r \cos \theta = a$
   (c) $r = a \csc \theta$ (d) $r(a \cos \theta + b \sin \theta) = c$

6. Show that each of the following is an equation for a circle by changing each to rectangular coordinates. Give the rectangular coordinates of the center and the radius of each.
   (a) $r = a$ (b) $r = a \cos \theta$
   (c) $r = a \sin \theta$ (d) $r = a \cos \theta + b \sin \theta$

7. Test for symmetry and graph each of the following in polar coordinates.
   (a) $r = 2 \sin \theta$ (b) $r = 1 - \cos \theta$
   (c) $r = 1 + \sin \theta$ (d) $r = 1 - \sin \theta$
   (e) $r = 1 + 2 \cos \theta$ (f) $r = 2 + \sin \theta$
   (g) $r = \sin 3\theta$ (h) $r = \cos 2\theta$
   (i) $r^2 = \cos 2\theta$ (j) $r^2 = \cos 3\theta$
   (k) $r = \cos 4\theta$ (l) $r = 1/\theta;\ \theta > 0$
   (m) $r^2 = \sin \theta$ (n) $r = 2 \sec \theta - 1$

8. Let $P_1(r_1, \theta_1)$ and $P_2(r_2, \theta_2)$ be two points in the polar plane. Show that the distance between them is given by the formula

$$d = \sqrt{r_1^2 + r_2^2 - 2r_1r_2 \cos (\theta_2 - \theta_1)}$$

   (Hint: Let (x_1, y_1) and (x_2, y_2) be the rectangular coordinates of P_1 and P_2. Substitute $x_1 = r_1 \cos \theta_1$, $y_1 = r_1 \sin \theta_1$, $x_2 = r_2 \cos \theta_2$, $y_2 = r_2 \sin \theta_2$ in the distance formula for rectangular coordinates and simplify.)

9. It can be shown that every conic which has a focus at the pole and axis on the polar axis or on the line through $\theta = \pi/2$ has an equation in polar coordinates of the form

$$r = \frac{a(1 + e)}{1 \pm e \cos \theta}$$

   or

$$r = \frac{a(1 + e)}{1 \pm e \sin \theta}$$

where e is the eccentricity of the conic and a is the distance from focus to vertex.

(a) Show that

$$r = \frac{1}{1 + \cos\theta}$$

is an equation of a parabola with focus at the origin by changing to rectangular coordinates.

(b) Show that

$$r = \frac{1}{1 - \left(\frac{1}{2}\right)\cos\theta}$$

is an equation of an ellipse with one focus at the origin and eccentricity 1/2 by changing to rectangular coordinates.

(c) Show that

$$r = \frac{1}{1 + 2\sin\theta}$$

is an equation of a hyperbola with one focus at the origin and eccentricity 2 by changing to rectangular coordinates.

10. (a) Show that substituting $\sin\theta$ for $\cos\theta$ in an equation $r = f(\cos\theta)$ in polar coordinates which involves only the cosine function has the effect of rotating the curve through $\theta = \pi/2$, and that consequently substituting $\cos\theta$ for $\sin\theta$ in $r = g(\sin\theta)$ rotates that curve through $-\pi/2$.
    (b) Show that substituting $-\sin\theta$ for $\sin\theta$ in an equation $r = f(\sin\theta)$ in polar coordinates has the effect of rotating the curve through $\theta = \pi$.
    (c) Show that substituting $-\cos\theta$ for $\cos\theta$ in an equation $r = f(\cos\theta)$ in polar coordinates has the effect of rotating the curve through $\theta = \pi$.

11. Find, in polar coordinates, equations for the following conics. (Hint: See problems 8 and 9.)
    (a) a circle with center at $(4, 0)$ and radius 4
    (b) a circle with center at $(3, \pi/4)$ and radius 1
    (c) a parabola with focus at the pole and vertex $(2, \pi)$
    (d) an ellipse with one focus at the pole, eccentricity 2/3 and one vertex at $(1, \pi)$

Review Exercise

1. Find an equation for the line
    (a) through $(3, 2)$ parallel to $x + 2y = 1$.
    (b) through $(1, 0)$ perpendicular to $2x - y + 7 = 0$.
    (c) with inclination $\pi/6$ and y-intercept 3.

2. Find an equation for the circle
   (a) with center at $(1, -2)$ passing through the point $(4, 2)$.
   (b) with center at $(1, 0)$ tangent to the y-axis.

3. Change $x^2 + y^2 + 2x - 5y + 4 = 0$ to standard form and determine if its graph is a circle, a point, or the empty set. If its graph is a circle, give the center and radius.

4. Find an equation for the parabola with focus $(1, 3)$ and directrix $y = 4$.

5. Put $y^2 + 6y + 7x + 5 = 0$ in standard form, find the vertex, the focus, the equation of the directrix, the focal width, and graph.

6. Find an equation for the ellipse with foci at $(3, -2)$ and $(5, -2)$ and eccentricity $1/2$.

7. Put $3x^2 + 4y^2 + 6x - 12y + 7 = 0$ in standard form, find the coordinates of the center, the foci, and vertices; find the eccentricity, and graph.

8. Find an equation for the hyperbola with foci at $(1, \pm 2)$ and eccentricity 2.

9. Put $2x^2 - 9y^2 + 4x + 20 = 0$ in standard form, find the coordinates of the center, the foci, and the vertices; find the eccentricity, and graph.

10. Simplify $3x^2 + 2y^2 + 6x - 10y - 2 = 0$ by a translation of axes. Draw both sets of axes and graph.

11. Simplify $2x^2 - 5xy + 2y^2 + 7 = 0$ by a rotation of axes. Draw both sets of axes and graph.

12. Identify each of the following conics. You may assume that none is a degenerate form.
    (a) $3x^2 - 2xy + y^2 + 7x - 2y - 14 = 0$
    (b) $2x^2 + 2xy + 3x - 4y + 10 = 0$
    (c) $2x^2 + 2y^2 + 3x - 2y = 0$
    (d) $x^2 + 2xy + y^2 + 7x - 3y + 4 = 0$
    (e) $x^2 - y^2 = 4$

13. (a) Change $r = 3 \sin \theta$ to rectangular coordinates.
    (b) Change $x^2 - y^2 = 1$ to polar coordinates.

14. Graph $r = 1 + \sin \theta$ in polar coordinates.

Bibliography

Coolidge, Julian L., *A History of the Conic Sections and Quadric Surfaces.* London: Oxford University Press, 1945.

Courant, Richard and Hubert Robbins. *What Is Mathematics?*, pp. 329–338; pp. 346–361. New York: Oxford University Press, 1941.

Eves, Howard, *An Introduction to the History of Mathematics*, Rev. Ed., pp. 281–290. New York: Holt, Rinehart and Winston, 1964.

Kasner, Edward and James Newman, *Mathematics and the Imagination*, pp. 96–107. New York: Simon and Schuster, 1940.

Kline, Morris, *Mathematics, a Cultural Approach*, Ch. 13. Reading, Massachusetts: Addison-Wesley Publishing Company, Inc., 1962.

———, *Mathematics in Western Culture*, Ch. 12. New York: Oxford University Press, 1953.

Van der Waerden, B. L., *Science Awakening*, pp. 237–261. New York: Oxford University Press, 1961.

Wolff, Peter, *Breakthroughs in Mathematics*, Ch. 3. New York: New American Library, 1963.

8

Analytic Geometry in Three Dimensions

8-1 Rectangular Coordinates in Three Dimensions and the Distance Formula

In studying analytic geometry in the plane, we set up a rectangular (or Cartesian) coordinate system in which the location of a point was given by its directed distance from two perpendicular lines. There was a one-to-one correspondence between points in the plane and ordered pairs of real numbers (x, y).

We now extend this idea to three dimensions. We can describe the location of a point in three dimensional space by giving its directed distances from three mutually perpendicular planes. For example, the position of a point in the classroom could be given by saying it was

2 feet from the front wall, 5 feet from an adjacent wall, and 6 feet from the floor.

We will set up our coordinate system in space by taking three mutually perpendicular lines which will be called the x-axis, the y-axis, and the z-axis. If we set these up as in Figure 8-1 with the positive y-axis to the right, the positive z-axis upward, and the positive x-axis coming out toward us perpendicular to the page, then the system is called a right-handed system. The intersection of these three lines is called the origin, and we select a scale on each axis as we did in the plane.

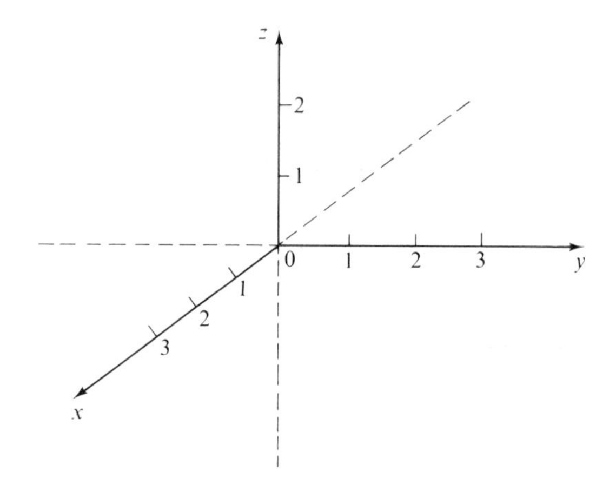

Figure 8-1

To draw these three dimensional diagrams on the two dimensional page we slant the x-axis and shorten the units on this axis slightly to give the illusion of depth.

The plane containing the x- and y-axes is called the xy plane; that containing the z- and y-axes is called the zy plane; and the plane containing the x- and z-axes is the xz plane. To locate a point P in space we drop perpendiculars from P to each of these three planes. The x-coordinate is the directed distance from the yz plane to the point, the y-coordinate the directed distance from the xz plane, and the z-coordinate the directed distance from the xy plane. We will agree to list these three coordinates in that order, (x, y, z).

The point with coordinates $(1, 2, 3)$, for example, is one unit in a positive direction from the yz plane, two units in a positive direction from the xz plane, and three units in a positive direction from the xy plane. The student will find it helpful in graphing these points to draw a box with the point at one of the corners as in Figure 8-2.

This method of assigning rectangular coordinates (a better name would be parallelepiped coordinates) to points in three dimensional space sets up a one-to-one correspondence between ordered triples of real numbers (x, y, z) and points in space. As before, because of this correspondence we will identify point and ordered triple and refer to "the point (x, y, z)."

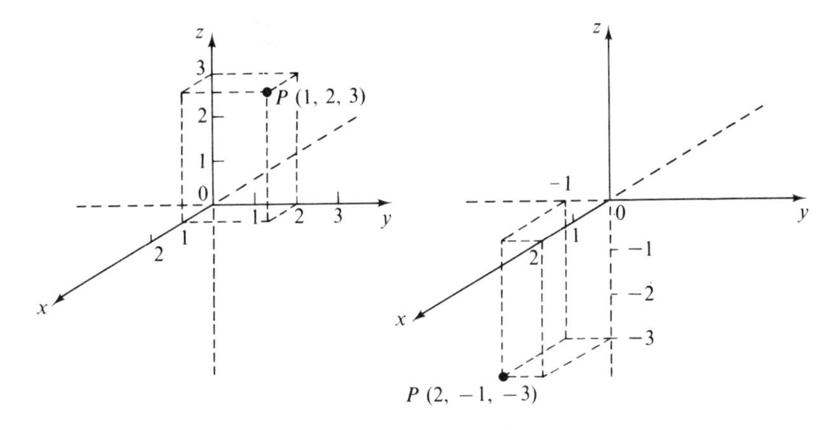

Figure 8-2

In this chapter we will be studying equations in x, y, and z just as in earlier chapters we studied equations in x and y. The graph of an equation in three variables is the set of all points (x, y, z) which satisfy the equation. The relationship between these equations and their graphs is the subject of three dimensional analytic geometry.

To find the distance between two points in space, we can use the Pythagorean Theorem as we did in the plane (Section 1-6). Let $P_1(x_1, y_1, z_1)$ and $P_2(x_2, y_2, z_2)$ be two points in space. (See Figure 8-3.)

If we sketch a box with sides parallel to the coordinate planes and with P_1 and P_2 at opposite corners, then we notice that P_1QP_2 is a right triangle with right angle at Q since P_2Q is perpendicular to the xy plane. Triangle P_1RQ is also a right triangle with right angle at R since P_1R and RQ are parallel to the y- and x-axes respectively. The coordinates of Q are (x_2, y_2, z_1) and those of R are (x_1, y_2, z_1). By the Pythagorean Theorem

$$(P_1P_2)^2 = (P_1Q)^2 + (QP_2)^2$$

and

$$(P_1Q)^2 = (P_1R)^2 + (RQ)^2$$

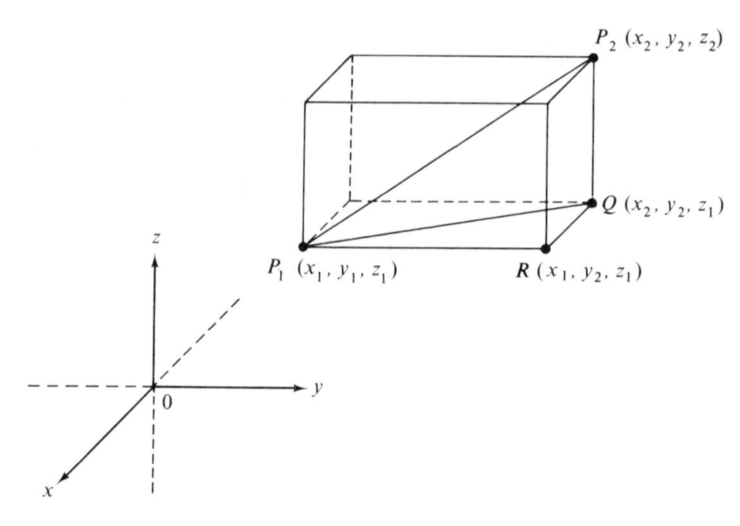

Figure 8-3

Thus,

$$(P_1P_2)^2 = (P_1R)^2 + (RQ)^2 + (QP_2)^2$$

But P_1 and R lie on a line parallel to the y-axis, thus $P_1R = |y_2 - y_1|$. Similarly $RQ = |x_2 - x_1|$ and $QP_2 = |z_2 - z_1|$.

Thus, we have

$$(P_1P_2)^2 = (|y_2 - y_1|)^2 + (|x_2 - x_1|)^2 + (|z_2 - z_1|)^2$$

or

$$P_1P_2 = \sqrt{(x_2 - x_1)^2 + (y_2 - y_1)^2 + (z_2 - z_1)^2} \qquad \textbf{(8-1)}$$

If P_1 and P_2 are in a plane parallel to one of the coordinate planes, for instance the xy plane, then their z-coordinates are equal and this formula reduces to the distance formula for the plane.

Example 8-1. Find the distance between the points (3, −2, 1) and (0, 4, −2).

Solution: By formula (8-1)

$$\begin{aligned} d &= \sqrt{(3-0)^2 + (-2-4)^2 + (1-(-2))^2} \\ &= \sqrt{9 + 36 + 9} \\ &= \sqrt{54} \\ &= 3\sqrt{6} \end{aligned}$$

Exercise 8-1

1. Locate the following points on the three dimensional rectangular coordinate system: $A(1, 0, 0)$; $B(0, 2, 1)$; $C(1, 3, 0)$; $D(3, 2, 1)$; $E(-1, 2, 1)$; $F(1, -1, 4)$; $G(1, 1, -2)$; $H(-1, -2, 5)$; $I(0, -2, -2)$; $J(-1, -2, -2)$; $K(3, -1, -1)$; $L(-1, 2, -3)$.

2. Give the coordinates of a typical point on the x-axis; on the y-axis; on the z-axis; on the xy plane; on the xz plane; on the yz plane.

3. Describe each of the following sets of points:
   (a) $\{(x, y, z) \mid z = 0\}$
   (b) $\{(x, y, z) \mid x = 0\}$
   (c) $\{(x, y, z) \mid x = 0 \text{ and } y = 0\}$
   (d) $\{(x, y, z) \mid z = 0 \text{ and } y = 0\}$

4. Find the distance between the following pairs of points:
   (a) $(0, 0, 0)$ and $(1, -2, 2)$ (b) $(3, 2, 0)$ and $(0, 0, 6)$
   (c) $(1, -2, -1)$ and $(3, 4, -2)$ (d) $(-1, 4, 6)$ and $(3, 1, 0)$
   (e) $(4, -1, -2)$ and $(-1, 1, -2)$ (f) $(2, 1, -1)$ and $(2, 1, 5)$

5. Show that the triangle whose vertices are $(2, 6, -3)$, $(-4, 3, -3)$, and $(-1, 3, 3)$ is isosceles.

6. What is the length of the diagonal of a cube if each edge is 5 inches long?

7. What is the length of the diagonal of a box whose dimensions are $2 \times 3 \times 4$?

8. (a) Find the coordinates of all the points on the z-axis that are 5 units from the point $(2, 1, 3)$. (Hint: Points on the z-axis have x- and y-coordinates equal to zero.)
   (b) Find the coördinates of all the points on the x-axis that are 5 units from the point $(2, 1, 3)$.
   (c) Find the coordinates of all the points on the y-axis that are 5 units from the point $(2, 1, 3)$.

9. Express as an equation the statement that the distance between the point (x, y, z) and the point $(0, 0, 0)$ is always equal to 2 units. Describe the graph of this equation.

10. Express as an equation the statement that the distance between the point (x, y, z) and the z-axis is always 2 units. Describe the graph of this equation.

8-2 Functions of Two Variables

In Chapter 2 we defined a function to be a set of ordered pairs (x, y) having the property that for every choice of x there corresponded one and only one y. The variable x was called the independent variable and

y was called the dependent variable. The set of all values assumed by x was called the domain of the function and the set of values taken on by y was called the range. Such a function is sometimes called a function of one variable, since there is only one independent variable x.

In this section we will define a function of *two* variables. Such a function will have two independent variables and one dependent variable. The domain of this function will be, not a set of numbers, but a set of ordered pairs of numbers. This function will assign to every ordered pair in its domain one and only one number. The student has seen many such functions. An example is the formula for the area of a rectangle, which is a function of the two variables, its length and its width.

Definition 8-1. A function of two variables is a set of ordered pairs of elements $((x, y), z)$ such that to every first element (x, y) there corresponds one and only one second element z.

We say that z is a function of the two variables x and y. The set of ordered pairs of numbers (x, y) is the domain of the function and the set of values assumed by z is its range. In the examples we will be studying, all of these numbers will be real numbers.

Example 8-2. The function $z = x^2 + y^2$ is a function of two variables. The domain is the set of all real number pairs, and the range is the set of nonnegative real numbers.

Example 8-3. The equation $z^2 = x^2 + y^2$ is not a function since to the ordered pair $(x, y) = (1, 0)$ there corresponds two values of z, $z = \pm 1$.

As before we name functions by the letters f, h, and so on. If z is a function of the variables x and y, we often write $z = f(x, y)$. If $z = f(x, y)$ is a function of two variables, then the graph of the function f is the set of all points (x, y, z) in space which satisfy the equation. The domain will be some subset of the xy plane. Since to every ordered pair (x, y) in the domain there corresponds one and only one value of z, then *every line perpendicular to the xy plane will intersect the graph of f in at most one point*. Usually the graph of f will be a *surface*.

Example 8-4. The graph of the function $z = \sqrt{1 - x^2 - y^2}$ is a hemisphere (Figure 8-4). Its domain is the set $\{(x, y) \mid x^2 \mid y^2 \leq 1\}$, and its range the set $\{z \mid 0 \leq z \leq 1\}$.

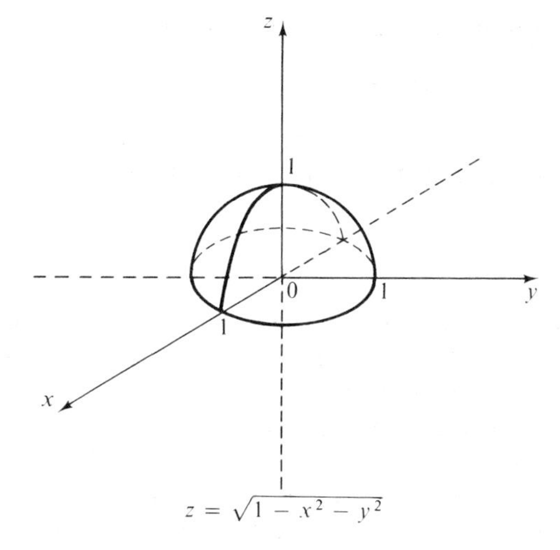

Figure 8-4

Exercise 8-2

1. How would you define a one-to-one function of two variables?

2. How would you define the inverse of a function of two variables? What would be the domain of such a function? The range?

3. Give the domain and range of each of the following functions:
   (a) $z = 1/(x^2 + y^2)$
   (b) $z = 2x + 3y$
   (c) $z = \sqrt{x^2 + y^2}$
   (d) $z = \sqrt{x} + \sqrt{y} + 1$

4. Show that the ordinary binary operations of addition, subtraction, multiplication, and division can be defined to be functions of two variables. Give the formulas representing each function.

5. Define a function of three independent variables x, y, and z. (See Definition 2-1 and Definition 8-1.) What is the domain of such a function? The range? Give an example of a function of three variables.

8-3 Planes and Lines

We saw in Chapter 7 that the graph of a first degree equation in two variables

$$Ax + By + C = 0$$

was a line provided A and B are not both zero.

The graph of a first degree equation in three variables

$$Ax + By + Cz + D = 0 \tag{8-2}$$

where A, B, and C are not all three zero, is a *plane*. Conversely every plane has an equation of this form. If $C \neq 0$, we can solve for z

$$z = \frac{-(Ax + By + D)}{C}$$

and we see that z is a function of the two variables x and y, since to every choice of (x, y) there corresponds a unique number z. If $C = 0$ then the plane is parallel to the z-axis.

To graph a plane it is helpful to find its *intercepts*. The x-intercept of a plane is the point of intersection of the plane and the x-axis. The y-intercept and z-intercept are defined similarly. To find the x-intercept we set $y = z = 0$ in Equation (8-2) and solve for x. The other intercepts are found by setting $x = z = 0$ and $x = y = 0$ respectively.

Example 8-5. Find the intercepts of the plane $2x + 3y + z = 6$ and sketch.

Solution: Setting $y = z = 0$ we find that $x = 3$ and the x-intercept is $(3, 0, 0)$. Similarly the other two intercepts are found to be $(0, 2, 0)$ and $(0, 0, 6)$. If we graph these three points and join them we get a triangular portion of the plane. (See Figure 8-5.)

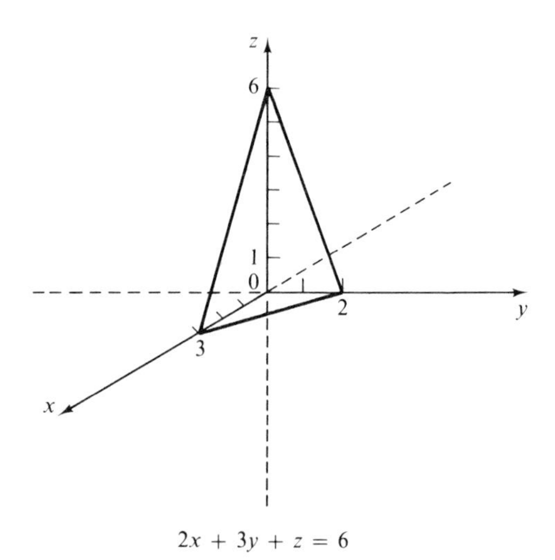

$2x + 3y + z = 6$

Figure 8-5

The lines joining the intercepts in Figure 8-5 are called the traces of the plane. The *traces* of a surface are the intersections of the surface and the coordinate planes. Sketching the traces of a surface is a tremendous help in visualizing these three dimensional objects. To find the trace of a surface in the xy plane, we set $z = 0$ in its equation. The traces in the yz and xz planes are found by setting $x = 0$ and $y = 0$ respectively.

Example 8-6. Graph the plane $3x + 2y = 6$ by finding the intercepts and sketching the traces.

Solution: The x-intercept is (2, 0, 0), the y-intercept is (0, 3, 0) and there is no z-intercept since the z term is missing and x and y cannot both be zero. We conclude that this plane must be parallel to the z-axis. To sketch it, we plot the two intercepts and join them to get the trace in the xy plane. To find the trace in the xz plane we set $y = 0$, getting the line $x = 2$, $y = 0$. The trace in the yz plane is the line $y = 3$, $x = 0$. A portion of this plane is pictured in Figure 8-6.

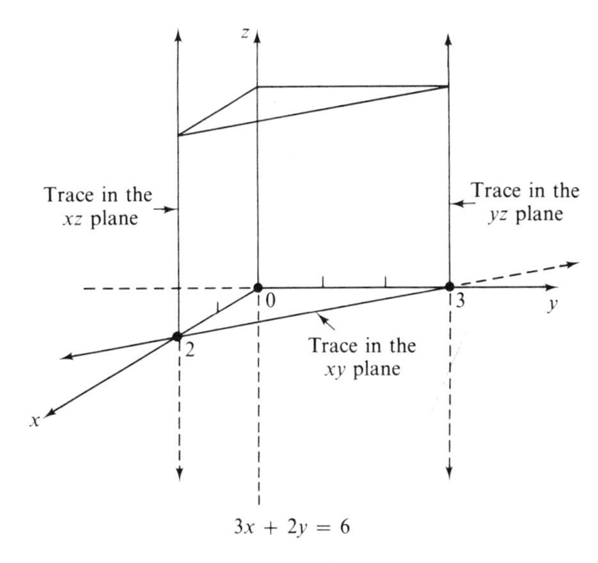

Figure 8-6

Example 8-7. Graph the plane $y = 3$.

Solution: In this case both the x and z terms are missing. Since y can never be zero, there is no x- or z-intercept. We conclude that the plane is parallel to both the x- and z-axes; thus, it must be parallel to the xz

plane. Its trace in the xy plane is $z = 0$, $y = 3$, and its trace in the yz plane is $x = 0$, $y = 3$. (See Figure 8-7.)

Clearly the equation of any plane parallel to the xz plane will be of the form $y = k$, where k is a constant. The equation of a plane parallel to the xy plane will have the form $z = k$, while one parallel to the yz plane will have the equation $x = k$.

On the other hand, if the z term is missing from the equation, we may conclude that the plane is parallel to the z-axis. Similarly, if the y term is missing the plane is parallel to the y-axis, and if the x term is missing the plane is parallel to the x-axis.

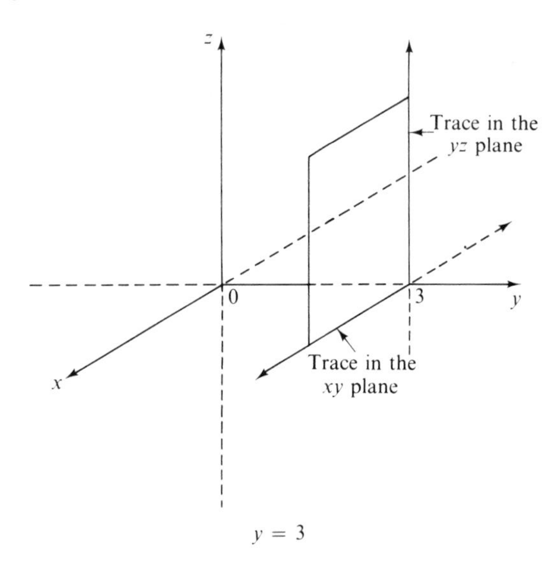

Figure 8-7

Since a line is the intersection of two nonparallel planes, the equation of a line in space will be given by *two* equations of the form of Equation (8-2) which hold simultaneously. Thus, the pair of equations

$$\begin{cases} x + 3y + z = 5 \\ x - y - z = 1 \end{cases}$$

represents a line in space. The student should not be misled into thinking that he can get *one* equation for the line by eliminating one of the unknowns from this pair of equations. If we eliminate z, for instance, we will get the equation

$$x + y = 3$$

which is the equation of still another *plane* passing through the line. Since infinitely many planes pass through a given line, a line can have infinitely many representations as a pair of first degree equations.

To graph a line in space, it will be helpful to find the points where the line intersects the coordinate planes. These are called the *piercing points* of the line. To find the point where the line intersects the xy plane, for example, we let $z = 0$ in both of the equations and solve for x and y. Similarly the piercing points in the xz and yz planes are found by setting $y = 0$ and $x = 0$ respectively.

Example 8-8. Find the piercing points for the line

$$\begin{cases} x + 3y + z = 5 \\ x - y - z = 1 \end{cases}$$

and sketch the line by graphing two of the piercing points and joining them.

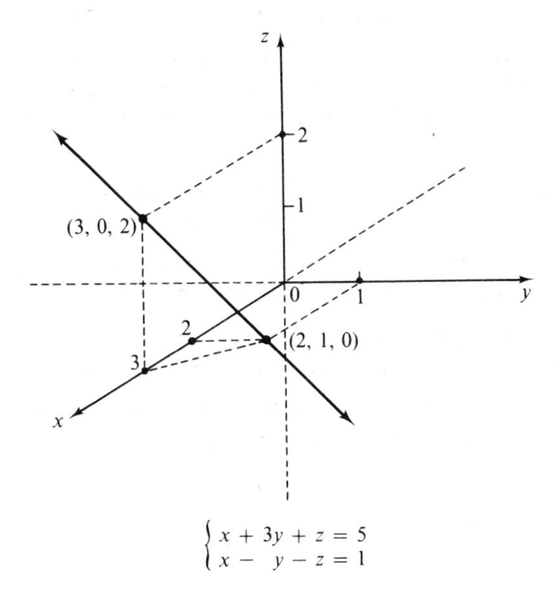

$$\begin{cases} x + 3y + z = 5 \\ x - y - z = 1 \end{cases}$$

Figure 8-8

Solution: Setting $z = 0$ in these equations gives us

$$\begin{cases} x + 3y = 5 \\ x - y = 1 \end{cases}$$

Solving simultaneously, we get $x = 2$, $y = 1$. The piercing point in the

xy plane thus is (2, 1, 0). Similarly setting $y = 0$ and $x = 0$ we find the piercing points (3, 0, 2) and (0, 3, −4). Plotting the first two of these gives us a sketch of the line. (See Figure 8-8.)

Exercise 8-3

1. Graph each of the following planes by finding their intercepts and sketching their traces:
   (a) $x + y + z = 3$
   (b) $2x + y + 4z = 8$
   (c) $x - y + z = 2$
   (d) $x + 2y - z = 4$
   (e) $2x + y = 6$
   (f) $y + 2z = 4$
   (g) $3x - z = 6$
   (h) $z = -2$
   (i) $x = 4$
   (j) $x - y = 0$
   (k) $y - z = 0$
   (l) $2x - y - z = 0$

2. Give an equation for a plane which is
   (a) parallel to the xy plane and 3 units above it.
   (b) parallel to the xz plane and passes through the point (2, −1, 3).
   (c) parallel to the yz plane and passes through the point (2, 1, 1).

3. What is the distance between two planes, one through (1, 2, −2) and the other through (−3, 4, 1), if both planes are perpendicular to the z-axis? If both are perpendicular to the y-axis? If both are perpendicular to the x-axis?

4. Express as an equation the statement that the point $P(x, y, z)$ is equidistant from the points (0, 0, 0) and (4, −2, 1). Describe the graph of this equation.

5. Express as an equation the statement that the point $P(x, y, z)$ is equidistant from the yz and xz planes. Describe the graph of this equation.

6. Express as an equation the statement that the point $P(x, y, z)$ is equidistant from the three coordinate planes. Describe the graph of this equation.

7. Find all of the piercing points for the following lines in space. Sketch the line by locating two of the piercing points and joining them.
   (a) $\begin{cases} x + y + z = 4 \\ 2x - y + z = 6 \end{cases}$
   (b) $\begin{cases} 2x - y + 2z = -1 \\ x - y - z = 5 \end{cases}$
   (c) $\begin{cases} x - y + 3z = 3 \\ 3x + y - z = 1 \end{cases}$
   (d) $\begin{cases} x + y = 6 \\ 2x - 3y + z = -8 \end{cases}$
   (e) $\begin{cases} z = -2 \\ x + y + 2z = 4 \end{cases}$
   (f) $\begin{cases} x = 1 \\ y = 2 \end{cases}$

8. Show that the three planes

$$x + y - z = 3$$
$$3x - y - 11z = 1$$
$$3x - 5y - 19z = -7$$

pass through the same line.

8-4 Cylinders

The graph of $x^2 + y^2 = 1$ in the plane is a circle; in three dimensions the graph of this equation is a right circular cylinder. (See Figure 8-9.)

Since z is missing from this equation, then no matter what value is assigned to z, we still have $x^2 + y^2 = 1$. Thus, the intersection of the surface with any plane $z = k$ parallel to the xy plane is

$$x^2 + y^2 = 1 \qquad z = k$$

The intersection of a surface and a plane is called the *cross section* of the surface in that plane. All of the cross sections of this cylinder in planes parallel to the xy plane are circles.

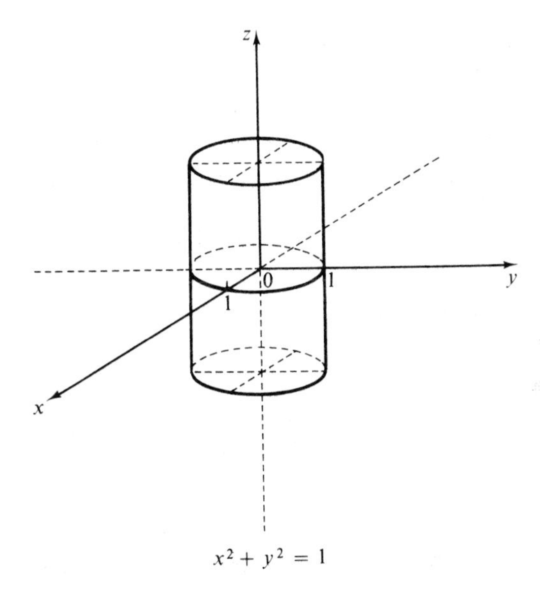

$x^2 + y^2 = 1$

Figure 8-9

In general, a *cylinder* is defined to be a surface that is generated by a line which moves along a plane curve in such a way that it always remains parallel to a fixed line which is not in the plane of the curve. The curve is called the *directrix* of the cylinder (this is not the same as the directrix of a parabola) and the moving line is called the *generator* of the cylinder. (See Figure 8-10.)

If the directrix is a line, then the cylinder is a plane. If the directrix is a circle and the generator is perpendicular to the plane of the circle, then the cylinder is a right circular cylinder.

If the directrix lies in one of the coordinate planes and the generator is perpendicular to this plane, then the equation of the cylinder is especially simple. In particular, suppose that the directrix lies in the xy plane and its equation in this plane is $f(x, y) = 0$; $z = 0$. Now for any point $P(x, y, z)$ on the cylinder, the point $Q(x, y, 0)$ will lie on the directrix and $Q(x, y, 0)$ will satisfy the equation $f(x, y) = 0$; $z = 0$. Thus, $P(x, y, z)$ lies on the cylinder if and only if $f(x, y) = 0$. We conclude that $f(x, y) = 0$ is an equation for the cylinder. In the same way it can be shown that $f(x, z) = 0$ is an equation of a cylinder with directrix $f(x, z) = 0$; $y = 0$ in the xz plane, and $f(y, z) = 0$ is an equation of a cylinder with directrix $f(y, z) = 0$; $x = 0$ in the yz plane. We conclude that in

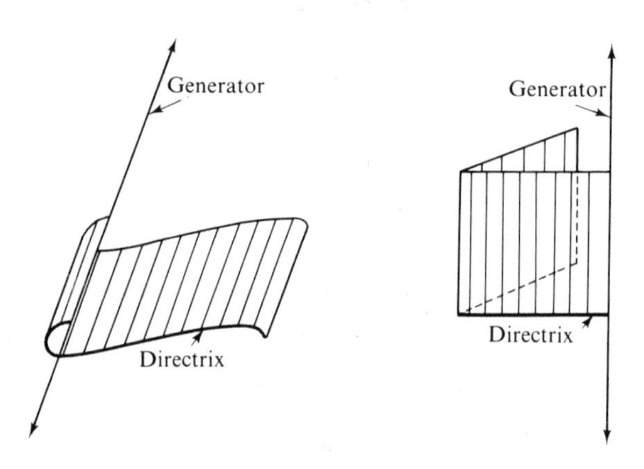

Figure 8-10

three dimensional space an equation in any two of the three variables x, y, z is a cylinder. The converse of this statement is not true. Cylinders whose generator is not perpendicular to one of the coordinate planes will have more complicated equations involving all three variables. (See Exercise 8-5, problem 7.)

Example 8-9. The graph of $y^2 = 2z$ is a *parabolic cylinder* with directrix in the yz plane. To sketch this surface we show its cross sections in the yz plane and some other plane parallel to it. [See Figure 8-11 (a).]

Example 8-10. The graph of $(x^2/a^2) - (y^2/b^2) = 1$ is a *hyperbolic cylinder* with directrix in the xy plane. [See Figure 8-11 (b).]

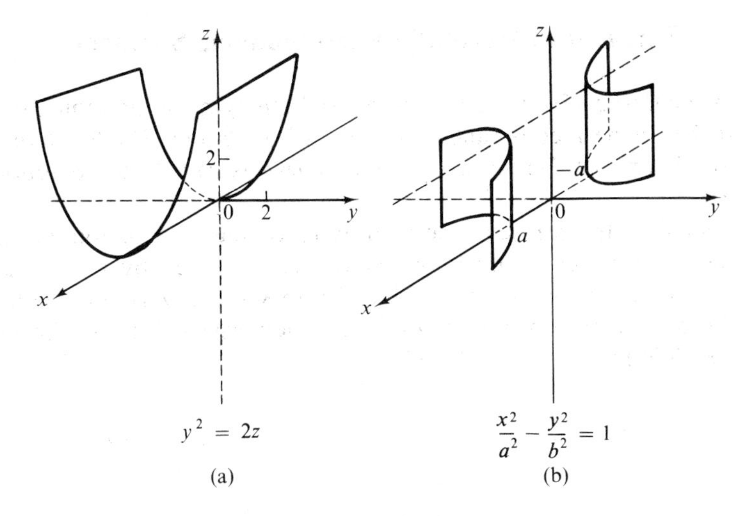

Figure 8-11

Exercise 8-4

1. Graph each of the following cylinders by sketching its cross sections in one of the coordinate planes and some other plane parallel to this coordinate plane.

   (a) $y = x$ (b) $x + z = 1$
   (c) $x^2 + z^2 = 4$ (d) $y^2 + z^2 = 1$
   (e) $x^2/4 + z^2 = 1$ (f) $z^2/4 + y^2 = 1$
   (g) $y^2 = 2x$ (h) $z^2 = 2x$
   (i) $x^2 - y^2 = 1$ (j) $zy = 1;\ z \geq 0;\ y \geq 0$
   (k) $y = \sin x;\ -\pi \leq x \leq \pi$ (l) $z = e^y$
   (m) $y = \ln x$ (n) $z = y^3$
   (o) $z = |y|$ (p) $z = [y]$

2. Find the equation of a cylinder whose directrix is a circle in the xz plane with center $(2, 0, -1)$ and radius 2.

3. Find the equation of a cylinder whose directrix is a line in the yz plane which passes through the points $(0, 1, -1)$ and $(0, 2, 3)$.

4. Find the equation of a cylinder whose directrix is an ellipse in the xy plane with center at the origin, one vertex at $(2, 0, 0)$ and minor axis two units in length.

5. Find the equation of a cylinder whose directrix is a parabola in the xz plane with vertex at the origin and focus $(1, 0, 0)$.

8-5 Surfaces of Revolution and Quadric Surfaces

If a plane curve is revolved about a fixed line lying in the same plane, the surface generated is called a *surface of revolution*. The fixed line is called the axis of the surface and the curve is called the *generating curve*.

A *sphere* is the surface generated by revolving a circle about a line through a diameter. A circular cylinder is generated by revolving a line about an axis parallel to it. (See Figure 8-12.) Clearly every cross section of a surface of revolution in a plane perpendicular to the axis will be a circle.

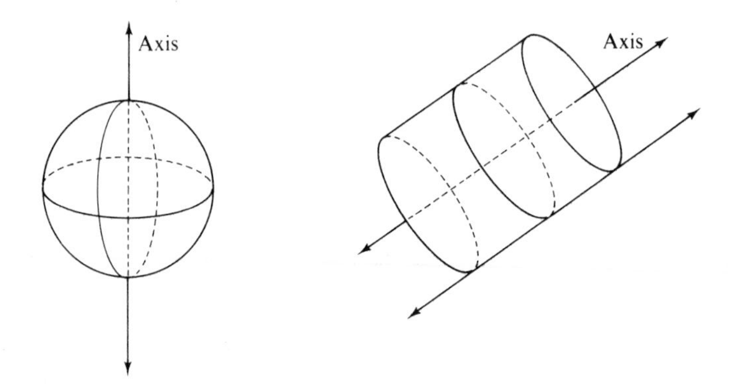

Figure 8-12

If the axis of a surface of revolution is taken to be one of the coordinate axes then an equation for the surface can be easily obtained from the two dimensional equation of the generating curve.

Suppose, for example, we let $z = f(y)$; $x = 0$ be the equation of a curve in the zy plane and revolve this curve about the y-axis. (See Figure 8-13.)

Let $P(x, y, z)$ be a point on the surface and consider the cross section of the surface in a plane through P perpendicular to the y-axis. This cross section will be a circle $x^2 + z^2 = r^2$; $y = y$, and will contain a point $P_0(0, y, z_0)$ of the generating curve. It is easy to see that the radius of this circle is z_0, so we have $x^2 + z^2 = z_0^2$. But P_0 is on the generating curve so y and z_0 satisfy the equation $z_0 = f(y)$. Thus, we have

$$x^2 + z^2 = [f(y)]^2$$

as the equation of the surface.

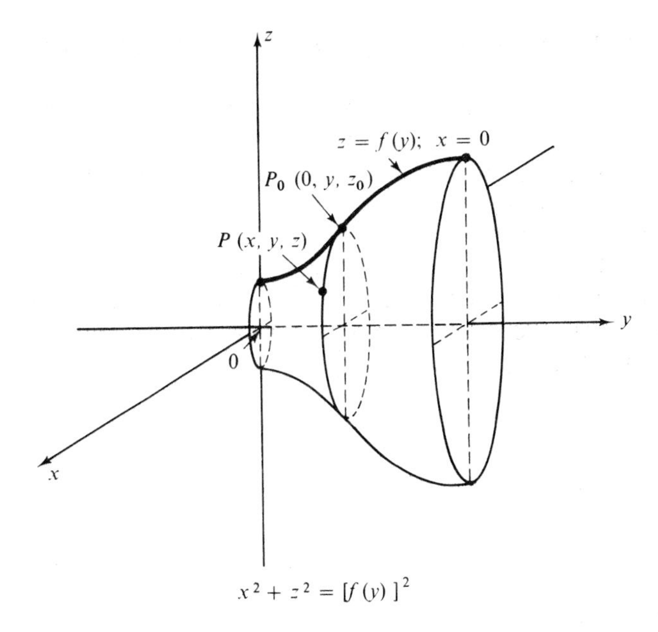

Figure 8-13

In the same way we could show that the equation of the surface generated by revolving this curve $z = f(y)$; $x = 0$ about the z-axis is

$$x^2 + y^2 = [g(z)]^2$$

where $y = g(z)$ is obtained by solving $z = f(y)$ for y.

In general, if the axis of the surface of revolution is

(1) the x-axis, then the equation will be

$$y^2 + z^2 = [f(x)]^2$$

(2) the y-axis, then the equation will be

$$x^2 + z^2 = [f(y)]^2$$

(3) the z-axis, then the equation will be

$$x^2 + y^2 = [f(z)]^2$$

where $f(x)$, $f(y)$, or $f(z)$ is found from the equation of the generating curve.

Example 8-11. Find the equation of the surface generated by revolving the parabola $z^2 = 4y$ about (a) the y-axis, (b) the z-axis.

Solution: (a) Since $z = f(y)$, $[f(y)]^2 = 4y$ and the equation of the surface is

$$x^2 + z^2 = 4y$$

when the parabola is revolved about the y-axis.

(b) On the other hand, $y = g(z) = z^2/4$, and the equation of the surface is

$$x^2 + y^2 = \left(\frac{z^2}{4}\right)^2$$

when the parabola is revolved about the z-axis. (See Figure 8-14.)

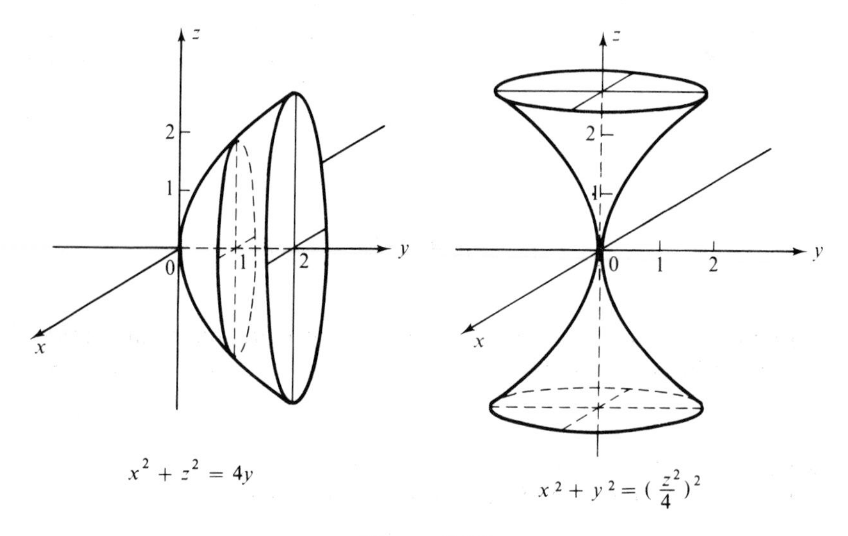

Figure 8-14

The first of these surfaces is called a *paraboloid of revolution.*

The paraboloid of revolution is an example of a class of surfaces called *quadric surfaces.* A quadric surface is the graph of a second degree equation in three variables. That is, its equation is of the general form

$$Ax^2 + By^2 + Cz^2 + Dxy + Exz + Fyz + Gx + Hy + Iz + J = 0$$

where A, B, C, D, E, and F are not all zero.

We have already studied some of these surfaces. The cylinders of Section 8-4 whose directrices were conics are quadric surfaces. It can be shown that every quadric surface which is not a cylinder or some degenerate form such as a plane, line, point, etc., can be classified as one of six general types. In describing each of these six types we will choose the coordinate axes so that the equation will be in its simplest form.

It should be noted that in the following equations our choice of the order in which the variables x, y, and z appear is purely arbitrary. Any permutation of these letters in the equation will result in a graph which is the same surface but has a different orientation on the axes. Thus, for example,

$$\frac{x^2}{a^2} + \frac{y^2}{b^2} = z$$

$$\frac{x^2}{a^2} + \frac{z^2}{c^2} = y$$

$$\frac{y^2}{b^2} + \frac{z^2}{c^2} = x$$

are all equations for the elliptic paraboloid.

1. *The Ellipsoid.* (Figure 8-15)

$$\frac{x^2}{a^2} + \frac{y^2}{b^2} + \frac{z^2}{c^2} = 1$$

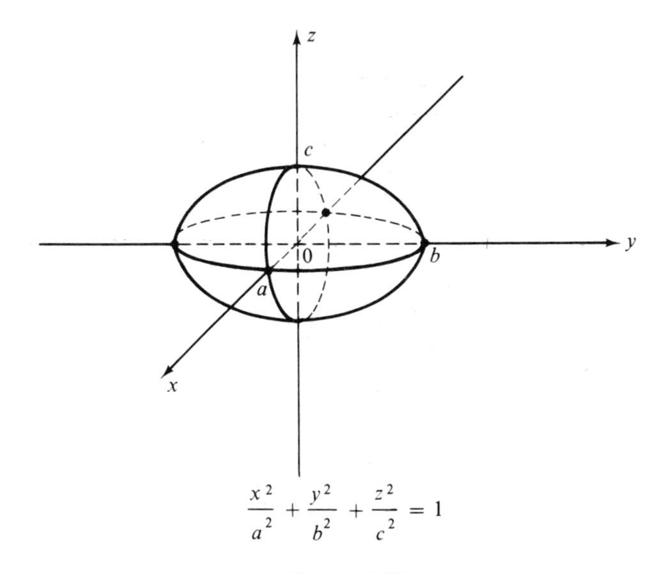

Figure 8-15

The traces of this surface are

$$\frac{x^2}{a^2} + \frac{y^2}{b^2} = 1 \qquad z = 0 \qquad \text{in the } xy \text{ plane}$$

$$\frac{y^2}{b^2} + \frac{z^2}{c^2} = 1 \qquad x = 0 \qquad \text{in the } yz \text{ plane}$$

$$\frac{x^2}{a^2} + \frac{z^2}{c^2} = 1 \qquad y = 0 \qquad \text{in the } xz \text{ plane}$$

These traces are ellipses if the constants are different, circles if they are equal. If $a = b = c$, the surface is a *sphere*.

If only two of the three constants are equal, then one trace is a circle, the other two are ellipses, and the surface is an *ellipsoid of revolution*, also called a *spheroid*, obtained when a two dimensional ellipse is revolved about one of its axes. (See Exercise 8-5, problem 3.)

2. *The Hyperboloid of One Sheet.* (Figure 8-16)

$$\frac{x^2}{a^2} + \frac{y^2}{b^2} - \frac{z^2}{c^2} = 1$$

The traces of this surface in the xz and yz planes are the hyperbolas

$$\frac{x^2}{a^2} - \frac{z^2}{c^2} = 1 \qquad y = 0$$

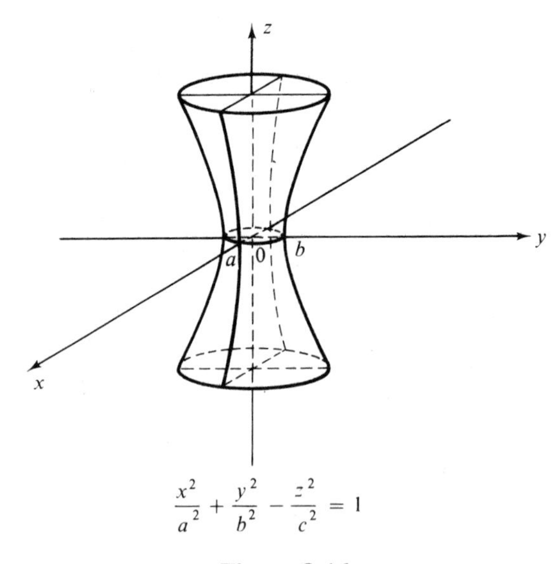

Figure 8-16

$$\frac{y^2}{b^2} - \frac{z^2}{c^2} = 1 \qquad x = 0$$

The trace in the xy plane is

$$\frac{x^2}{a^2} + \frac{y^2}{b^2} = 1 \qquad z = 0$$

which is an ellipse if $a \neq b$, a circle if $a = b$. If $a = b$ the surface is a hyperboloid of revolution obtained by revolving a hyperbola about its conjugate axis. (See Exercise 8-5, problem 4.)

3. *The Hyperboloid of Two Sheets.* (Figure 8-17)

$$\frac{y^2}{b^2} - \frac{x^2}{a^2} - \frac{z^2}{c^2} = 1$$

The traces of this surface in the xy and yz planes are hyperbolas. There is no trace in the xz plane, since

$$-\frac{x^2}{a^2} - \frac{z^2}{c^2} = 1 \qquad y = 0$$

has no real solutions. However, for $|k| > b$ the cross sections of the surface in the planes $y = k$ are the curves

$$\frac{x^2}{a^2} + \frac{z^2}{c^2} = \frac{k^2}{b^2} - 1 \qquad y = k$$

Since $k^2/b^2 - 1 > 0$, this is an ellipse if $a \neq c$, a circle otherwise. If $a = c$, the surface is a hyperboloid of revolution generated by revolving a hyperbola about its transverse axis. (See Exercise 8-5, problem 4.)

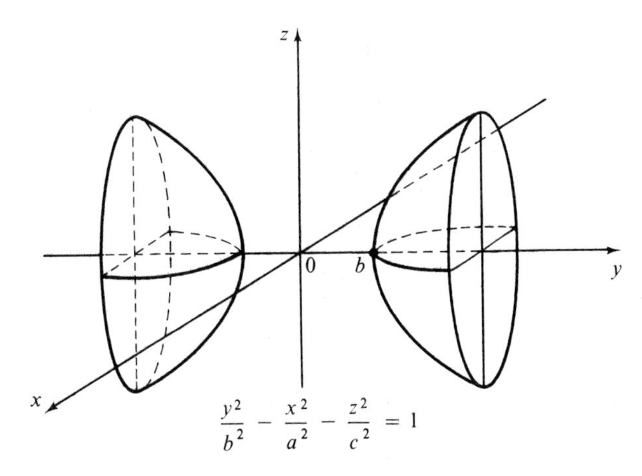

Figure 8-17

4. *The Elliptic Paraboloid.* (Figure 8-18)

$$\frac{x^2}{a^2}+\frac{y^2}{b^2}=z$$

The traces of this surface in the xz and yz planes are the parabolas

$$\frac{x^2}{a^2}=z \qquad y=0$$

and

$$\frac{y^2}{b^2}=z \qquad x=0$$

The trace in the xy plane is the single point (0, 0, 0). For $k>0$, the cross sections in the planes $z=k$ are the curves

$$\frac{x^2}{a^2}+\frac{y^2}{b^2}=k \qquad z=k$$

These are ellipses if $a \neq b$, circles otherwise. If $a=b$ the surface is a paraboloid of revolution generated by revolving a parabola about its axis.

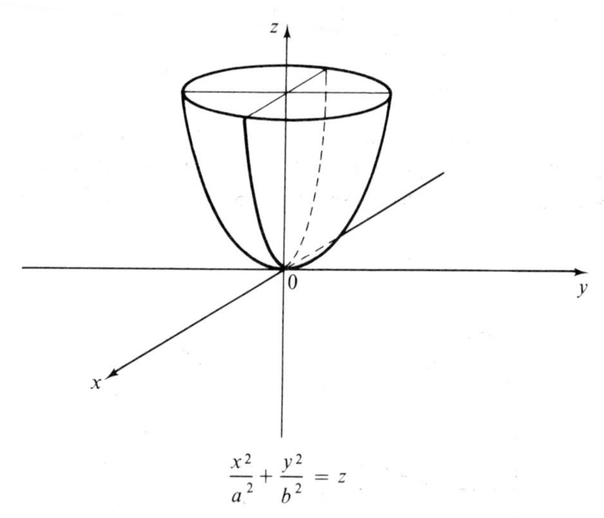

Figure 8-18

5. *The Hyperbolic Paraboloid.* (Figure 8-19)

$$\frac{y^2}{b^2}-\frac{x^2}{a^2}=z$$

The trace of this surface in the xy plane is a pair of intersecting lines

$$\begin{cases} ay - bx = 0 \\ ay + bx = 0 \end{cases} \qquad z = 0$$

The cross sections in the planes $z = k$ will be the hyperbolas

$$\frac{y^2}{b^2} - \frac{x^2}{a^2} = k \qquad z = k$$

The transverse axis of these hyperbolas will lie on the y-axis if $k > 0$; on the x-axis if $k < 0$.

The trace in the yz plane is the parabola

$$\frac{y^2}{b^2} = z \qquad x = 0$$

which opens upward.

The trace in the xz plane is the parabola

$$-\frac{x^2}{a^2} = z \qquad y = 0$$

which opens downward.

This rather complicated surface is shaped like a saddle or a mountain pass.

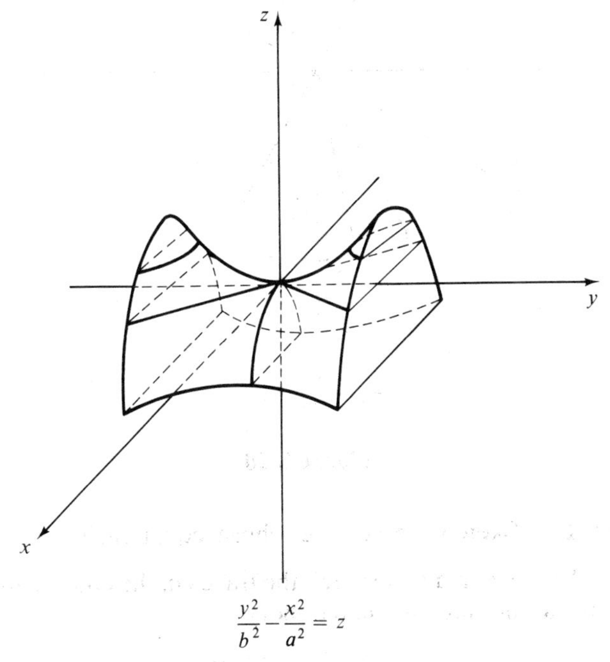

$$\frac{y^2}{b^2} - \frac{x^2}{a^2} = z$$

Figure 8-19

6. *The Elliptic Cone.* (Figure 8-20)

$$\frac{x^2}{a^2} + \frac{y^2}{b^2} = z^2$$

The trace of this surface in the xz or yz planes is a pair of intersecting lines. The cross sections in planes $x = k$ or $y = k$ are hyperbolas.

The trace in the xy plane is the point (0, 0, 0). Cross sections in the planes $z = k$ are the curves

$$\frac{x^2}{a^2} + \frac{y^2}{b^2} = k^2 \qquad z = k$$

which are ellipses if $a \neq b$, circles otherwise. If $a = b$ this surface is called simply a *cone*. It is a surface of revolution obtained by revolving a line through the origin about the z-axis. (See Exercise 8-5, problem 1.)

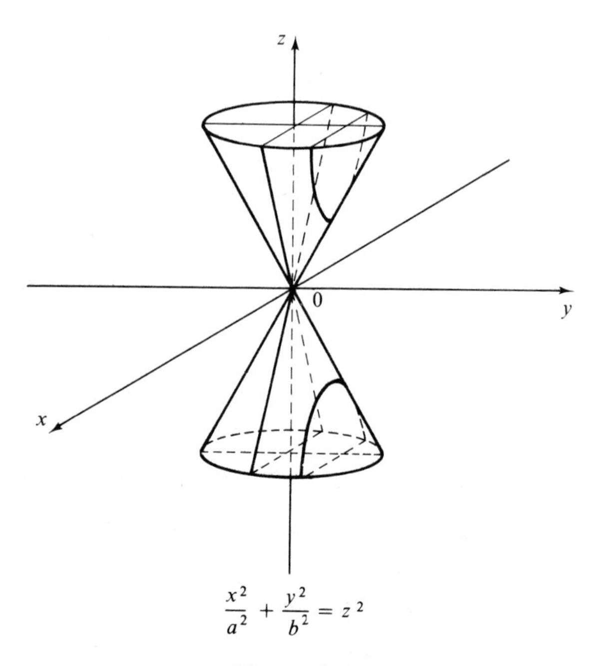

Figure 8-20

Example 8-12. Sketch the surface whose equation is $y = 2x^2 + z^2$.

Solution: First we find and sketch the traces in the coordinate planes. In the xy plane the trace is the parabola

$$y = 2x^2 \qquad z = 0$$

and in the zy plane the parabola

$$y = z^2 \qquad x = 0$$

In the xz plane the trace is the single point (0, 0, 0), so we find the cross section in some plane parallel to the xz plane, for example $y = 2$. This is the ellipse

$$2x^2 + z^2 = 2 \qquad y = 2$$

The surface is an elliptic paraboloid. (See Figure 8-21.)

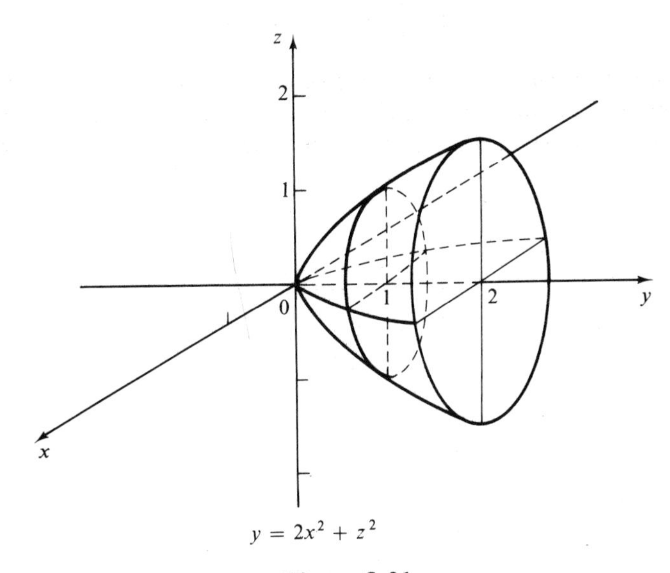

Figure 8-21

Example 8-13. Sketch the surface whose equation is $x^2 - 2y^2 + z^2 = 4$.

Solution: The trace in the xy plane is the hyperbola

$$\frac{x^2}{4} - \frac{y^2}{2} = 1 \qquad z = 0$$

In the yz plane it is the hyperbola

$$\frac{z^2}{4} - \frac{y^2}{2} = 1 \qquad x = 0$$

and in the xz plane it is the circle

$$x^2 + z^2 = 4 \qquad y = 0$$

In this example we sketch only the portion of the surface that lies

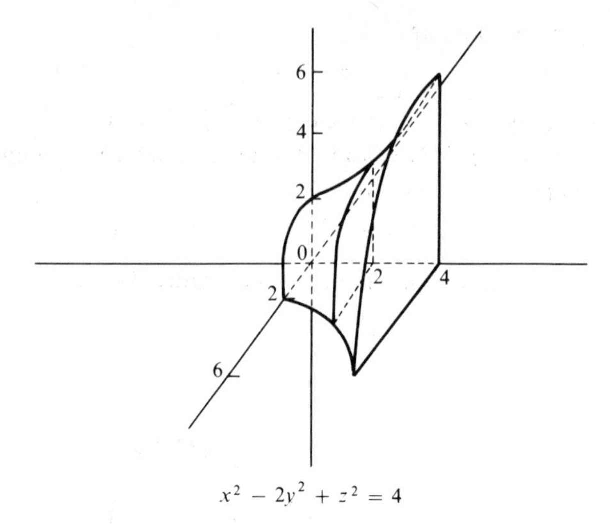

$x^2 - 2y^2 + z^2 = 4$

Figure 8-22

in the octant $x \geq 0$; $y \geq 0$; $z \geq 0$. Because of its symmetry the remainder of the surface can then be visualized. The surface is a hyperboloid of one sheet. (See Figure 8-22.)

Exercise 8-5

1. Find the equation of the surface of revolution generated when the line $z = my$, $m \neq 0$, is rotated about the z-axis. This surface is a *cone*. Sketch this surface.

2. Find the equation of the surface of revolution generated when the circle $y^2 + z^2 = r^2$ is rotated about the z-axis, about the y-axis. This surface is the *sphere*. Sketch this surface.

3. Find the equation of the surface of revolution generated when the ellipse

$$\frac{y^2}{a^2} + \frac{z^2}{b^2} = 1 \qquad a > b$$

is rotated
(a) about its major axis. This is called a *prolate spheroid.*
(b) about its minor axis. This is called an *oblate spheroid.*
Sketch both of these surfaces.

4. Find the equation of the surface of revolution generated when the hyperbola

$$\frac{y^2}{a^2} - \frac{z^2}{b^2} = 1$$

is rotated

(a) about its conjugate axis. This is the *hyperboloid of one sheet.*
(b) about its transverse axis. This is the *hyperboloid of two sheets.*
Sketch both of these surfaces.

5. Find the equation of the surface of revolution generated when the circle $(y-2)^2+z^2=1$ is rotated about the z-axis. This surface is called a *torus* and it is shaped like a doughnut. Sketch this surface.

6. In each of the following find an equation for the surface of revolution generated by revolving the given plane curve about the indicated axis. Sketch each surface.
   (a) $y+z=1$ about the z-axis
   (b) $y+z=1$ about the y-axis
   (c) $z=e^y$ about the y-axis
   (d) $z=\sin y$ about the y-axis
   (e) $y=x^3$ about the x-axis

7. Show that the graph of $x^2=y-z$ is a cylinder whose generating line is not perpendicular to the directrix by sketching the surface. Find the equation of the directrix. What is the inclination of the generating line?

8. Identify each of the following surfaces. Find the traces of each in the coordinate planes. Sketch the surface by first drawing the traces and, if necessary, cross sections in planes parallel to the coordinate planes.
   (a) $x^2+y^2+z^2=9$ (b) $x^2+4y^2+z^2=4$
   (c) $x^2+4y^2-z^2=4$ (d) $2x^2+z^2-y=0$
   (e) $x^2+4y^2-z^2=0$ (f) $x^2-4y^2-z^2=4$
   (g) $z=y^2-x^2$

9. Each of the following equations has as its graph a degenerate form of a quadric surface. Describe the graph of each.
   (a) $z^2=1$ (b) $x^2-y^2=0$
   (c) $x^2=0$ (d) $x^2+y^2=0$
   (e) $x^2+y^2+z^2=0$ (f) $x^2+y^2+z^2+1=0$

10. Describe each of the following surfaces:
   (a) $x^2+y^2=1$ (b) $x^2+y^2=x$
   (c) $x^2+y^2=z$ (d) $x^2+y^2=z^2$
   (e) $x^2+y^2=z^2+1$ (f) $x^2+y^2=z^2-1$
   (g) $x^2-y^2=1$ (h) $x^2-y^2=z$
   (i) $x^2-y^2=z^2$ (j) $x^2-y^2=z^2+1$
   (k) $x^2-y^2=z^2-1$ (l) $x^2+y^2=1-z^2$

8-6 Cylindrical and Spherical Coordinates

The system of rectangular coordinates we have been using is not the only way to locate points in three dimensional space. Two other useful systems are the cylindrical and spherical coordinate systems. These systems are extremely useful in calculus in problems such as finding the volume or the mass of certain solids.

The cylindrical coordinate system is simply an extension of polar coordinates in the plane. To find the cylindrical coordinates of a point P whose rectangular coordinates are (x, y, z), we drop a perpendicular from P to the xy plane. The point Q at the foot of the perpendicular has polar coordinates (r, θ) in the xy plane. The cylindrical coordinates of P are taken to be (r, θ, z), where z is the same as in rectangular coordinates. (See Figure 8-23.)

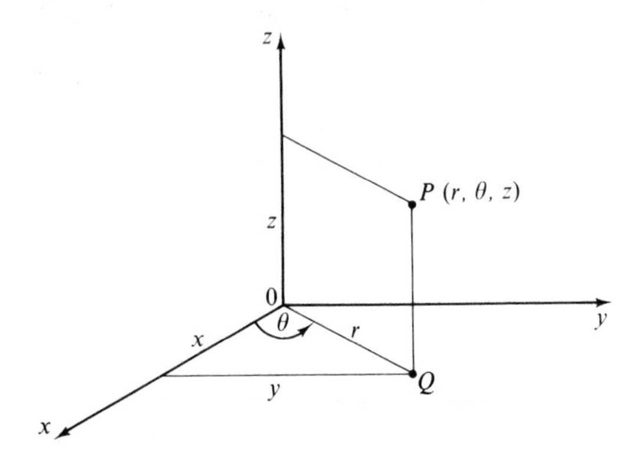

Figure 8-23

The relationship between rectangular coordinates (x, y, z) and cylindrical coordinates (r, θ, z) is given by the equations

$$x = r\cos\theta$$
$$y = r\sin\theta \qquad \textbf{(8-3)}$$
$$z = z$$

or

$$r^2 = x^2 + y^2$$
$$\tan\theta = \frac{y}{x} \qquad \textbf{(8-4)}$$
$$z = z$$

Just as in polar coordinates, a point may have many different representations in cylindrical coordinates. The point whose rectangular coordinates are $(1, 1, 1)$ has cylindrical coordinates $(\sqrt{2}, \pi/4, 1)$ or $(\sqrt{2}, -7\pi/4, 1)$ or $(-\sqrt{2}, 5\pi/4, 1)$ and so on.

The equation of the circular cylinder $x^2 + y^2 = k^2$ is particularly simple in cylindrical coordinates, and this accounts for the name given to this coordinate system. Since $x^2 + y^2 = r^2$, the equation of the cylinder becomes simply $r = k$.

To find the spherical coordinates of a point P whose rectangular coordinates are (x, y, z), we again drop a perpendicular from P to a point Q in the xy plane. (See Figure 8-24.) The spherical coordinates of P are (ρ, θ, ϕ) where ρ is the distance from P to the origin; θ is the measure of the angle from the positive x-axis to the ray OQ, as in cylindrical coordinates; and ϕ is the measure of the angle from the positive z-axis to the ray OP. Unlike polar and cylindrical coordinates, there are some restrictions on the values these coordinates may assume. We have

$$\rho \geq 0$$
$$0 \leq \theta < 2\pi$$
$$0 \leq \phi \leq \pi$$

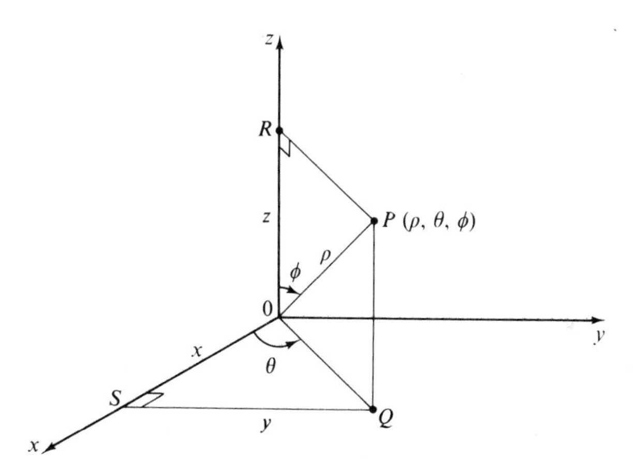

Figure 8-24

As a result of these restrictions there is a one-to-one correspondence between points in space and the number triples which are their spherical coordinates, with the exception of points on the z-axis.

The relationship between rectangular coordinates (x, y, z) and spherical coordinates (ρ, θ, ϕ) is easily derived. From Figure 8-24, we see that triangle OPR is a right triangle with right angle at R. Thus, $\cos \phi = OR/OP = z/\rho$ and we have $z = \rho \cos \phi$.

Triangle OQS is a right triangle also with right angle at S, thus $\cos \theta = x/OQ$ and $\sin \theta = y/OQ$. From this we get

$$x = OQ \cos \theta \qquad \textbf{(8-5)}$$
$$y = OQ \sin \theta$$

But $OQ = RP$ since they are opposite sides of a parallelogram and from triangle OPR we have $\sin \phi = RP/\rho$. Thus, $OQ = RP = \rho \sin \phi$. Substituting this in Equations (8-5) we have $x = \rho \sin \phi \cos \theta$; $y = \rho \sin \phi \sin \theta$. The equations giving x, y, and z in terms of ρ, θ, and ϕ, thus are

$$\begin{aligned} x &= \rho \sin \phi \cos \theta \\ y &= \rho \sin \phi \sin \theta \\ z &= \rho \cos \phi \end{aligned} \tag{8-6}$$

To find equations giving ρ, θ, and ϕ in terms of x, y, and z, first we observe that since ρ is the distance from P to the origin, then $\rho = \sqrt{x^2+y^2+z^2}$. The coordinate θ is the same as in cylindrical coordinates, so $\tan \theta = y/x$, where $0 \le \theta < 2\pi$. From the last of the three equations in (8-6) we get $\cos \phi = z/\rho = z/\sqrt{x^2+y^2+z^2}$. Thus,

$$\begin{aligned} \rho &= \sqrt{x^2+y^2+z^2} \\ \theta &= \arctan \frac{y}{x} \qquad 0 \le \theta < 2\pi \\ \phi &= \arccos z/\sqrt{x^2+y^2+z^2} \qquad 0 \le \phi \le \pi \end{aligned} \tag{8-7}$$

The equation of the sphere $x^2 + y^2 + z^2 = k^2$ is especially simple in spherical coordinates, and this is the reason for the name given to this coordinate system. Since $\rho^2 = x^2 + y^2 + z^2$, the equation of the sphere is simply $\rho = |k|$.

Exercise 8-6

1. (a) Locate each of the following points in a cylindrical coordinate system: $A(1, 0, 0)$; $B(2, \pi/2, 1)$; $C(-2, \pi/2, 1)$; $D(1, \pi/4, -1)$; $E(0, \pi, 2)$.
   (b) Locate each of the following points in a spherical coordinate system: $A(1, \pi/2, \pi)$; $B(1, \pi, 0)$; $C(0, \pi, \pi)$; $D(2, 0, \pi/2)$; $E(2, \pi/2, \pi/2)$.

2. Show that points on the z-axis can have more than one set of spherical coordinates.

3. Find (i) cylindrical and (ii) spherical coordinates for the points whose rectangular coordinates are:
   (a) $(1, 0, 0)$ (b) $(1/\sqrt{2}, 1/\sqrt{2}, 1)$
   (c) $(1, 1, 1)$ (d) $(0, 0, 2)$
   (e) $(0, 2, 0)$ (f) $(1, -1, 1)$
   (g) $(\sqrt{3}/4, 3/4, 1/2)$ (h) $(1/2\sqrt{2}, 1/2\sqrt{2}, \sqrt{3}/2)$

4. Find rectangular coordinates for the points whose cylindrical coordinates are:
   (a) $(2, \pi/2, -1)$ (b) $(2, \pi/3, 2)$
   (c) $(1, 0, 1)$ (d) $(1, 3\pi/4, -1)$

5. Find rectangular coordinates for the points whose spherical coordinates are:
   (a) $(5, \pi/2, \pi/2)$ (b) $(1, \pi, \pi/2)$
   (c) $(2, \pi, \pi/6)$ (d) $(1, \pi/4, \pi/4)$

6. Describe the graph of (a) $r = 0$, (b) $\theta = k$, (c) $z = k$, where k is a constant, in cylindrical coordinates.

7. Describe the graph of (a) $\rho = 0$, (b) $\theta = k$ $(0 \le k < 2\pi)$, (c) $\phi = k$ $(0 \le k \le \pi)$ in spherical coordinates.

8. Change the equation of the cone $x^2 + y^2 = 2z^2$ to (a) cylindrical, (b) spherical coordinates.

9. Change the equation of the cylinder $x^2 + y^2 = 4$ to (a) cylindrical, (b) spherical coordinates.

10. Change the equation of the sphere $x^2 + y^2 + z^2 = 1$ to (a) cylindrical, (b) spherical coordinates.

11. Change the equation of the paraboloid $x^2 + y^2 + z - 4 = 0$ to (a) cylindrical, (b) spherical coordinates.

12. Change the equation of the ellipsoid $x^2 + y^2 + 2z^2 = 4$ to (a) cylindrical, (b) spherical coordinates.

13. (a) Derive equations expressing the cylindrical coordinates r, θ, and z in terms of the spherical coordinates ρ, θ, and ϕ.
    (b) Derive equations expressing the spherical coordinates ρ, θ, and ϕ in terms of the cylindrical coordinates r, θ, and z.

Review Exercise

1. Give the formula for the distance between two points in (a) three dimensional space, (b) two dimensional space (a plane), (c) one dimensional space (a line).

2. Is the triangle whose vertices are $(1, 1, 1)$, $(2, -1, 0)$, and $(3, 0, 2)$ isosceles? Is it equilateral?

3. (a) Define a function of one variable. (b) Define a function of two variables.

4. Find the intercepts and the equations of the traces of $3x + 2y - z = 6$ and graph.

5. Show that the three planes

$$x + y + z = 4$$
$$2x - y + z = 6$$
$$x + 4y + 2z = 6$$

pass through the same line.

6. Find an equation for the cylinder whose directrix is a circle in the yz plane with center (0, 1, 2) and radius 3. Graph.

7. Find an equation for the surface of revolution generated when the ellipse $2y^2 + 3z^2 = 6$; $x = 0$ is rotated about (a) its major axis, (b) its minor axis.

8. Identify each of the following surfaces:
(a) $2x^2 + 3y^2 + 6z^2 = 6$ (b) $2x^2 - 3y^2 - 6z^2 = 6$
(c) $2x^2 - 3y^2 = 6$ (d) $2x^2 - 3y^2 = z$
(e) $2x^2 + 3y^2 = z^2$

9. Give the equations which express the rectangular coordinates x, y, and z in terms of (a) cylindrical coordinates r, θ, and z; (b) spherical coordinates ρ, θ, and ϕ.

10. Change the equation of the hyperboloid $x^2 + y^2 - z^2 = 4$ to (a) cylindrical, (b) spherical coordinates.

Bibliography

Albert, Adrian, *Solid Analytic Geometry*. New York: McGraw-Hill, 1949.

Coolidge, Julian L., *A History of the Conic Sections and Quadric Surfaces*. London: Oxford University Press, 1945.

Kline, Morris, *Mathematics in Western Culture*, Ch. 12. New York: Oxford University Press, 1953.

Van der Waerden, B. L., *Science Awakening*, Ch. 7. New York: Oxford University Press, 1961.

Appendix

Table 1

Logarithms of Numbers

N	0	1	2	3	4	5	6	7	8	9
10	0000	0043	0086	0128	0170	0212	0253	0294	0334	0374
11	0414	0453	0492	0531	0569	0607	0645	0682	0719	0755
12	0792	0828	0864	0899	0934	0969	1004	1038	1072	1106
13	1139	1173	1206	1239	1271	1303	1335	1367	1399	1430
14	1461	1492	1523	1553	1584	1614	1644	1673	1703	1732
15	1761	1790	1818	1847	1875	1903	1931	1959	1987	2014
16	2041	2068	2095	2122	2148	2175	2201	2227	2253	2279
17	2304	2330	2355	2380	2405	2430	2455	2480	2504	2529
18	2553	2577	2601	2625	2648	2672	2695	2718	2742	2765
19	2788	2810	2833	2856	2878	2900	2923	2945	2967	2989
20	3010	3032	3054	3075	3096	3118	3139	3160	3181	3201
21	3222	3243	3263	3284	3304	3324	3345	3365	3385	3404
22	3424	3444	3464	3483	3502	3522	3541	3560	3579	3598
23	3617	3636	3655	3674	3692	3711	3729	3747	3766	3784
24	3802	3820	3838	3856	3874	3892	3909	3927	3945	3962
25	3979	3997	4014	4031	4048	4065	4082	4099	4116	4133
26	4150	4166	4183	4200	4216	4232	4249	4265	4281	4298
27	4314	4330	4346	4362	4378	4393	4409	4425	4440	4456
28	4472	4487	4502	4518	4533	4548	4564	4579	4594	4609
29	4624	4639	4654	4669	4683	4698	4713	4728	4742	4757
30	4771	4786	4800	4814	4829	4843	4857	4871	4886	4900
31	4914	4928	4942	4955	4969	4983	4997	5011	5024	5038
32	5051	5065	5079	5092	5105	5119	5132	5145	5159	5172
33	5185	5198	5211	5224	5237	5250	5263	5276	5289	5302
34	5315	5328	5340	5353	5366	5378	5391	5403	5416	5428
35	5441	5453	5465	5478	5490	5502	5514	5527	5539	5551
36	5563	5575	5587	5599	5611	5623	5635	5647	5658	5670
37	5682	5694	5705	5717	5729	5740	5752	5763	5775	5786
38	5798	5809	5821	5832	5843	5855	5866	5877	5888	5899
39	5911	5922	5933	5944	5955	5966	5977	5988	5999	6010
40	6021	6031	6042	6053	6064	6075	6085	6096	6107	6117
41	6128	6138	6149	6160	6170	6180	6191	6201	6212	6222
42	6232	6243	6253	6263	6274	6284	6294	6304	6314	6325
43	6335	6345	6355	6365	6375	6385	6395	6405	6415	6425
44	6435	6444	6454	6464	6474	6484	6493	6503	6513	6522
45	6532	6542	6551	6561	6571	6580	6590	6599	6609	6618
46	6628	6637	6646	6656	6665	6675	6684	6693	6702	6712
47	6721	6730	6739	6749	6758	6767	6776	6785	6794	6803
48	6812	6821	6830	6839	6848	6857	6866	6875	6884	6893
49	6902	6911	6920	6928	6937	6946	6955	6964	6972	6981
50	6990	6998	7007	7016	7024	7033	7042	7050	7059	7067
51	7076	7084	7093	7101	7110	7118	7126	7135	7143	7152
52	7160	7168	7177	7185	7193	7202	7210	7218	7226	7235
53	7243	7251	7259	7267	7275	7284	7292	7300	7308	7316
54	7324	7332	7340	7348	7356	7364	7372	7380	7388	7396

Table 1

Logarithms of Numbers (continued)

N	0	1	2	3	4	5	6	7	8	9
55	7404	7412	7419	7427	7435	7443	7451	7459	7466	7474
56	7482	7490	7497	7505	7513	7520	7528	7536	7543	7551
57	7559	7566	7574	7582	7589	7597	7604	7612	7619	7627
58	7634	7642	7649	7657	7664	7672	7679	7686	7694	7701
59	7709	7716	7723	7731	7738	7745	7752	7760	7767	7774
60	7782	7789	7796	7803	7810	7818	7825	7832	7839	7846
61	7853	7860	7868	7875	7882	7889	7896	7903	7910	7917
62	7924	7931	7938	7945	7952	7959	7966	7973	7980	7987
63	7993	8000	8007	8014	8021	8028	8035	8041	8048	8055
64	8062	8069	8075	8082	8089	8096	8102	8109	8116	8122
65	8129	8136	8142	8149	8156	8162	8169	8176	8182	8189
66	8195	8202	8209	8215	8222	8228	8235	8241	8248	8254
67	8261	8267	8274	8280	8287	8293	8299	8306	8312	8319
68	8325	8331	8338	8344	8351	8357	8363	8370	8376	8382
69	8388	8395	8401	8407	8414	8420	8426	8432	8439	8445
70	8451	8457	8463	8470	8476	8482	8488	8494	8500	8506
71	8513	8519	8525	8531	8537	8543	8549	8555	8561	8567
72	8573	8579	8585	8591	8597	8603	8609	8615	8621	8627
73	8633	8639	8645	8651	8657	8663	8669	8675	8681	8686
74	8692	8698	8704	8710	8716	8722	8727	8733	8739	8745
75	8751	8756	8762	8768	8774	8779	8785	8791	8797	8802
76	8808	8814	8820	8825	8831	8837	8842	8848	8854	8859
77	8865	8871	8876	8882	8887	8893	8899	8904	8910	8915
78	8921	8927	8932	8938	8943	8949	8954	8960	8965	8971
79	8976	8982	8987	8993	8998	9004	9009	9015	9020	9025
80	9031	9036	9042	9047	9053	9058	9063	9069	9074	9079
81	9085	9090	9096	9101	9106	9112	9117	9122	9128	9133
82	9138	9143	9149	9154	9159	9165	9170	9175	9180	9186
83	9191	9196	9201	9206	9212	9217	9222	9227	9232	9238
84	9243	9248	9253	9258	9263	9269	9274	9279	9284	9289
85	9294	9299	9304	9309	9315	9320	9325	9330	9335	9340
86	9345	9350	9355	9360	9365	9370	9375	9380	9385	9390
87	9395	9400	9405	9410	9415	9420	9425	9430	9435	9440
88	9445	9450	9455	9460	9465	9469	9474	9479	9484	9489
89	9494	9499	9504	9509	9513	9518	9523	9528	9533	9538
90	9542	9547	9552	9557	9562	9566	9571	9576	9581	9586
91	9590	9595	9600	9605	9609	9614	9619	9624	9628	9633
92	9638	9643	9647	9652	9657	9661	9666	9671	9675	9680
93	9685	9689	9694	9699	9703	9708	9713	9717	9722	9727
94	9731	9736	9741	9745	9750	9754	9759	9763	9768	9773
95	9777	9782	9786	9791	9795	9800	9805	9809	9814	9818
96	9823	9827	9832	9836	9841	9845	9850	9854	9859	9863
97	9868	9872	9877	9881	9886	9890	9894	9899	9903	9908
98	9912	9917	9921	9926	9930	9934	9939	9943	9948	9952
99	9956	9961	9965	9969	9974	9978	9983	9987	9991	9996

Table 2

Values of the Trigonometric Functions

Angle θ									
Degrees	Radians	$\sin\theta$	$\cos\theta$	$\tan\theta$	$\cot\theta$	$\sec\theta$	$\csc\theta$		
0° 00′	.0000	.0000	1.0000	.0000	No value	1.000	No value	1.5708	90° 00′
10	029	029	000	029	343.8	000	343.8	679	50
20	058	058	000	058	171.9	000	171.9	650	40
30	087	087	1.0000	087	114.6	000	114.6	621	30
40	116	116	.9999	116	85.94	000	85.95	592	20
50	145	145	999	145	68.75	000	68.76	563	10
1° 00′	.0175	.0175	.9998	.0175	57.29	1.000	57.30	1.5533	89° 00′
10	204	204	998	204	49.10	000	49.11	504	50
20	233	233	997	233	42.96	000	42.98	475	40
30	262	262	997	262	38.19	000	38.20	446	30
40	291	291	996	291	34.37	000	34.38	417	20
50	320	320	995	320	31.24	001	31.26	388	10
2° 00′	.0349	.0349	.9994	.0349	28.64	1.001	28.65	1.5359	88° 00′
10	378	378	993	378	26.43	001	26.45	330	50
20	407	407	992	407	24.54	001	24.56	301	40
30	436	436	990	437	22.90	001	22.93	272	30
40	465	465	989	466	21.47	001	21.49	243	20
50	495	494	988	495	20.21	001	20.23	213	10
3° 00′	.0524	.0523	.9986	.0524	19.08	1.001	19.11	1.5184	87° 00′
10	553	552	985	553	18.07	002	18.10	155	50
20	582	581	983	582	17.17	002	17.20	126	40
30	611	610	981	612	16.35	002	16.38	097	30
40	640	640	980	641	15.60	002	15.64	068	20
50	669	669	978	670	14.92	002	14.96	039	10
4° 00′	.0698	.0698	.9976	.0699	14.30	1.002	14.34	1.5010	86° 00′
10	727	727	974	729	13.73	003	13.76	981	50
20	756	756	971	758	13.20	003	13.23	952	40
30	785	785	969	787	12.71	003	12.75	923	30
40	814	814	967	816	12.25	003	12.29	893	20
50	844	843	964	846	11.83	004	11.87	864	10
5° 00′	.0873	.0872	.9962	.0875	11.43	1.004	11.47	1.4835	85° 00′
10	902	901	959	904	11.06	004	11.10	806	50
20	931	929	957	934	10.71	004	10.76	777	40
30	960	958	954	963	10.39	005	10.43	748	30
40	.0989	.0987	951	.0992	10.08	005	10.13	719	20
50	.1018	.1016	948	.1022	9.788	005	9.839	690	10
6° 00′	.1047	.1045	.9945	.1051	9.514	1.006	9.567	1.4661	84° 00′
10	076	074	942	080	9.255	006	9.309	632	50
20	105	103	939	110	9.010	006	9.065	603	40
30	134	132	936	139	8.777	006	8.834	573	30
40	164	161	932	169	8.556	007	8.614	544	20
50	193	190	929	198	8.345	007	8.405	515	10
7° 00′	.1222	.1219	.9925	.1228	8.144	1.008	8.206	1.4486	83° 00′
10	251	248	922	257	7.953	008	8.016	457	50
20	280	276	918	287	7.770	008	7.834	428	40
30	309	305	914	317	7.596	009	7.661	399	30
40	338	334	911	346	7.429	009	7.496	370	20
50	367	363	907	376	7.269	009	7.337	341	10
8° 00′	.1396	.1392	.9903	.1405	7.115	1.010	7.185	1.4312	82° 00′
		$\cos\theta$	$\sin\theta$	$\cot\theta$	$\tan\theta$	$\csc\theta$	$\sec\theta$	Radians	Degrees
								Angle θ	

Table 2

Values of the Trigonometric Functions (continued)

Angle θ									
Degrees	Radians	$\sin\theta$	$\cos\theta$	$\tan\theta$	$\cot\theta$	$\sec\theta$	$\csc\theta$		
8° 00′	.1396	.1392	.9903	.1405	7.115	1.010	7.185	1.4312	82° 00′
10	425	421	899	435	6.968	010	7.040	283	50
20	454	449	894	465	827	011	6.900	254	40
30	484	478	890	495	691	011	765	224	30
40	513	507	886	524	561	012	636	195	20
50	542	536	881	554	435	012	512	166	10
9° 00′	.1571	.1564	.9877	.1584	6.314	1.012	6.392	1.4137	81° 00′
10	600	593	872	614	197	013	277	108	50
20	629	622	868	644	6.084	013	166	079	40
30	658	650	863	673	5.976	014	6.059	050	30
40	687	679	858	703	871	014	5.955	1.4021	20
50	716	708	853	733	769	015	855	1.3992	10
10° 00′	.1745	.1736	.9848	.1763	5.671	1.015	5.759	1.3963	80° 00′
10	774	765	843	793	576	016	665	934	50
20	804	794	838	823	485	016	575	904	40
30	833	822	833	853	396	017	487	875	30
40	862	851	827	883	309	018	403	846	20
50	891	880	822	914	226	018	320	817	10
11° 00′	.1920	.1908	.9816	.1944	5.145	1.019	5.241	1.3788	79° 00′
10	949	937	811	.1974	5.066	019	164	759	50
20	.1978	965	805	.2004	4.989	020	089	730	40
30	.2007	.1994	799	035	915	020	5.016	701	30
40	036	.2022	793	065	843	021	4.945	672	20
50	065	051	787	095	773	022	876	643	10
12° 00′	.2094	.2079	.9781	.2126	4.705	1.022	4.810	1.3614	78° 00′
10	123	108	775	156	638	023	745	584	50
20	153	136	769	186	574	024	682	555	40
30	182	164	763	217	511	024	620	526	30
40	211	193	757	247	449	025	560	497	20
50	240	221	750	278	390	026	502	468	10
13° 00′	.2269	.2250	.9744	.2309	4.331	1.026	4.445	1.3439	77° 00′
10	298	278	737	339	275	027	390	410	50
20	327	306	730	370	219	028	336	381	40
30	356	334	724	401	165	028	284	352	30
40	385	363	717	432	113	029	232	323	20
50	414	391	710	462	061	030	182	294	10
14° 00′	.2443	.2419	.9703	.2493	4.011	1.031	4.134	1.3265	76° 00′
10	473	447	696	524	3.962	031	086	235	50
20	502	476	689	555	914	032	4.039	206	40
30	531	504	681	586	867	033	3.994	177	30
40	560	532	674	617	821	034	950	148	20
50	589	560	667	648	776	034	906	119	10
15° 00′	.2618	.2588	.9659	.2679	3.732	1.035	3.864	1.3090	75° 00′
10	647	616	652	711	689	036	822	061	50
20	676	644	644	742	647	037	782	032	40
30	705	672	636	773	606	038	742	1.3003	30
40	734	700	628	805	566	039	703	1.2974	20
50	763	728	621	836	526	039	665	945	10
16° 00′	.2793	.2756	.9613	.2867	3.487	1.040	3.628	1.2915	74° 00′
		$\cos\theta$	$\sin\theta$	$\cot\theta$	$\tan\theta$	$\csc\theta$	$\sec\theta$	Radians	Degrees
								Angle θ	

Table 2

Values of the Trigonometric Functions (continued)

Angle θ									
Degrees	Radians	sin θ	cos θ	tan θ	cot θ	sec θ	csc θ		
16° 00′	.2793	.2756	.9613	.2867	3.487	1.040	3.628	1.2915	74° 00′
10	822	784	605	899	450	041	592	886	50
20	851	812	596	931	412	042	556	857	40
30	880	840	588	962	376	043	521	828	30
40	909	868	580	.2994	340	044	487	799	20
50	938	896	572	.3026	305	045	453	770	10
17° 00′	.2967	.2924	.9563	.3057	3.271	1.046	3.420	1.2741	73° 00′
10	.2996	952	555	089	237	047	388	712	50
20	.3025	.2979	546	121	204	048	357	683	40
30	054	.3007	537	153	172	048	326	654	30
40	083	035	528	185	140	049	295	625	20
50	113	062	520	217	108	050	265	595	10
18° 00′	.3142	.3090	.9511	.3249	3.078	1.051	3.236	1.2566	72° 00′
10	171	118	502	281	047	052	207	537	50
20	200	145	492	314	3.018	053	179	508	40
30	229	173	483	346	2.989	054	152	479	30
40	258	201	474	378	960	056	124	450	20
50	287	228	465	411	932	057	098	421	10
19° 00′	.3316	.3256	.9455	.3443	2.904	1.058	3.072	1.2392	71° 00′
10	345	283	446	476	877	059	046	363	50
20	374	311	436	508	850	060	3.021	334	40
30	403	338	426	541	824	061	2.996	305	30
40	432	365	417	574	798	062	971	275	20
50	462	393	407	607	773	063	947	246	10
20° 00′	.3491	.3420	.9397	.3640	2.747	1.064	2.924	1.2217	70° 00′
10	520	448	387	673	723	065	901	188	50
20	549	475	377	706	699	066	878	159	40
30	578	502	367	739	675	068	855	130	30
40	607	529	356	772	651	069	833	101	20
50	636	557	346	805	628	070	812	072	10
21° 00′	.3665	.3584	.9336	.3839	2.605	1.071	2.790	1.2043	69° 00′
10	694	611	325	872	583	072	769	1.2014	50
20	723	638	315	906	560	074	749	1.1985	40
30	752	665	304	939	539	075	729	956	30
40	782	692	293	.3973	517	076	709	926	20
50	811	719	283	.4006	496	077	689	897	10
22° 00′	.3840	.3746	.9272	.4040	2.475	1.079	2.669	1.1868	68° 00′
10	869	773	261	074	455	080	650	839	50
20	898	800	250	108	434	081	632	810	40
30	927	827	239	142	414	082	613	781	30
40	956	854	228	176	394	084	595	752	20
50	985	881	216	210	375	085	577	723	10
23° 00′	.4014	.3907	.9205	.4245	2.356	1.086	2.559	1.1694	67° 00′
10	043	934	194	279	337	088	542	665	50
20	072	961	182	314	318	089	525	636	40
30	102	.3987	171	348	300	090	508	606	30
40	131	.4014	159	383	282	092	491	577	20
50	160	041	147	417	264	093	475	548	10
24° 00′	.4189	.4067	.9135	.4452	2.246	1.095	2.459	1.1519	66° 00′
		cos θ	sin θ	cot θ	tan θ	csc θ	sec θ	Radians	Degrees
								Angle θ	

Table 2

Values of the Trigonometric Functions (continued)

Angle θ									
Degrees	Radians	sin θ	cos θ	tan θ	cot θ	sec θ	csc θ		
24° 00′	.4189	.4067	.9135	.4452	2.246	1.095	2.459	1.1519	66° 00′
10	218	094	124	487	229	096	443	490	50
20	247	120	112	522	211	097	427	461	40
30	276	147	100	557	194	099	411	432	30
40	305	173	088	592	177	100	396	403	20
50	334	200	075	628	161	102	381	374	10
25° 00′	.4363	.4226	.9063	.4663	2.145	1.103	2.366	1.1345	65° 00′
10	392	253	051	699	128	105	352	316	50
20	422	279	038	734	112	106	337	286	40
30	451	305	026	770	097	108	323	257	30
40	480	331	013	806	081	109	309	228	20
50	509	358	.9001	841	066	111	295	199	10
26° 00′	.4538	.4384	.8988	.4877	2.050	1.113	2.281	1.1170	64° 00′
10	567	410	975	913	035	114	268	141	50
20	596	436	962	950	020	116	254	112	40
30	625	462	949	.4986	2.006	117	241	083	30
40	654	488	936	.5022	1.991	119	228	054	20
50	683	514	923	059	977	121	215	1.1025	10
27° 00′	.4712	.4540	.8910	.5095	1.963	1.122	2.203	1.0996	63° 00′
10	741	566	897	132	949	124	190	966	50
20	771	592	884	169	935	126	178	937	40
30	800	617	870	206	921	127	166	908	30
40	829	643	857	243	907	129	154	879	20
50	858	669	843	280	894	131	142	850	10
28° 00′	.4887	.4695	.8829	.5317	1.881	1.133	2.130	1.0821	62° 00′
10	916	720	816	354	868	134	118	792	50
20	945	746	802	392	855	136	107	763	40
30	.4974	772	788	430	842	138	096	734	30
40	.5003	797	774	467	829	140	085	705	20
50	032	823	760	505	816	142	074	676	10
29° 00′	.5061	.4848	.8746	.5543	1.804	1.143	2.063	1.0647	61° 00′
10	091	874	732	581	792	145	052	617	50
20	120	899	718	619	780	147	041	588	40
30	149	924	704	658	767	149	031	559	30
40	178	950	689	696	756	151	020	530	20
50	207	.4975	675	735	744	153	010	501	10
30° 00′	.5236	.5000	.8660	.5774	1.732	1.155	2.000	1.0472	60° 00′
10	265	025	646	812	720	157	1.990	443	50
20	294	050	631	851	709	159	980	414	40
30	323	075	616	890	698	161	970	385	30
40	352	100	601	930	686	163	961	356	20
50	381	125	587	.5969	675	165	951	327	10
31° 00′	.5411	.5150	.8572	.6009	1.664	1.167	1.942	1.0297	59° 00′
10	440	175	557	048	653	169	932	268	50
20	469	200	542	088	643	171	923	239	40
30	498	225	526	128	632	173	914	210	30
40	527	250	511	168	621	175	905	181	20
50	556	275	496	208	611	177	896	152	10
32° 00′	.5585	.5299	.8480	.6249	1.600	1.179	1.887	1.0123	58° 00′
		cos θ	sin θ	cot θ	tan θ	csc θ	sec θ	Radians	Degrees
								Angle θ	

Table 2

Values of the Trigonometric Functions (continued)

Angle θ									
Degrees	Radians	sin θ	cos θ	tan θ	cot θ	sec θ	csc θ		
32° 00′	.5585	.5299	.8480	.6249	1.600	1.179	1.887	1.0123	58° 00′
10	614	324	465	289	590	181	878	094	50
20	643	348	450	330	580	184	870	065	40
30	672	373	434	371	570	186	861	036	30
40	701	398	418	412	560	188	853	1.0007	20
50	730	422	403	453	550	190	844	.9977	10
33° 00′	.5760	.5446	.8387	.6494	1.540	1.192	1.836	.9948	57° 00′
10	789	471	371	536	530	195	828	919	50
20	818	495	355	577	520	197	820	890	40
30	847	519	339	619	511	199	812	861	30
40	876	544	323	661	501	202	804	832	20
50	905	568	307	703	492	204	796	803	10
34° 00′	.5934	.5592	.8290	.6745	1.483	1.206	1.788	.9774	56° 00′
10	963	616	274	787	473	209	781	745	50
20	.5992	640	258	830	464	211	773	716	40
30	.6021	664	241	873	455	213	766	687	30
40	050	688	225	916	446	216	758	657	20
50	080	712	208	.6959	437	218	751	628	10
35° 00′	.6109	.5736	.8192	.7002	1.428	1.221	1.743	.9599	55° 00′
10	138	760	175	046	419	223	736	570	50
20	167	783	158	089	411	226	729	541	40
30	196	807	141	133	402	228	722	512	30
40	225	831	124	177	393	231	715	483	20
50	254	854	107	221	385	233	708	454	10
36° 00′	.6283	.5878	.8090	.7265	1.376	1.236	1.701	.9425	54° 00′
10	312	901	073	310	368	239	695	396	50
20	341	925	056	355	360	241	688	367	40
30	370	948	039	400	351	244	681	338	30
40	400	972	021	445	343	247	675	308	20
50	429	.5995	.8004	490	335	249	668	279	10
37° 00′	.6458	.6018	.7986	.7536	1.327	1.252	1.662	.9250	53° 00′
10	487	041	969	581	319	255	655	221	50
20	516	065	951	627	311	258	649	192	40
30	545	088	934	673	303	260	643	163	30
40	574	111	916	720	295	263	636	134	20
50	603	134	898	766	288	266	630	105	10
38° 00′	.6632	.6157	.7880	.7813	1.280	1.269	1.624	.9076	52° 00′
10	661	180	862	860	272	272	618	047	50
20	690	202	844	907	265	275	612	.9018	40
30	720	225	826	.7954	257	278	606	.8988	30
40	749	248	808	.8002	250	281	601	959	20
50	778	271	790	050	242	284	595	930	10
39° 00′	.6807	.6293	.7771	.8098	1.235	1.287	1.589	.8901	51° 00′
10	836	316	753	146	228	290	583	872	50
20	865	338	735	195	220	293	578	843	40
30	894	361	715	243	213	296	572	814	30
40	923	383	698	292	206	299	567	785	20
50	952	406	679	342	199	302	561	756	10
40° 00′	.6981	.6428	.7660	.8391	1.192	1.305	1.556	.8727	50° 00′
		cos θ	sin θ	cot θ	tan θ	csc θ	sec θ	Radians	Degrees
								Angle θ	

Table 2

Values of the Trigonometric Functions (continued)

Angle θ									
Degrees	Radians	sin θ	cos θ	tan θ	cot θ	sec θ	csc θ		
40° 00′	.6981	.6428	.7660	.8391	1.192	1.305	1.556	.8727	50° 00′
10	.7010	450	642	441	185	309	550	698	50
20	039	472	623	491	178	312	545	668	40
30	069	494	604	541	171	315	540	639	30
40	098	517	585	591	164	318	535	610	20
50	127	539	566	642	157	322	529	581	10
41° 00′	.7156	.6561	.7547	.8693	1.150	1.325	1.524	.8552	49° 00′
10	185	583	528	744	144	328	519	523	50
20	214	604	509	796	137	332	514	494	40
30	243	626	490	847	130	335	509	465	30
40	272	648	470	899	124	339	504	436	20
50	301	670	451	.8952	117	342	499	407	10
42° 00′	.7330	.6691	.7431	.9004	1.111	1.346	1.494	.8378	48° 00′
10	359	713	412	057	104	349	490	348	50
20	389	734	392	110	098	353	485	319	40
30	418	756	373	163	091	356	480	290	30
40	447	777	353	217	085	360	476	261	20
50	476	799	333	271	079	364	471	232	10
43° 00′	.7505	.6820	.7314	.9325	1.072	1.367	1.466	.8203	47° 00′
10	534	841	294	380	066	371	462	174	50
20	563	862	274	435	060	375	457	145	40
30	592	884	254	490	054	379	453	116	30
40	621	905	234	545	048	382	448	087	20
50	650	926	214	601	042	386	444	058	10
44° 00′	.7679	.6947	.7193	.9657	1.036	1.390	1.440	.8029	46° 00′
10	709	967	173	713	030	394	435	.7999	50
20	738	.6988	153	770	024	398	431	970	40
30	767	.7009	133	827	018	402	427	941	30
40	796	030	112	884	012	406	423	912	20
50	825	050	092	.9942	006	410	418	883	10
45° 00′	.7854	.7071	.7071	1.000	1.000	1.414	1.414	.7854	45° 00′
		cos θ	sin θ	cot θ	tan θ	csc θ	sec θ	Radians	Degrees
								Angle θ	

Answers to Selected Problems

Chapter 1

Exercise 1-2

1. (a) $B \subset A$; (b) $B \not\subset C$; (c) $\phi \subset B$; (d) $1 \notin C$; (e) $2 \in B$; (f) $\{2\} \subset B$; (g) $\phi \in \{\phi\}$, or $\phi \subset \{\phi\}$; (h) $A \subseteq A$

3. (a) $\{x \mid x$ is a state outside the continental United States$\}$
   (b) $\{x \mid x$ is a counting number less than $5\}$
   (c) $\{x \mid x$ is a color in the American flag$\}$
   (d) $\{x \mid x$ is a counting number between 4 and $101\}$
   (e) $\{x \mid x$ is a counting number and a multiple of $3\}$

5. (a) U; (b) ϕ; (c) U; (d) A; (e) A; (f) ϕ; (g) U; (h) ϕ; (i) A

7. (a) $\{x \mid x$ is a consonant in the alphabet$\}$; (b) ϕ; (c) C; (d) U; (e) $\{x \mid x$ is a letter of the alphabet and $x \neq a, b, c, d, e, x, y, z\}$

Exercise 1-3

1. .666 . . . ; .142857 . . . ; .25; .1875; .0; 1.57079 . . . ; 2.82842 . . . ; .41421 . . . ; 6.28318 . . .

3. (a) $\{1, 2, 3\}$; (b) $\{0, 1, 2, 3\}$; (c) $\{\ldots -3, -2, -1, 0, 1, 2, 3\}$

5. $(1 + \sqrt{2}) + (1 - \sqrt{2}) = 2$; $(\sqrt{2})(\sqrt{2}) = 2$

7. $i^4 = 1$; $i^5 = i$; $i^6 = -1$; $i^7 = -i$; $i^8 = 1$; $i^{29} = i$; $i^{107} = -i$

Exercise 1-4

1. (a) $<$; (b) $>$; (c) $\leq$ or $\geq$; (d) $>$; (e) $>$; (f) $<$; (g) $<$; (h) $>$; (i) $>$; (j) $<$

3. (a) $3 < x < 10$; (b) $-1 < y < 5$; (c) $10 < x < 100$; (d) $0 \leq z \leq 1$; (e) $-10 < n \leq 14$; (f) $0 < x < 10$; (g) $-5 < x < 0$; (h) $-2 < x \leq 0$

5. (a) $\{x \mid -4 < x < 1\}$; (b) ϕ; (c) $\{x \mid x \leq 10\}$; (d) $\{x \mid x > -4\}$; (e) $\{x \mid x > 0\}$

7. $(1/a) < -1$; $0 < (1/b) < 1$; $0 < -a < 1$; $-1 < -b < -2$; $a < [(a + b)/2] < b$; $0 < 1 + a < 1$; $2 < 1 + b < 3$

Exercise 1-5

1. 8; 1; 1; 0; 5; 1/2; 1

3. (a) $\{5, 3\}$; (b) $\{7, -7\}$; (c) $\{-1, -5\}$; (d) $\{2.5, 1.5\}$; (e) $\{-5, 3\}$; (f) $\{2, -2\}$; (g) $\{2.5, -1.5\}$; (h) $\{\sqrt{2} + 2, \sqrt{2} - 2\}$; (i) $\{1/4, -5/4\}$; (j) $\{1, -2\}$

5. Not always. $|a - b| \geq |a| - |b|$

7. (a) $\{x \mid x > 6 \text{ or } x < 4\}$
   (b) $\{x \mid x \geq 0 \text{ or } x \leq -4\}$
   (c) $\{x \mid x > 1 \text{ or } x < -1\}$
   (d) $\{x \mid x > -3 \text{ or } x < -7\}$
   (e) $\{x \mid x \geq 5 \text{ or } x \leq -3\}$
   (f) $\{x \mid x \geq 4 \text{ or } x \leq -2\}$
   (g) $\{x \mid x > 5/4 \text{ or } x < -1/4\}$
   (h) $\{x \mid x > 99 \text{ or } x < -101\}$

9. (a) By Theorem1-1, if $|x - a| < r$, then $-r < x - a$ and $x - a < r$. Adding a to both sides of these inequalities gives $a - r < x$ and $x < a + r$. Thus $a - r < x < a + r$.
   (b) Use Theorem 1-2

11. (i) If $a > 0$, then $-a < 0$ and $|a| = a$, $|-a| = -(-a) = a$. If $a < 0$, $-a > 0$ and $|a| = -a$, $|-a| = -a$.
    (ii) Since $|a \cdot b| = |a| \cdot |b|$, then $|-a| = |(-1)(a)| = |-1| \cdot |a| = 1 \cdot |a| = |a|$.

13. Converse of Theorem 1-2. If $x - a > r$ or $x - a < -r$ and $r \geq 0$, then $|x - a| > r$.
    Case 1. $x - a > r$. Since $r \geq 0$, $x - a > 0$ and $x - a = |x - a|$. Thus in this case $|x - a| > r$.
    Case 2. $x - a < -r$. Then $-(x - a) > r$ and $-(x - a) > 0$. Then $-(x - a) = |-(x - a)| = |x - a|$ by problem 11, and in this case also $|x - a| > r$.

15. The set of all real numbers; the set of all real numbers except a.

Exercise 1-6

3. (a) (1/2, 1/2); (−1/2, 1/2); (−1/2, −1/2); (1/2, −1/2)
   (b) (1, 1); (2, 1); (2, 2); (1, 2)
   (c) (3, 8); (−2, 8); (−2, 4); (3, 4)

5. All three sets of points are collinear.

9. (a) $x = \pm 2\sqrt{2}$; (b) $y = \pm\sqrt{3}$

11. $x^2 + y^2 = 1$

13. (a) exterior; (b) interior; (c) on C; (d) exterior

Review Exercise

1. (a) $\{2, 4, 6\}$; (b) U; (c) ϕ; (d) $\{4, 6\}$; (e) U; (f) $\{1, 3, 4, 5, 6, 7\}$; (g) $\{1, 2, 3, 5, 7\}$

3. $(A \cap B) \cup (A \cap C) \cup (B \cap C)$

5. $\{-2, -1, 0, 1, 2, 3, 4, 5\}$; $\{x \mid x \text{ is an integer and } x > 5 \text{ or } x \leq -3\}$

7. $\sqrt{37}$

9. $(x-1)^2+(y+2)^2=1$

Chapter 2

Exercise 2-3

1. (a) t independent, s dependent; (b) r independent, V dependent; (c) y independent, x dependent; (d) d independent, C dependent

3. (a) Domain $\{1, 2, 3, 4\}$; Range $\{1\}$
   (b) Domain $\{1, 2, 3, 4\}$; Range $\{1, 2, 3, 4\}$
   (c) Domain $\{1\}$; Range $\{1, 2, 3, 4\}$
   (a) and (b) are functions.

5. (a) Domain $\{x \mid x \text{ is a real number}\}$; Range $\{y \mid -1 \le y \le 1\}$
   (b) Domain $\{x \mid x \text{ is a real number}\}$; Range $\{y \mid y \ge 0\}$
   (c) Domain $\{x \mid x \text{ is a real number}\}$; Range $\{-2, 2\}$
   (d) Domain $\{x \mid -1 \le x \le 1\}$; Range $\{y \mid -1 \le y \le 1\}$
   (a), (b), and (c) are functions.

7. $p = a/V$; $V = a/p$. Both domain and range are the set of all positive real numbers.

9. No, because for each choice of x there are many corresponding values of y.

11.

s	60	120	170	220	270
t	1	2	3	4	5

Exercise 2-5

1. (a) -1; (b) 4; (c) 3; (d) undefined; (e) $3x-7$; (f) p^2-2p+1; (g) $3x^2-1$; (h) x^2+2x+1; (i) $3/x-1$; (j) $3x^2-6x+2$

3. Domain $\{x \mid x \text{ is a real number}\}$; Range $\{y \mid y \ge 0\}$

5. Domain $\{x \mid x \text{ is a real number}\}$; Range $\{6\}$

7. Domain $\{x \mid x \text{ is a real number}\}$; Range $\{y \mid y \ge 0\}$

9. Any function of the form $y = kx$, where k is a constant; $y = x^2$

11. Domain and range are both the set of all real numbers. $I(0) = 0$; $I(2) = 2$; $I(-1) = -1$.

13. Domain $\{x \mid x \text{ is a real number}\}$; range $\{-1, 0, 1\}$. $\text{sgn}(-5) = -1$; $\text{sgn}(14) = 1$; $\text{sgn}(0) = 0$.

15. It is not a function since $1/2 = 2/4$, but $f(1/2) = 2$; $f(2/4) = 4$. Its range is $\{x \mid x \neq 0\}$.

17. $f(a) = 1$; $g(b) = 4$; $f(b) \cdot g(c) = 12$; $g(d) \div f(d) = 2$; $f(c) + g(c) = 9$.

Exercise 2-6

1. $x + y = 3$

3. $y = \sqrt{1 - x^2}$

5. $y = 1/x$

7. $x = t - 2$; $y = t^2 - 11$

9. $x = t + 1$; $y = -t/2$

Exercise 2-7

1. Symmetric with respect to the y-axis.

3. Symmetric with respect to the origin.

5. Symmetric with respect to the origin.

7. Symmetric with respect to the y-axis.

9. even

11. even

Exercise 2-8

1. (a) $x = y - 1$; (b) $x = y/3$; (c) $x = (y - 3)/2$; (d) $x = -(y - 7)/3$; (e) $x = 1/y$; (f) $x = 1/y + 1$

5. (a) does not have an inverse.
(b) g^{-1}

4	1
5	2
6	3

(c) h^{-1}

1	1
2	2
3	3

7. $(f^{-1})^{-1} = f$

9. Hint: Consider three cases: $x > 0$, $x = 0$, and $x < 0$.

11. $g^{-1}(x) = x^2 - 1$; its domain is $\{x \mid x \geq 0\}$

Exercise 2-9

1. $(f + g)(x) = \sqrt{x} + 1/x$; $D_{f+g} = \{x \mid x > 0\}$
$(f - g)(x) = \sqrt{x} - 1/x$; $D_{f-g} = \{x \mid x > 0\}$

$(f \cdot g)(x) = \sqrt{x}/x;\ D_{f \cdot g} = \{x \mid x > 0\}$
$(g/f)(x) = 1/x\sqrt{x};\ D_{g/f} = \{x \mid x > 0\}$

3. $f+g$

1	−1
2	−1
3	−1

$f-g$

1	−1
2	3
3	−7

$f \cdot g$

1	0
2	−2
3	−12

f/g

2	−1/2
3	−4/3

5. 1/3; 1; 4; −3

7. $(g \circ f)(x) = \sqrt{x+2} - 1$; Domain $\{x \mid x \geq -2\}$
   $(f \circ g)(x) = \sqrt{x+1}$; Domain $\{x \mid x \geq -1\}$

9. $(g \circ f)(x) = 2/x + 1$; Domain $\{x \mid x \neq 0\}$
   $(f \circ g)(x) = 1/(2x+1)$; Domain $\{x \mid x \neq -1/2\}$

11. $(g \circ f)(x) = 3x^2 + 6x - 22$; Domain $\{x \mid x$ is a real number$\}$
    $(f \circ g)(x) = 9x^2 - 8$; Domain $\{x \mid x$ is a real number$\}$

13. $(g \circ f)(x) = 1/(x^2+1)^3$; Domain $\{x \mid x$ is a real number$\}$
    $(f \circ g)(x) = 1/(x^6+1)$; Domain $\{x \mid x$ is a real number$\}$

15. $(g \circ f)(x) = \sqrt{(x+1)/(x-1)}$; Domain $\{x \mid x > 1$ or $x \leq -1\}$
    $(f \circ g)(x) = (\sqrt{x}+1)/(\sqrt{x}-1)$; Domain $\{x \mid x > 0$ and $x \neq 1\}$

17. $(f \circ g)(x) = (g \circ f)(x) = (x+1)/(x+2)$; Domain $\{x \mid x \neq -2$ and $x \neq -1\}$

19. $(g \circ f)(x) = 1/x$; Domain $\{x \mid x \neq 0\}$
    $(f \circ g)(x) = 1/(4x)$; Domain $\{x \mid x \neq 0\}$

21. $(f \circ g)(x) = (g \circ f)(x) = x$; Domain $\{x \mid x \neq 0\}$

23. $(I \circ f)(x) = f(x)$; $(f \circ I)(x) = f(x)$. $I \circ g = g \circ I = g$ and will be defined on the domain of g.

25. $f^{-1}(x) = (x-b)/a$

27. $f^{+} + f^{-} = f$

Review Exercise

1. (a) and (d) are functions.

3. (a) Domain $\{1, 2\}$; Range $\{1, 3\}$
   (b) Domain $\{1, 0\}$; Range $\{1\}$
   (c) Domain $\{1\}$; Range $\{1, 2\}$
   (d) Domain $\{0, 4\}$; Range $\{0, 4\}$
   (a), (b), and (d) are functions. (a) and (d) are one-to-one. The inverse of function (a) is

1	1
3	2

; of function (d) is

4	0
0	4

.

5. $A = 2w^2$

7. (a) Domain $\{x \mid -1 \le x \le 1\}$; Range $\{y \mid -1 \le y \le 1\}$
   (b) Domain $\{x \mid x \ge 0\}$; Range $\{y \mid y \text{ is a real number}\}$
   (c) Domain $\{x \mid x \text{ is a real number}\}$; Range $\{y \mid y \ge 2\}$
   (d) Domain $\{x \mid x \ne 0\}$; Range $\{y \mid y \ne 0\}$
   (c) and (d) are functions.

9. (a) Symmetric with respect to the y-axis
   (b) Symmetric with respect to the y-axis
   (c) Symmetric with respect to the origin
   (d) No symmetry

11.

$f+g$	
1	−1
2	−1
3	−1

$f-g$	
1	−1
2	3
3	−7

$f \cdot g$	
1	0
2	−2
3	−12

f/g	
2	−1/2
3	−4/3

13. No, they do not have the same domain.

Chapter 3

Exercise 3-2

1. No. A vertical line does not represent a function.

3. It is a measure of the inclination of the line.

5. They are perpendicular lines.

7. (a) (1, 0), up; (b) (3/2, 37/4), down; (c) (0, 5), up; (d) (0, −2), up.

9. No.

11. (a) $\frac{3 \pm i\sqrt{47}}{4}$; (b) $\frac{-5 \pm \sqrt{21}}{2}$; (c) $\frac{-1 \pm i\sqrt{11}}{6}$; (d) −2, −2
    (e) $\pm\sqrt{6}/2$; (f) 0, −7

13. $k > 2$ or $k < -2$; $k = \pm 2$; $-2 < k < 2$

15. (a) 2, 3; (b) $\pm\sqrt{6}/2$; (c) ± 1; (d) 1, 1

19. −2, 1; −3, −3; the circle would not touch the x-axis.

Exercise 3-3

1. (a) 3; (b) 8; (c) 4; (d) 5; (e) 1; (f) 1; (g) 0; (h) no degree

3. none; no; yes

5. (a) $f = -3 \cdot I \cdot I + 7 \cdot I - 1$

(b) $f = I \cdot I \cdot I + 2 \cdot I \cdot I - I + 2$
(c) $f = I \cdot I \cdot I \cdot I \cdot I + I \cdot I \cdot I - I$
(d) $f = I \cdot I \cdot I \cdot I - 1$

9. (a) F; (b) F; (c) T (provided $f + g \neq 0$); (d) F; (e) T; (f) T; (g) T; (h) F

Exercise 3-4

1. (a) -2; (b) -3; (c) 54; (d) 2; (e) -357; (f) 21

3. $x^3 - x$; $2x^3 - 2x$

5. (a) $-1, -1, 3, -4$; 4 roots
(b) $1/2, -3/2, -3/2, -1, -1, -1$; 6 roots
(c) $0, 0, 0, -\sqrt{3}, \sqrt{3}$; 5 roots
(d) $-i, -i, i, i, -1, -1$; 6 roots

7. $\dfrac{-1 \pm i\sqrt{39}}{2}$

9. If $P(x) = x^n - a^n$; $P(a) = a^n - a^n = 0$, and by the Factor Theorem, $x - a$ is a factor of $P(x)$.

11. $k = -92$

13. $Q_1 = (1/2)x^2 + (5/4)x - 9/8$; $R = -1/8$
$Q_2 = x^2 + (5/2)x - 9/4$; $R = -1/8$
The remainders are the same, but $Q_2 = 2Q_1$.

Exercise 3-5

1. If $a + bi$ and $a - bi$ are roots, then $[x - (a + bi)]$ and $[x - (a - bi)]$ are factors of the polynomial. $[x - (a + bi)][x - (a - bi)] = (x - a - bi)(x - a + bi) = (x - a)^2 - (bi)^2 = x^2 - 2ax + a^2 + b^2$.

3. If $a + \sqrt{b}$ and $a - \sqrt{b}$ are roots, then $[x - (a + \sqrt{b})]$ and $[x - (a - \sqrt{b})]$ are factors of the polynomial. $[x - (a + \sqrt{b})][x - (a - \sqrt{b})] = (x - a - \sqrt{b})(x - a + \sqrt{b}) = (x - a)^2 - (\sqrt{b})^2 = x^2 - 2ax + a^2 - b^2$.

5. $1 - i, -3$

7. $1 + \sqrt{3}, -1$

9. (a) No. Theorem 3-5 does not apply because the coefficients are not real numbers.
(b) Yes. All of the coefficients are real numbers.

11. (a) $x^3 - 3x^2 + x + 1 = 0$
(b) $x^2 - (2 + \sqrt{2})x + (1 + \sqrt{2}) = 0$

13. $x^3 - 3x^2 + 3x - 1$; one answer to the second part is $x^4 - 2(i + \sqrt{2})x^3 + (1 + 4\sqrt{2}i)x^2 + 2(\sqrt{2} - 2i)x - 2$

Exercise 3-6

1. $-74; -14; 1; -35$

3. (a) $3x^3 + 9x^2 + 20x + 70; R : 209$
   (b) $3x^3 - 6x^2 + 5x; R : -1$
   (c) $6x^4 + 4x^3 - 2x^2; R : 7$
   (d) $x^4 + 2x^3 + 4x^2 + 8x + 16; R : 64$
   (e) $2x - 2\sqrt{2} - 3; R : 3\sqrt{2}$
   (f) $-x^6 - 2x^5 - 4x^4 - 6x^3 - 5x^2 - 10x - 21; R : -42$
   (g) $2x^2 - 7x + 35; R : -75$
   (h) $x^5 - x^4 + x^3 - 2x^2 + 2x - 2; R : 3$

5.

x	0	1	2	3	-1	-2	-3
y	10	9	18	73	9	18	73

7. (a) 12; (b) 0; (c) -3; (d) 1

Exercise 3-7

1. (a) $\pm 21; \pm 1; \pm 3; \pm 7; \pm 1/3; \pm 7/3$
   (b) $\pm 12; \pm 1; \pm 3; \pm 4; \pm 2; \pm 6$
   (c) $\pm 9; \pm 1; \pm 3; \pm 9/4; \pm 1/4; \pm 3/4; \pm 9/2; \pm 3/2; \pm 1/2$
   (d) $\pm 24; \pm 1; \pm 12; \pm 2; \pm 3; \pm 8; \pm 4; \pm 6; \pm 1/2; \pm 3/2$

7. (a) $2, \pm i\sqrt{3}$
   (b) $-1/2, \pm i(2\sqrt{2})$

Exercise 3-8

1. $u = 0; l = -3$

3. $u = 1; l = -1$

5. $u = 1; l = -1$

7. $2/3; -1/2, \pm i\sqrt{6}$

9. $4, -1/2, \pm i/\sqrt{3}$

Exercise 3-9

1. By Descartes' Rule of Signs there is exactly one positive root and one negative root. Since the degree of the equation is 4, two roots must be complex.

3. (a) $x^n + a = 0$ has one negative root, no positive roots, and no zero roots.
   (b) $x^n - a = 0$ has one positive root, no negative roots, and no zero roots.

5.

0	+	−	i
0	3	1	0
0	1	1	2

7.

0	+	−	i
0	3	0	2
0	1	0	4

9.

0	+	−	i
2	1	2	2
2	1	0	4

Exercise 3-10

1. (a) $4x; R: -3x + 3$
   (b) $x^2 + 2x + 4; R: 7x + 40$
   (c) $3x^2 - 2x + 2; R: -18x + 9$
   (d) $x^3 - 7x^2 + 4x - 18; R: 16x - 83$

3. 2, 2, 5

5. $5, 3 \pm \sqrt{2}$

7. 3, 3, −4

Exercise 3-11

1. 3.18

3. −1.24

5. 1.442

Exercise 3-12

1. (a) $\{x \mid x \neq 0 \text{ and } x \neq 1\}$
   (b) $\{x \mid x \neq -3 \text{ and } x \neq 1\}$
   (c) $\{x \mid x \text{ is a real number}\}$
   (d) $\{x \mid x \neq 0 \text{ and } x \neq -3\}$
   (e) $\{x \mid x \neq 1/2\}$

3. No. The domain of f is $\{x \mid x \neq 0\}$. The domain of g is the set of all real numbers.

5. (a) Domain $\{x \mid x \neq 1 \text{ and } x \neq -1\}$; vertical asymptotes: $x = 1$; $x = -1$; horizontal asymptote: $y = 0$; symmetric with respect to the origin.
   (b) Domain $\{x \mid x \neq -1\}$; vertical asymptote: $x = -1$; horizontal asymptote: $y = 2$.

(c) Domain $\{x \mid x \neq -1 \text{ and } x \neq 1\}$; vertical asymptotes: $x = 1$ and $x = -1$; horizontal asymptote: $y = 1$; symmetric with respect to the y-axis.
(d) Domain $\{x \mid x \text{ is a real number}\}$; no vertical asymptotes; horizontal asymptote: $y = 0$; symmetric with respect to the y-axis.
(e) Domain $\{x \mid x \neq 2 \text{ and } x \neq -2\}$; vertical asymptotes: $x = 2$ and $x = -2$; horizontal asymptote: $y = 3$; symmetric with respect to the y-axis.

7. $y = \dfrac{3x + 1}{x + 1}$

Review Exercise

1. $(g \circ f)(x) = cax + (cb + d)$; $(f + g)(x) = (a + c)x + (b + d)$
3. $k < 9/4$; $k = 9/4$; $k > 9/4$
5. $0, 0, -1/2, 3, 3, -3, 1$; 7 roots
7. $x^5 - x^4 + 8x^2 - 9x - 15 = 0$
9. (a) $\pm 15; \pm 1; \pm 3; \pm 5; \pm 15/2; \pm 1/2; \pm 3/2; \pm 5/2$
   (b) $u = 3$; $l = -1$
   (c) $3, -1/2, \pm i\sqrt{5}$
11. $\pm 12; \pm 1; \pm 3; \pm 4; \pm 6; \pm 2; \pm 1/2; \pm 3/2$; Roots: $-3/2, 1, 1, \pm 2i$
13. (a) $\{x \mid x \neq -1\}$
    (b) vertical asymptote: $x = -1$; horizontal asymptote: $y = 2$

Chapter 4

Exercise 4-1

1. (a) 5^5; (b) 3^4; (c) $1/7^3$; (d) 2^3; (e) $1/3^5$; (f) $2^7/5^4$
3. (a) 8; (b) 1; (c) -1; (d) 1/3; (e) -2; (f) 3; (g) 1/4; (h) 5/2; (i) 8; (j) 8/27
5. (a) 9.75; (b) 10 5/8; (c) 1111.11; (d) 412.87

Exercise 4-2

1.

x	2	1	0	-1	-2
y	9	3	1	1/3	1/9

5. If a were negative, then the function would not be defined for values of x such as 1/2, 1/4, etc. If a were zero, then the function would be undefined for $x = 0$.

7. (a) $0 < x < 1$; (b) $4 < x < 5$; (c) $-3 < x < -2$; (d) $-4 < x < -3$; (e) $2 < x < 3$; (f) $-2 < x < -1$

Exercise 4-4

3. (a) $\log_4 16 = 2$; (b) $\log_2(1/2) = -1$; (c) $\log_5 125 = 3$; (d) $\log_2 32 = 5$; (e) $\log_{27}(1/3) = -1/3$; (f) $\log_{64} 2 = 1/6$; (g) $\log_{10}(.1) = -1$; (h) $\log_{16} 64 = 3/2$

5. (a) 3; (b) -1; (c) 5; (d) 10; (e) 3; (f) 4.2; (g) 1; (h) 0; (i) -1; (j) x

7. Let $\log_a M = x$ and $\log_a N = y$. Then $a^x = M$ and $a^y = N$. $M/N = a^x/a^y = a^{x-y}$. Thus $\log_a(M/N) = x - y = \log_a M - \log_a N$.

9. (a) $2 \log_a x + 2 \log_a y$; (b) $\log_a x + \log_a y - \log_a u - \log_a v$;
   (c) $(1/2)(\log_a x + \log_a y + \log_a z)$; (d) $(1/3)\log_a(x + y)$;
   (e) $(1/2[2 \log_a x - (1/2)\log_a y + (2/3)\log_a z]$;
   (f) $(3/2)\log_a x - (2/3)\log_a y$; (g) $(1/2)[\log_{10} s + \log_{10}(s - a) + \log_{10}(s - b) + \log_{10}(s - c)]$; (h) $\ln 2 + \ln \pi + (1/2)(\ln l - \ln g)$

11. (a) 1.3862; (b) $-.4055$; (c) 1.7917; (d) -1.0986; (e) 1.5041; (f) $-.1179$

13. (a) $x = \dfrac{\log 10}{\log 5}$; (b) $x = \dfrac{\log 2}{2 \log 3}$; (c) $x = \dfrac{\log 3 - \log 2}{\log 2}$; (d) $x = \dfrac{\log 5}{\log 3 - \log 5}$

15. Let $M = \log_b a$; $N = \log_c a$; $P = \log_c b$. Then $b^M = a$; $c^N = a$, and $c^P = b$. Substituting gives $(c^P)^M = a = c^N$ or $PM = N$. Thus $M = N/P$.

17. Let $a = \log_N b$. Then $N^a = b$ and $(1/N)^a = 1/b$. Thus $\log_{(1/N)}(1/b) = a$.

Exercise 4-5

1. (a) 2.371×10^3; (b) 1.24×10^{-2}; (c) 1.0×10^{-5}; (d) 1.07×10^1; (e) 2.43×10^0; (f) 5.41×10^8

3. (a) 0.1875; (b) 2.1847; (c) 7.9196 − 10; (d) 4.0414; (e) 1.4116; (f) 8.0934 − 10

5. (a) 524.8; (b) .3326; (c) 7.633; (d) 63.08; (e) 1746.; (f) .005155

7. (a) 2.6355; (b) 3; (c) 1.016; (d) 1.723

9. (a) $\ln 10 = 2.3026$; (b) $\ln 1.02 = .0198$; (c) $\ln 2 = .6931$; (d) $\ln .00722 = -4.9309$

11. Domain $\{x \mid x > 1\}$; yes

Exercise 4-6

1. .445

3. 40.4

5. 5.76

7. 11.83

9. 1.47

11. .00230

13. 4.45

15. 8.829

17. 8.29

19. 15.4 sq. in.

21. 6730 years

23. Approximately 5 minutes

Review Exercise

1. (a) y/x^3; (b) $a^{(7/2)}b^{(5/3)}$

3. Domain $\{x \mid x \text{ is a real number}\}$; range $\{x \mid x > 0\}$; $x = \log_2 y$

5. (a) $3^2 = 9$; (b) $8^{(-1/3)} = 1/2$

9. (a) $x = \dfrac{\log 7}{\log 5 - 2 \log 7}$; (b) $x = 9$

11. (a) 7.5128 − 10; (b) 74.56

13. −1.5041

Chapter 5

Exercise 5-1

1. (a) IV; (b) II; (c) IV; (d) III; (e) IV; (f) III; (g) III; (h) II

3. (a) $180° \pm k \cdot 360°$; $k = 0, 1, 2, 3, \ldots$
   (b) $270° \pm k \cdot 360°$; $k = 0, 1, 2, 3, \ldots$
   (c) $0° \pm k \cdot 360°$; $k = 0, 1, 2, 3, \ldots$

5. (a) $\pi/6$; (b) $-\pi$; (c) $5\pi/4$; (d) $13\pi/6$; (e) $-5\pi/2$; (f) $\pi/3$; (g) $-\pi/4$; (h) $\pi/10$

7. (a) $3\pi/2$; (b) $13\pi/8$; (c) $\pi/4$; (d) $2\pi/3$; (e) 0; (f) $\pi/2$

9. 288 deg/sec; 1.6 rad/sec

Exercise 5-2

1. Domain of $\sin x$: $\{x \mid x \text{ is a real number}\}$
   Domain of $\cos x$: $\{x \mid x \text{ is a real number}\}$

Domain of tan x : $\{x \mid x \neq \pm(2n+1)(\pi/2);\ n = 0, 1, 2, \ldots\}$
Domain of csc x : $\{x \mid x \neq \pm n\pi;\ n = 0, 1, 2, \ldots\}$
Domain of sec x : $\{x \mid x \neq \pm(2n+1)(\pi/2);\ n = 0, 1, 2, \ldots\}$
Domain of cot x : $\{x \mid x \neq \pm n\pi;\ n = 0, 1, 2, \ldots\}$

3. (a) I; (b) III; (c) IV; (d) II; (e) III; (f) II

5. (a) sin, csc, +; (b) tan, cot, +; (c) cos, sec, +; (d) cos, sec, −; (e) tan, cot, −; (f) sin, csc, −

7. Quadrant II: $\sin\theta = \sqrt{11}/6$; $\tan\theta = -\sqrt{11}/5$
Quadrant III: $\sin\theta = -\sqrt{11}/6$; $\tan\theta = \sqrt{11}/5$

9. $\sin\theta = -3/\sqrt{10}$; $\cos\theta = -1/\sqrt{10}$; $\csc\theta = -\sqrt{10}/3$; $\sec\theta = -\sqrt{10}$; $\cot\theta = 1/3$

11.

θ	$\sin\theta$	$\cos\theta$	$\tan\theta$	$\csc\theta$	$\sec\theta$	$\cot\theta$
0	0	1	0	und	1	und
$\pi/2$	1	0	und	1	und	0
π	0	−1	0	und	−1	und
$3\pi/2$	−1	0	und	−1	und	0

13. Because $\sec\theta = r/x$ and $r \geq x$.

15. Because $\sin\theta \leq 1$ and $\csc\theta = 1/\sin\theta \geq 1$

Exercise 5-3

1. $\sin A = \cos B = 4/5$; $\cos A = \sin B = 3/5$; $\tan A = \cot B = 4/3$; $\csc A = \sec B = 5/4$; $\sec A = \csc B = 5/3$; $\cot A = \tan B = 3/4$

3. $\csc\theta = 3/2$; $\cos\theta = \sqrt{5}/3$

5. (a) $-1/2$; (b) 0; (c) 1; (d) $\sqrt{3}/2$; (e) $2\sqrt{3}$

7. (a) $\sqrt{3}/2, -1/2, -\sqrt{3}$; (b) $-\sqrt{3}/2, -1/2, \sqrt{3}$; (c) $-1/\sqrt{2}, 1/\sqrt{2}, -1$; (d) $-1/2, \sqrt{3}/2, -1/\sqrt{3}$; (e) $1/2, -\sqrt{3}/2, -1/\sqrt{3}$; (f) $-1/\sqrt{2}, -1/\sqrt{2}, 1$; (g) $-\sqrt{3}/2, 1/2, -\sqrt{3}$

9. (a) $\pi/4$; (b) $\pi/3$; (c) $\pi/4$; (d) $\pi/3$; (e) $\pi/6$; (f) $\pi/2$;

11. (a) $3\pi/4, 7\pi/4$; (b) $\pi/3, 5\pi/3$; (c) $5\pi/4, 7\pi/4$; (d) $5\pi/6, 11\pi/6$

Exercise 5-4

1. (a) $\sin\pi$; (b) $\sin 140°$; (c) $\cos 2\pi/3$; (d) $\cos 159°$; (e) $\tan 2\pi/7$; (f) $\tan 2\pi/3$; (g) $-\csc 120°$; (h) $\sec 341°$; (i) $\sin(16 - 4\pi) \approx \sin 3.43$; (j) $\cos(21 - 6\pi) \approx \cos 2.15$

3. (a) −1; (b) 0; (c) $\sqrt{3}$; (d) $-1/\sqrt{2}$; (e) $\sqrt{3}$; (f) −1

7. If $f(x) = k$ for all x then $f(x + p) = f(x) = k$ for all $p > 0$ and for all x

and f is periodic. Since this is true for all $p > 0$, there is no smallest positive number p for which this is true, therefore f does not have a period.

Exercise 5-5

1.

Function	Domain	Range	Period
sine	$\{x \mid x$ is a real number$\}$	$\{x \mid -1 \leq x \leq 1\}$	2π
cosine	$\{x \mid x$ is a real number$\}$	$\{x \mid -1 \leq x \leq 1\}$	2π
tangent	$\{x \mid x \neq \pm(2n+1)(\pi/2);\ n = 0, 1, 2, \ldots\}$	$\{x \mid x$ is a real number$\}$	π
cosecant	$\{x \mid x \neq \pm n\pi;\ n = 0, 1, 2, \ldots\}$	$\{x \mid x \geq 1$ or $x \leq -1\}$	2π
secant	$\{x \mid x \neq \pm(2n+1)(\pi/2);\ n = 0, 1, 2, \ldots\}$	$\{x \mid x \geq 1$ or $x \leq -1\}$	2π
cotangent	$\{x \mid x \neq \pm n\pi;\ n = 0, 1, 2, \ldots\}$	$\{x \mid x$ is a real number$\}$	π

7. The graph of $y = -f(x)$ is the reflection of the graph of $y = f(x)$ on the x-axis.

9.

	Amplitude	Period	Phase Shift
(a)	3	π	0
(b)	2	$2\pi/3$	0
(c)	None	$\pi/2$	0
(d)	2	4π	2π
(e)	4	4π	0
(f)	1/2	2π	1
(g)	1	2π	$-\pi/4$
(h)	2	2π	$\pi/3$

11. They are the same function.

13. $y = \sin^2 x + \cos^2 x = 1$ for all x

Exercise 5-6

1. Period 2π

3. Period 2π

5. Not periodic

7. 4π

9. 4π

11. $(f+g)(x+p)=f(x+p)+g(x+p)=f(x)+g(x)=(f+g)(x)$
    $(f\cdot g)(x+p)=f(x+p)\cdot g(x+p)=f(x)\cdot g(x)=(f\cdot g)(x)$

Exercise 5-8

1. (a) $3\pi/4$; (b) $\pi/4$; (c) $\pi/6$; (d) $-\pi/3$; (e) $-\pi/6$; (f) $2\pi/3$; (g) $-\pi/4$; (h) $\pi/3$; (i) $5\pi/6$; (j) $\pi/3$
3. (a) $\sqrt{15}/4$; (b) $1/\sqrt{10}$; (c) $\sqrt{5}/2$; (d) $\sqrt{15}/4$; (e) $1/\sqrt{10}$; (f) $-\sqrt{5}/2$
5. (a) $-\pi/2+\pi=\pi/2$; (b) $\pi/2+(-\pi/4)=\pi/4$; (c) $2\pi/3+(-\pi/3)=\pi/3$; (d) $-\pi/4+\pi=3\pi/4$
9. Domain $\{x \mid x \text{ is a real number}\}$; Range $\{y \mid -\pi/2 < y < \pi/2\}$
11. (a) $x=0$; (b) $x=5/2$; (c) $x=\sqrt{5}/3-2$
19. Set $\arcsin x=\theta$, then $\sin\theta=x$. If $0<x<1$, then $0<\theta<\pi/2$ and $\cos(\pi/2-\theta)=x$. Thus $\arccos x=\pi/2-\theta$. If $x=0$ or 1, the equation is clearly true.
21. (a) 2; (b) $2\pi/5$; (c) $1/4$; (d) $-\pi/4$; (e) 0; (f) $\pi/4$; (g) 0; (h) $-\pi/2$

Exercise 5-9

1. $\tan\theta=\dfrac{\sin\theta}{\cos\theta}$ is defined for $\{\theta \mid \theta\neq\pm(2n+1)(\pi/2);\ n=0,1,2,\ldots\}$

   $\cot\theta=\dfrac{\cos\theta}{\sin\theta}$ is defined for $\{\theta \mid \theta\neq\pm n\pi;\ n=0,1,2,\ldots\}$

3. (a) $\sin\theta=\pm\sqrt{1-\cos^2\theta}$; $\cos\theta=\cos\theta$; $\tan\theta=\dfrac{\pm\sqrt{1-\cos^2\theta}}{\cos\theta}$;

   $\csc\theta=\dfrac{\pm1}{\sqrt{1-\cos^2\theta}}$; $\sec\theta=\dfrac{1}{\cos\theta}$; $\cot\theta=\dfrac{\cos\theta}{\pm\sqrt{1-\cos^2\theta}}$

   (b) $\sin\theta=\dfrac{\tan\theta}{\pm\sqrt{\tan^2\theta+1}}$; $\cos\theta=\dfrac{1}{\pm\sqrt{\tan^2\theta+1}}$; $\tan\theta$ $\tan\theta$;

   $\csc\theta=\dfrac{\pm\sqrt{\tan^2\theta+1}}{\tan\theta}$; $\sec\theta=\pm\sqrt{\tan^2\theta+1}$; $\cot\theta=\dfrac{1}{\tan\theta}$

   (c) $\sin\theta=\dfrac{1}{\csc\theta}$; $\cos\theta=\dfrac{\pm\sqrt{\csc^2\theta-1}}{\csc\theta}$; $\tan\theta=\dfrac{1}{\pm\sqrt{\csc^2\theta-1}}$

   $\csc\theta=\csc\theta$; $\sec\theta=\dfrac{\csc\theta}{\pm\sqrt{\csc^2\theta-1}}$; $\cot\theta=\pm\sqrt{\csc^2\theta-1}$

   (d) $\sin\theta=\dfrac{\pm\sqrt{\sec^2\theta-1}}{\sec\theta}$; $\cos\theta=\dfrac{1}{\sec\theta}$; $\tan\theta=\pm\sqrt{\sec^2\theta-1}$

   $\csc\theta=\dfrac{\sec\theta}{\pm\sqrt{\sec^2\theta-1}}$; $\sec\theta=\sec\theta$; $\cot\theta=\dfrac{1}{\pm\sqrt{\sec^2\theta-1}}$

(e) $\sin\theta = \dfrac{1}{\pm\sqrt{1+\cot^2\theta}}$; $\cos\theta = \dfrac{\cot\theta}{\pm\sqrt{1+\cot^2\theta}}$; $\tan\theta = \dfrac{1}{\cot\theta}$

$\csc\theta = \pm\sqrt{1+\cot^2\theta}$; $\sec\theta = \dfrac{\pm\sqrt{1+\cot^2\theta}}{\cot\theta}$; $\cot\theta = \cot\theta$

Exercise 5-10

3. (a) $\cot(\alpha+\beta) = \dfrac{\cot\alpha\cot\beta - 1}{\cot\beta + \cot\alpha}$

   (b) $\cot(\alpha-\beta) = \dfrac{\cot\alpha\cot\beta + 1}{\cot\beta - \cot\alpha}$

5. (a) $(1/2)(\sin 3x + \sin x)$; (b) $(1/2)(\cos 4x + \cos 2x)$; (c) $(1/2)(\cos x - \cos 5x)$; (d) $(1/2)(\sin 5\theta - \sin 3\theta)$

7. (a) $5\sin(x+\alpha)$; $\alpha = \arctan(3/4)$
   (b) $\sqrt{2}\sin(x+\alpha)$; $\alpha = \pi/4$
   (c) $2\sin(x+\alpha)$; $\alpha = \pi/3$
   (d) $\sqrt{13}\sin(x-\alpha)$; $\alpha = \arctan(-3/2)$
   (e) $2\sin(x-\alpha)$; $\alpha = -\pi/6$
   (f) $\sqrt{5}\sin(x+\alpha)$; $\alpha = \arctan 2$

9. $\sin(5\pi/12) = \dfrac{\sqrt{2+\sqrt{3}}}{2}$; $\cos(5\pi/13) = \dfrac{\sqrt{2-\sqrt{3}}}{2}$; $\tan(5\pi/12) = 2+\sqrt{3}$

11. $\sin(\pi/8) = \dfrac{\sqrt{2-\sqrt{2}}}{2}$; $\cos(\pi/8) = \dfrac{\sqrt{2+\sqrt{2}}}{2}$; $\tan(\pi/8) = \sqrt{2}-1$

17. $\sin(\theta/2) = 1/\sqrt{5}$; $\cos(\theta/2) = 2/\sqrt{5}$; $\tan(\theta/2) = 1/2$; $\sin 2\theta = 24/25$; $\cos 2\theta = -7/25$; $\tan 2\theta = -24/7$

Exercise 5-11

1. (a) $5(\cos 2\pi k + i\sin 2\pi k)$; k an integer
   (b) $3[\cos(\pi+2\pi k) + i\sin(\pi+2\pi k)]$; k an integer
   (c) $\sqrt{2}[\cos(3\pi/4+2\pi k) + i\sin(3\pi/4+2\pi k)]$; k an integer
   (d) $2[\cos(\pi/3+2\pi k) + i\sin(\pi/3+2\pi k)]$; k an integer
   (e) $\cos(\pi/2+2\pi k) + i\sin(\pi/2+2\pi k)$; k an integer
   (f) $2[\cos(3\pi/2+2\pi k) + i\sin(3\pi/2+2\pi k)]$; k an integer

3. $\pm 3, \pm 3i$

5. $-1, (1 \pm i\sqrt{3})/2$

7. $\pm(1+i)/\sqrt{2}$

11. -1

Exercise 5-12

1. $\pi + 2\pi k$; k an integer
3. $\pi/6 + 2\pi k$; $5\pi/6 + 2\pi k$; k an integer
5. $\pi/4 + (\pi/2)k$; k an integer
7. $\pi/2$, $3\pi/2$
9. $\pi/3$, $2\pi/3$, $4\pi/3$, $5\pi/3$
11. 0, $4\pi/3$
13. $\pi/2$, $7\pi/6$, $3\pi/2$, $11\pi/6$
15. $2\pi/3$
17. $\pi/18$, $13\pi/18$, $25\pi/18$, $11\pi/18$, $23\pi/18$, $35\pi/18$
19. $3\pi/8$, $7\pi/8$, $11\pi/8$, $15\pi/8$
21. 0
23. π
25. $\pi/2$, $3\pi/2$, $\pi/6$, $5\pi/6$

Exercise 5-13

3. Domain $\{x \mid x \text{ is a real number}\}$, range $\{x \mid x \text{ is a real number}\}$. No; yes.
7. $x = \ln(y + \sqrt{y+1})$

Review Exercise

1. (a) $4\pi/3$, $\pi/12$, π, $31\pi/18$, $-\pi/3$
   (b) 150°, 60°, −180°, 135°, $360/\pi$
3. $\sin\theta = -3/\sqrt{10}$; $\cos\theta = -1/\sqrt{10}$
5. (a) $\cos(\pi/10)$; (b) $\sin(\pi/10)$; (c) $\cot(\pi/14)$; (d) $-\sec(\pi/8)$
7.

	Amplitude	Period	Phase shift
(a)	None	$\pi/3$	$\pi/3$
(b)	1	π	$-1/2$
(c)	1/2	π	0
(d)	None	$\pi/2$	$\pi/4$
(e)	3	3π	$3\pi/4$
(f)	1	4π	$2\pi/3$

9. 2π

11. (a) $-\pi/2$; (b) $\pi/4$; (c) $-\pi/6$; (d) $\pi/6$; (e) $5\pi/6$; (f) $\pi/3$

15. $2\sqrt{2}\sin(x + \alpha)$; $\alpha = \pi/4$

19. $\pi/2$, $7\pi/6$, $11\pi/6$

Chapter 6

Exercise 6-2

1. (a) sin 40°; (b) −cos 65°; (c) −tan 56°; (d) csc 35°; (e) −cot 75°; (f) −sec 72°; (g) −sin 52°; (h) tan 48°

3. sin 294° = −sin 66° = −.9135; cos 294° = .4067; tan 294° = −2.246; csc 294° = −1.095; sec 294° = 2.459; cot 294° = −.4452

5. (a) .7986; (b) .1736; (c) 1.483; (d) −.4452; (e) −1.113; (f) −.9703; (g) −28.65; (h) 1.428

7. (a) 64°; (b) 8°9′; (c) 40°30′; (d) 15°20′; (e) 19°56′; (f) 35°24′; (g) 50°27′; (h) 44°4′

9. (a) 170°; (b) 121°; (c) 113°30′; (d) 137°20′

Exercise 6-3

1. $\alpha = 33°40'$; $\beta = 56°20'$; $c = 45.9$

3. $\alpha = 30°56'$; $\beta = 59°4'$; $b = 844.7$

5. $\alpha = 41°20'$; $a = 149$; $b = 169$

7. 79 sq. in.

9. 72°

11. approximately 28″

13. 3°00′

15. 43 sq. in.

Exercise 6-4

5. (a) $\alpha = 16°$; $\beta = 145°$; $\gamma = 19°$
   (b) no solution
   (c) $\gamma = 80°$; $a = 8.6$; $b = 15$
   (d) $c = 8.1$; $\alpha = 46°$; $\beta = 102°$
   (e) $a = 244$; $\beta = 21°50'$; $\gamma = 94°$
   (f) $\gamma = 73°$; $a = 2.8$; $b = 11.5$
   (g) $\gamma = 32°$; $a = 48$; $c = 31$
   (h) $\alpha = 15°$; $\beta = 31°$; $\gamma = 134°$

7. (a) $\beta = 13°$; $\gamma = 151°$; $c = 53$
   (b) No solution
   (c) $\beta = 41°$; $\gamma = 123°$; $c = 30$
   $\beta' = 139°$; $\gamma' = 25°$; $c' = 15$
   (d) No solution
   (e) $\beta = 29°$; $\gamma = 116°$; $c = 47$
   (f) $\beta = 49°$; $\gamma = 96°$; $c = 33$
   $\beta' = 131°$; $\gamma' = 14°$; $c' = 8$

9. 24°

11. 274 square units

Review Exercise

3. 167 square inches

5. (a) 28.9; (b) 29°; (c) 10

7. No, because 6.0 + 8.0 is not greater than 15.

Chapter 7

Exercise 7-2

1. (a) 1/4; (b) −6; (c) undefined; (d) 0; (e) −4/5; (f) 1/5; (g) undefined; (h) 4

3. (a) (3, 1); (b) (2, 4); (c) (1, 1); (d) (1, 0); (e) (2, −5); (f) (4, −2)

5. $m = 1/2$

7. (a) $x = 7$; (b) $y = 1$; (c) $y = 6/5$; (d) $y = -4$

9. The slopes of P_1P_2 and P_3P_4 are $-1/4$; the slopes of P_2P_3 and P_4P_1 are 3.

Exercise 7-3

1. (a) $2x + y - 2 = 0$; (b) $x - 2y + 7 = 0$; (c) $3x - 2y - 4 = 0$; (d) $y = -1$; (e) $5x + y - 14 = 0$; (f) $x = 1$; (g) $y = -1$; (h) $24x - 5y - 2 = 0$; (i) $x = -3$; (j) $y = -1$; (k) $x = 4$; (l) $x + 3y - 15 = 0$

3. (b), (d), (h) are parallel; (f) and (c) are parallel; (a) is perpendicular to (b), (d) and (h); (g) is perpendicular to (f) and (c); (e) and (i) are perpendicular.

5. (a) $k = -5/2$; (b) $k = 10$

7. Hint: Use the two point form of the line.

9. $3x - 4y + 26 = 0$

Exercise 7-4

1. (a) $(x-4)^2 + (y+2)^2 = 1$; (b) $(x-1)^2 + y^2 = 1/4$;
   (c) $(x-1)^2 + (y-1)^2 = 18$; (d) $(x-3)^2 + (y+1)^2 = 13$;
   (e) $(x-3)^2 + (y+1)^2 = 1$; (f) $(x-3)^2 + (y+1)^2 = 9$;
   (g) $(x-2)^2 + y^2 = 4$; (h) $(x-1)^2 + (y-1)^2 = 1$, or
   $(x-5)^2 + (y-5)^2 = 25$

3. (a) $(x+2)^2 + (y+2)^2 = 16$; $C(-2, -2)$; $r = 4$
   (b) $(x-3)^2 + (y+1)^2 = 0$; a single point $(3, -1)$
   (c) $(x+1)^2 + (y-4)^2 = 15$; $C(-1, 4)$; $r = \sqrt{15}$
   (d) $(x+3/2)^2 + (y-1)^2 = 9/4$; $C(-3/2, 1)$; $r = 3/2$
   (e) $(x-5/2)^2 + (y-7/2)^2 = -3/2$; the empty set
   (f) $(x+1)^2 + y^2 = 7/2$; $C(-1, 0)$; $r = \sqrt{7/2}$
   (g) $(x-7/6)^2 + (y-1/3)^2 = 5/36$; $C(7/6, 1/3)$; $r = \sqrt{5}/6$
   (h) $(x-2/5)^2 + (y+2)^2 = 9$; $C(2/5, -2)$; $r = 3$
   (i) $x^2 + (y+1/5)^2 = 6/25$; $C(0, -1/5)$; $r = \sqrt{6}/5$

5. Hint: Complete the square and consider the constant term.

Exercise 7-6

1. (a) $y^2 = 12x$; (b) $x^2 = -4y$; (c) $y^2 = -4x$; (d) $x^2 = 2y$; (e) $y^2 = -32x$; (f) $x^2 = -12(y-3)$; (g) $(y-5)^2 = 8(x+3)$; (h) $(y-3)^2 = -4(x-1)$; (i) $(x-2)^2 = -8(y-4)$; (j) $(x-4)^2 = 20(y+2)$

3. (a) $y^2 = 2(x+1/2)$; $V(-1/2, 0)$; $F(0, 0)$; focal width $= 2$; opens right; (b) $(x-1)^2 = -8(y+1)$; $V(1, -1)$; $F(1, -3)$; focal width $= 8$; opens down; (c) $(y+3)^2 = -4(x-2)$; $V(2, -3)$; $F(1, -3)$; focal width $= 4$; opens left; (d) $(y-7/2)^2 = -2(x-49/8)$; $V(49/8, 7/2)$; $F(45/8, 7/2)$; focal width $= 2$; opens left; (e) $(x+3/2)^2 = 7(y+1/28)$; $V(-3/2, -1/28$; $F(-3/2, 12/7)$; focal width $= 7$; opens up; (f) $(x+1/4)^2 = (1/2)(y+1/8)$; $V(-1/4, -1/8)$: $F(-1/4, 0)$; focal width $= 1/2$; opens up; (g) $(x-5/4)^2 = (-1/2)(y-25/8)$; $V(5/4, 25/8)$; $F(5/4, 3)$; focal width $= 1/2$; opens down; (h) $(y-1/3)^2 = (4/3)(x-1/6)$; $V(1/6, 1/3)$; $F(1/2, 1/3)$; focal width $= 4/3$; opens right

5. (7-7) $F(h+a, k)$; $D: x = h-a$; Axis : $y = k$
   (7-8) $F(h-a, k)$; $D: x = h+a$; Axis : $y = k$
   (7-9) $F(h, k+a)$; $D: y = k-a$; Axis : $x = h$
   (7-10) $F(h, k-a)$; $D: y = k+a$; Axis : $x = h$

7. $(y-1)^2 = 2(x+3/2)$; $(y-1)^2 = -2(x+1/2)$; $(x+1)^2 = 2(y-1/2)$; $(x+1)^2 = -2(y-3/2)$

9. $x^2 = 8(y+1)$; $x^2 = -8(y+5)$

11. $10\sqrt{5}$. The equation for the parabola is $x^2 = -25y$.

Exercise 7-7

1. (a) $C(0, 0)$; $V(0, \pm 3)$; $F(0, \pm 2\sqrt{2})$; $e = 2\sqrt{2}/3$
   (b) $C(0, 0)$; $V(\pm 3, 0)$; $F(\pm 2\sqrt{2}, 0)$; $e = 2\sqrt{2}/3$
   (c) $C(0, 0)$; $V(\pm 5, 0)$; $F(\pm 1, 0)$; $e = 1/\sqrt{5}$
   (d) $C(1, 0)$; $V(1 \pm \sqrt{10}, 0)$; $F(4, 0), (-2, 0)$; $e = 3/\sqrt{10}$
   (e) $C(-2, -3)$; $V(-2, 2), (-2, -8)$; $F(-2, 1), (-2, -7)$; $e = 4/5$
   (f) $C(1, -2)$; $V(5, -2), (-3, -2)$; $F(4, -2), (-2, -2)$; $e = 3/4$
   (g) $C(-1, 0)$; $V(-1, \pm 4)$; $F(-1, \pm 2\sqrt{3})$; $e = \sqrt{3}/2$
   (h) $C(0, -3)$; $V(0, 10), (0, -16)$; $F(0, 9), (0, -15)$; $e = 12/13$

3. (7-13) $F(h \pm c, k)$; $V(h \pm a, k)$
   (7-14) $F(h, k \pm c)$; $V(h, k \pm a)$

5. $$\left[\frac{(x - c)^2}{a^2}\right] + \left(\frac{y^2}{b^2}\right) = 1; \left[\frac{(x + c)^2}{a^2}\right] + \left(\frac{y^2}{b^2}\right) = 1$$

$$\left[\frac{(y - c)^2}{a^2}\right] + \left(\frac{x^2}{b^2}\right) = 1; \left[\frac{(y + c)^2}{a^2}\right] + \left(\frac{x^2}{b^2}\right) = 1$$

7. $4b^2c/a$

9. Nearly circular

11. $1/\sqrt{2}$

Exercise 7-8

1. (a) $C(0, 0)$; $V(0, \pm 3)$; $F(0, \pm\sqrt{10})$; $e = \sqrt{10}/3$
   (b) $C(0, 0)$; $V(\pm 1, 0)$; $F(\pm\sqrt{10}, 0)$; $e = \sqrt{10}$
   (c) $C(0, 0)$; $V(\pm 2, 0)$; $F(\pm 3, 0)$; $e = 3/2$
   (d) $C(1, 0)$; $V(1 \pm 2\sqrt{2}, 0)$; $F(4, 0), (-2, 0)$; $e = 3/2\sqrt{2}$
   (e) $C(-3, -2)$; $V(-3, 0), (-3, -4)$; $F(-3, -2 \pm 2\sqrt{2})$, $e = \sqrt{2}$
   (f) $C(1, -2)$; $V(3, -2), (-1, -2)$; $F(1 \pm \sqrt{13}, -2)$; $e = \sqrt{13}/2$
   (g) $C(-1, -2)$; $V(-1, -1), (-1, -3)$; $F(-1, -5), (-1, 1)$; $e = 3$
   (h) $C(-3, 0)$; $V(-3, \pm 2)$; $F(-3, \pm 4)$; $e = 2$

3. (a) $x^2/4 - y^2/3 = 1$
   (b) $y^2/10 - x^2/20 = 1$
   (c) $y^2/(1/4) - x^2/(1/8) = 1$
   (d) $x^2/(3/2) - y^2 = 1$
   (e) $(x + 2)^2/4 - y^2 = 1$
   (f) $(x - 1)^2 - (y + 2)^2 = 1$
   (g) $(y - 1)^2 - x^2/2 = 1$
   (h) $(y - 1)^2/5 - x^2/4 = 1$
   (i) $(x - 1)^2/9 - (y + 3)^2/4 = 1$
   (j) $(x - 1)^2/(14/3) - y^2/14 = 1$

5. (a) $y^2/16 - x^2/9 = 1$; (b) $y^2 - x^2 = 1$

9. $xy = 1$

Exercise 7-9

1. $X^2 + Y^2 = 6$

3. $Y^2 = -X$

5. $X^2 - 2Y^2 = 9$

7. $X^2/9 + Y^2/4 = 1$

Exercise 7-10

1. (a) circle; (b) hyperbola; (c) ellipse; (d) hyperbola; (e) parabola; (f) parabola

3. (a) $x = X/\sqrt{2} - Y/\sqrt{2}$; $y = X/\sqrt{2} + Y/\sqrt{2}$
(b) $x = 3X/\sqrt{10} - Y/\sqrt{10}$; $y = X/\sqrt{10} + 3Y/\sqrt{10}$
(c) $x = 2X/\sqrt{5} - Y/\sqrt{5}$; $y = X/\sqrt{5} + 2Y/\sqrt{5}$
(d) $x = X/\sqrt{2} - Y/\sqrt{2}$; $y = X/\sqrt{2} + Y/\sqrt{2}$
(e) $x = X/\sqrt{10} - 3Y/\sqrt{10}$; $y = 3X/\sqrt{10} + Y/\sqrt{10}$

5. (a) hyperbola; (b) ellipse; (c) hyperbola: (d) parabola; (e) hyperbola; (f) parabola

7. If $A = C = \pm B/2$, then $B^2 - 4AC = B^2 - 4(B^2/4) = 0$.

Exercise 7-11

3. (a) $r \cos \theta = 2$; (b) $r \sin \theta = -1$; (c) $r = 2$; (d) $r^2 \sin 2\theta = 2$; (e) $r \sin^2\theta = 2 \cos \theta$; (f) $r^2(2 \cos^2\theta + 3 \sin^2\theta) = 6$; (g) $2r = \cos \theta$; (h) $r^2 = \theta$

5. (a) $y = (\tan a)x$; $a \neq \pi/2$; $x = 0$ if $a = \pi/2$
(b) $x = a$; (c) $y = a$; (d) $ax + by = c$

7. (a) circle with center (0, 1) and radius 1; symmetric with respect to $\theta = \pi/2$.
(b) cardioid; symmetric with respect to the polar axis.
(c) cardioid
(d) cardioid
(e) limaçon; symmetric with respect to the polar axis.
(f) limaçon
(g) three-leaved rose; symmetric with respect to $\theta = \pi/2$.
(h) four-leaved rose; symmetric with respect to the polar axis and the pole.
(i) lemniscate; symmetric with respect to the polar axis, the pole, and the line $\theta = \pi/2$.
(j) symmetric with respect to the polar axis, the pole, and the line $\theta = \pi/2$.
(k) eight-leaved rose; symmetric with respect to the pole and the polar axis.

(l) spiral
(m) symmetric with respect to the pole.
(n) symmetric with respect to the polar axis.

9. (a) $y^2 = -2(x - 1/2)$
(b) $(x - 2/3)^2/(16/9) + y^2/(4/3) = 1$
(c) $(y + 2/3)^2/(1/9) - x^2/(1/3) = 1$

11. (a) $r = 8 \cos \theta$
(b) $(r^2 - (6r/\sqrt{2})\cos\theta + (6r/\sqrt{2})\sin \theta + 8 = 0$
(c) $r = \dfrac{4}{1 - \cos \theta}$
(d) $r = \dfrac{5}{3 - 2 \cos \theta}$

Review Exercise

1. (a) $x + 2y - 7 = 0$; (b) $x + 2y - 1 = 0$; (c) $x - 2y + 6 = 0$
3. $(x + 1)^2 + (y - 5/2)^2 = 13/4$; $C(-1, 5/2)$; $r = \sqrt{13}/2$
5. $(y + 3)^2 = -7(x - 4/7)$; $V(4/7, -3)$; $F(-33/28, -3)$; $D : x = 65/28$; focal width = 7; opens left.
7. $(x + 1)^2/(5/3) + (y - 3/2)^2/(5/4) = 1$; $C(-1, 3/2)$; $F(-1 \pm \sqrt{5}/2\sqrt{3}, 3/2)$; $V(-1 \pm \sqrt{5/3}, 3/2)$; $e = 1/2$
9. $y^2/2 - (x + 1)^2/9 = 1$; $C(-1, 0)$; $F(-1, \pm\sqrt{11})$; $V(-1, \pm\sqrt{2})$; $e = \sqrt{11/2}$
11. $X^2/14 - Y^2/(14/9) = 1$
13. (a) $x^2 + y^2 - 3y = 0$; (b) $r^2\cos 2\theta = 1$

Chapter 8

Exercise 8-1

3. (a) the xy plane; (b) the yz plane; (c) the z-axis; (d) the x-axis
7. $\sqrt{29}$
9. $x^2 + y^2 + z^2 = 4$; a sphere with center at the origin and radius 2.

Exercise 8-2

1. A function of two variables is one-to-one if to every element z in the range there corresponds one and only one ordered pair (x, y) in the domain.
3. (a) Domain $\{(x, y) \mid (x, y) \neq (0, 0)\}$; Range $\{z \mid z > 0\}$

(b) Domain $\{(x, y) \mid x$ and y are real numbers$\}$; Range $\{z \mid z$ is a real number$\}$
(c) Domain $\{(x, y) \mid x$ and y are real numbers$\}$; Range $\{z \mid z \geq 0\}$
(d) Domain $\{(x, y) \mid x \geq 0$ and $y \geq 0\}$; Range $\{z \mid z \geq 1\}$

5. A function of three variables is a set of ordered pairs $[(x, y, z), w]$ of elements such that to every first element (x, y, z) there corresponds one and only one second element, w. The domain is a set of ordered triples of numbers; the range is a set of numbers. An example is the formula for the volume of a box, $V = l \cdot w \cdot h$.

Exercise 8-3

1. (a) (3, 0, 0), (0, 3, 0), (0, 0, 3); (b) (4, 0, 0), (0, 8, 0), (0, 0, 2); (c) (2, 0, 0), (0, −2, 0), (0, 0, 2); (d) (4, 0, 0), (0, 2, 0), (0, 0, −4); (e) (3, 0, 0), (0, 6, 0), no z-intercept; (f) no x-intercept, (0, 4, 0), (0, 0, 2); (g) (2, 0, 0), no y-intercept, (0, 0, −6); (h) no x- or y-intercept, (0, 0, −2); (i) (4, 0, 0), no y- or z-intercept; (j) (0, 0, 0) is the only intercept; (k) (0, 0, 0) is the only intercept; (l) (0, 0, 0) is the only intercept

3. 3, 2, 4

5. $x = \pm y$; a pair of perpendicular planes, each making a 45° angle with the xz and yz planes

7. (a) (0, −1, 5), (2, 0, 2), (10/3, 2/3, 0)
(b) (0, −3, −2), (9/4, 0, −11/4), (−6, −11, 0)
(c) (0, 3, 2), (3/5, 0, 4/5), (1, −2, 0)
(d) (0, 6, 10), (6, 0, −20), (2, 4, 0)
(e) (0, 8, −2), (8, 0, −2), does not pierce the xy plane
(f) does not pierce the zx or zy planes, (1, 2, 0)

Exercise 8-4

3. $4y - z - 5 = 0$

5. $z^2 = 4x$

Exercise 8-5

1. $x^2 + y^2 = (z/m)^2$

3. (a) $x^2/b^2 + y^2/a^2 + z^2/b^2 = 1$; (b) $x^2/a^2 + y^2/a^2 + z^2/b^2 = 1$

5. $x^2 + y^2 + z^2 \pm 4\sqrt{1 - z^2} - 5 = 0$

7. $x^2 = y$, $\pi/4$

9. (a) two parallel planes, $z = 1$, $z = -1$
(b) two intersecting planes, $x = y$, $x = -y$

(c) the yz plane
(d) the z-axis
(e) the origin
(f) the empty set

Exercise 8-6

3. (a) $(1, 0, 0)$, $(1, 0, \pi/2)$; (b) $(1, \pi/4, 1)$, $(\sqrt{2}, \pi/4, \pi/4)$; (c) $(\sqrt{2}, \pi/4, 1)$, $(\sqrt{3}, \pi/4, \arccos 1/\sqrt{3})$; (d) $(0, \theta, 2)$, $(2, \theta, 0)$; (e) $(2, \pi/2, 0)$, $(2, \pi/2, \pi/2)$; (f) $(\sqrt{2}, 3\pi/4, 1)$, $(\sqrt{3}, 3\pi/4, \arccos 1/\sqrt{3})$; (g) $(\sqrt{3}/2, \pi/3, 1/2)$, $(1, \pi/3, \pi/3)$; (h) $(1/2, \pi/4, \sqrt{3}/2)$, $(1, \pi/4, \pi/6)$

5. (a) $(0, 5, 0)$; (b) $(-1, 0, 0)$; (c) $(-1, 0, \sqrt{3})$; (d) $(1/2, 1/2, 1/\sqrt{2})$

7. (a) the origin; (b) a half plane whose boundary is the z-axis; (c) If $k = 0$, the positive z-azis; if $0 < k < \pi/2$, the upper half of a cone whose axis is the z-axis; if $k = \pi/2$, the xy plane; if $\pi/2 < k < \pi$, the lower half of a cone; if $k = \pi$, the negative z-axis.

9. (a) $r = 2$; (b) $\rho^2\sin^2\phi = 4$

11. (a) $r^2 + z - 4 = 0$; (b) $\rho^2\sin^2\phi + \rho \cos \phi - 4 = 0$

13. (a) $r = \rho \sin \phi;\ \theta = \theta;\ z = \rho \cos \phi$

    (b) $\rho = \sqrt{r^2 + z^2};\ \theta = \theta;\ \phi = \arccos \dfrac{z}{\sqrt{r^2 + z^2}}$

Review Exercise

1. (a) $d = \sqrt{(x_1 - x_2)^2 + (y_1 - y_2)^2 + (z_1 - z_2)^2}$
   (b) $d = \sqrt{(x_1 - x_2)^2 + (y_1 - y_2)^2}$
   (c) $d = |x_1 - x_2|$

5. The line $x + y + z = 4$; $2x - y + z = 6$ has the piercing points $(10/3, 2/3, 0)$, $(2, 0, 2)$, and $(0, -1, 5)$. The line $x + y + z = 4$; $x + 4y + 2z = 6$ has the piercing points $(2, 2, 0)$, $(2, 0, 2)$, and $(0, -1, 5)$. All three planes pass through the line through $(2, 0, 2)$ and $(0, -1, 5)$.

7. (a) $x^2/2 + y^2/3 + z^2/2 = 1$; (b) $x^2/3 + y^2/3 + z^2/2 = 1$

Index